Real Analysis
with
Point-Set Topology

MONOGRAPHS AND TEXTBOOKS IN PURE AND APPLIED MATHEMATICS

1. *K. Yano,* Integral Formulas in Riemannian Geometry (1970)*(out of print)*
2. *S. Kobayashi,* Hyperbolic Manifolds and Holomorphic Mappings (1970) *(out of print)*
3. *V. S. Vladimirov,* Equations of Mathematical Physics (A. Jeffrey, editor; A. Littlewood, translator) (1970) *(out of print)*
4. *B. N. Pshenichnyi,* Necessary Conditions for an Extremum (L. Neustadt, translation editor; K. Makowski, translator) (1971)
5. *L. Narici, E. Beckenstein, and G. Bachman,* Functional Analysis and Valuation Theory (1971)
6. *D. S. Passman,* Infinite Group Rings (1971)
7. *L. Dornhoff,* Group Representation Theory (in two parts). Part A: Ordinary Representation Theory. Part B: Modular Representation Theory (1971, 1972)
8. *W. Boothby and G. L. Weiss (eds.),* Symmetric Spaces: Short Courses Presented at Washington University (1972)
9. *Y. Matsushima,* Differentiable Manifolds (E. T. Kobayashi, translator) (1972)
10. *L. E. Ward, Jr.,* Topology: An Outline for a First Course (1972) *(out of print)*
11. *A. Babakhanian,* Cohomological Methods in Group Theory (1972)
12. *R. Gilmer,* Multiplicative Ideal Theory (1972)
13. *J. Yeh,* Stochastic Processes and the Wiener Integral (1973) *(out of print)*
14. *J. Barros-Neto,* Introduction to the Theory of Distributions (1973) *(out of print)*
15. *R. Larsen,* Functional Analysis: An Introduction (1973) *(out of print)*
16. *K. Yano and S. Ishihara,* Tangent and Cotangent Bundles: Differential Geometry (1973) *(out of print)*
17. *C. Procesi,* Rings with Polynomial Identities (1973)
18. *R. Hermann,* Geometry, Physics, and Systems (1973)
19. *N. R. Wallach,* Harmonic Analysis on Homogeneous Spaces (1973) *(out of print)*
20. *J. Dieudonné,* Introduction to the Theory of Formal Groups (1973)
21. *I. Vaisman,* Cohomology and Differential Forms (1973)
22. *B.-Y. Chen,* Geometry of Submanifolds (1973)
23. *M. Marcus,* Finite Dimensional Multilinear Algebra (in two parts) (1973, 1975)
24. *R. Larsen,* Banach Algebras: An Introduction (1973)
25. *R. O. Kujala and A. L. Vitter (eds.),* Value Distribution Theory: Part A; Part B: Deficit and Bezout Estimates by Wilhelm Stoll (1973)
26. *K. B. Stolarsky,* Algebraic Numbers and Diophantine Approximation (1974)
27. *A. R. Magid,* The Separable Galois Theory of Commutative Rings (1974)
28. *B. R. McDonald,* Finite Rings with Identity (1974)
29. *J. Satake,* Linear Algebra (S. Koh, T. A. Akiba, and S. Ihara, translators) (1975)

30. *J. S. Golan,* Localization of Noncommutative Rings (1975)
31. *G. Klambauer,* Mathematical Analysis (1975)
32. *M. K. Agoston,* Algebraic Topology: A First Course (1976)
33. *K. R. Goodearl,* Ring Theory: Nonsingular Rings and Modules (1976)
34. *L. E. Mansfield,* Linear Algebra with Geometric Applications: Selected Topics (1976)
35. *N. J. Pullman,* Matrix Theory and Its Applications (1976)
36. *B. R. McDonald,* Geometric Algebra Over Local Rings (1976)
37. *C. W. Groetsch,* Generalized Inverses of Linear Operators: Representation and Approximation (1977)
38. *J. E. Kuczkowski and J. L. Gersting,* Abstract Algebra: A First Look (1977)
39. *C. O. Christenson and W. L. Voxman,* Aspects of Topology (1977)
40. *M. Nagata,* Field Theory (1977)
41. *R. L. Long,* Algebraic Number Theory (1977)
42. *W. F. Pfeffer,* Integrals and Measures (1977)
43. *R. L. Wheeden and A. Zygmund,* Measure and Integral: An Introduction to Real Analysis (1977)
44. *J. H. Curtiss,* Introduction to Functions of a Complex Variable (1978)
45. *K. Hrbacek and T. Jech,* Introduction to Set Theory (1978)
46. *W. S. Massey,* Homology and Cohomology Theory (1978)
47. *M. Marcus,* Introduction to Modern Algebra (1978)
48. *E. C. Young,* Vector and Tensor Analysis (1978)
49. *S. B. Nadler, Jr.,* Hyperspaces of Sets (1978)
50. *S. K. Segal,* Topics in Group Rings (1978)
51. *A. C. M. van Rooij,* Non-Archimedean Functional Analysis (1978)
54. *L. Corwin and R. Szczarba,* Calculus in Vector Spaces (1979)
53. *C. Sadosky,* Interpolation of Operators and Singular Integrals: An Introduction to Harmonic Analysis (1979)
54. *J. Cronin,* Differential Equations: Introduction and Quantitative Theory (1980)
55. *C. W. Groetsch,* Elements of Applicable Functional Analysis (1980)
56. *I. Vaisman,* Foundations of Three-Dimensional Euclidean Geometry (1980)
57. *H. I. Freedman,* Deterministic Mathematical Models in Population Ecology (1980)
58. *S. B. Chae,* Lebesgue Integration (1980)
59. *C. S. Rees, S. M. Shah, and C. V. Stanojević,* Theory and Applications of Fourier Analysis (1981)
60. *L. Nachbin,* Introduction to Functional Analysis: Banach Spaces and Differential Calculus (R. M. Aron, translator) (1981)
61. *G. Orzech and M. Orzech,* Plane Algebraic Curves: An Introduction Via Valuations (1981)
62. *R. Johnsonbaugh and W. E. Pfaffenberger,* Foundations of Mathematical Analysis (1981)
63. *W. L. Voxman and R. H. Goetschel,* Advanced Calculus: An Introduction to Modern Analysis (1981)
64. *L. J. Corwin and R. H. Szcarba,* Multivariable Calculus (1982)
65. *V. I. Istrăţescu,* Introduction to Linear Operator Theory (1981)
66. *R. D. Järvinen,* Finite and Infinite Dimensional Linear Spaces: A Comparative Study in Algebraic and Analytic Settings (1981)

67. *J. K. Beem and P. E. Ehrlich,* Global Lorentzian Geometry (1981)
68. *D. L. Armacost,* The Structure of Locally Compact Abelian Groups (1981)
69. *J. W. Brewer and M. K. Smith, eds.,* Emmy Noether: A Tribute to Her Lif and Work (1981)
70. *K. H. Kim,* Boolean Matrix Theory and Applications (1982)
71. *T. W. Wieting,* The Mathematical Theory of Chromatic Plane Ornaments (1982)
72. *D. B. Gauld,* Differential Topology: An Introduction (1982)
73. *R. L. Faber,* Foundations of Euclidean and Non-Euclidean Geometry (198
74. *M. Carmeli,* Statistical Theory and Random Matrices (1983)
75. *J. H. Carruth, J. A. Hildebrant, and R. J. Koch,* The Theory of Topological Semigroups (1983)
76. *R. L. Faber,* Differential Geometry and Relativity Theory: An Introduction (1983)
77. *S. Barnett,* Polynomials and Linear Control Systems (1983)
78. *G. Karpilovsky,* Commutative Group Algebras (1983)
79. *F. Van Oystaeyen and A. Verschoren,* Relative Invariants of Rings: The Commutative Theory (1983)
80. *I. Vaisman,* A First Course in Differential Geometry (1984)
81. *G. W. Swan,* Applications of Optimal Control Theory in Biomedicine (198
82. *T. Petrie and J. D. Randall,* Transformation Groups on Manifolds (1984)
83. *K. Goebel and S. Reich,* Uniform Convexity, Hyperbolic Geometry, and Nonexpansive Mappings (1984)
84. *T. Albu and C. Năstăsescu,* Relative Finiteness in Module Theory (1984)
85. *K. Hrbacek and T. Jech,* Introduction to Set Theory, Second Edition, Revised and Expanded (1984)
86. *F. Van Oystaeyen and A. Verschoren,* Relative Invariants of Rings: The Noncommutative Theory (1984)
87. *B. R. McDonald,* Linear Algebra Over Commutative Rings (1984)
88. *M. Namba,* Geometry of Projective Algebraic Curves (1984)
89. *G. F. Webb,* Theory of Nonlinear Age-Dependent Population Dynamics (1985)
90. *M. R. Bremner, R. V. Moody, and J. Patera,* Tables of Dominant Weight Multiplicities for Representations of Simple Lie Algebras (1985)
91. *A. E. Fekete,* Real Linear Algebra (1985)
92. *S. B. Chae,* Holomorphy and Calculus in Normed Spaces (1985)
93. *A. J. Jerri,* Introduction to Integral Equations with Applications (1985)
94. *G. Karpilovsky,* Projective Representations of Finite Groups (1985)
95. *L. Narici and E. Beckenstein,* Topological Vector Spaces (1985)
96. *J. Weeks,* The Shape of Space: How to Visualize Surfaces and Three-Dimensional Manifolds (1985)
97. *P. R. Gribik and K. O. Kortanek,* Extremal Methods of Operations Researc (1985)
98. *J.-A. Chao and W. A. Woyczynski, eds.,* Probability Theory and Harmonic Analysis (1986)
99. *G. D. Crown, M. H. Fenrick, and R. J. Valenza,* Abstract Algebra (1986)
100. *J. H. Carruth, J. A. Hildebrant, and R. J. Koch,* The Theory of Topological Semigroups, Volume 2 (1986)

101. *R. S. Doran and V. A. Belfi,* Characterizations of C*-Algebras: The Gelfand-Naimark Theorems (1986)
102. *M. W. Jeter,* Mathematical Programming: An Introduction to Optimization (1986)
103. *M. Altman,* A Unified Theory of Nonlinear Operator and Evolution Equations with Applications: A New Approach to Nonlinear Partial Differentia Equations (1986)
104. *A. Verschoren,* Relative Invariants of Sheaves (1987)
105. *R. A. Usmani,* Applied Linear Algebra (1987)
106. *P. Blass and J. Lang,* Zariski Surfaces and Differential Equations in Characteristic $p > 0$ (1987)
107. *J. A. Reneke, R. E. Fennell, and R. B. Minton.* Structured Hereditary Systems (1987)
108. *H. Busemann and B. B. Phadke,* Spaces with Distinguished Geodesics (1987)
109. *R. Harte,* Invertibility and Singularity for Bounded Linear Operators (1987)
110. *G. S. Ladde, V. Lakshmikantham, and B. G. Zhang,* Oscillation Theory of Differential Equations with Deviating Arguments (1987)
111. *L. Dudkin, I. Rabinovich, and I. Vakhutinsky,* Iterative Aggregation Theory: Mathematical Methods of Coordinating Detailed and Aggregate Problems in Large Control Systems (1987)
112. *T. Okubo,* Differential Geometry (1987)
113. *D. L. Stancl and M. L. Stancl,* Real Analysis with Point-Set Topology (1987)
114. *H. Strade and R. Farnsteiner,* Modular Lie Algebras and Their Representations (1987)
115. *T. C. Gard,* Stochastic Differential Equations (1987)

Other Volumes in Preparation

Real Analysis with Point-Set Topology

DONALD L. STANCL
MILDRED L. STANCL

ST. ANSELM COLLEGE
MANCHESTER, NEW HAMPSHIRE

MARCEL DEKKER, INC. NEW YORK AND BASEL

Library of Congress Cataloging-in-Publication Data

Stancl, Donald L.
Real analysis with point-set topology.

(Monographs and textbooks in pure and applied mathematics ; 113)
Bibliography: p.
Includes index.
1. Functions of real variables. 2. Mathematical analysis. 3. Topology. I. Stancl, Mildred L. II. Point-set topology. III. Series: Monographs and textbooks in pure and applied mathematics ; v. 113.
QA331.5.S73 1987 515.8 87-3465
ISBN 0-8247-7790-5

MARCEL DEKKER, INC.
270 Madison Avenue, New York, New York 10016

Current printing (last digit):
10 9 8 7 6 5 4 3 2 1

PRINTED IN THE UNITED STATES OF AMERICA

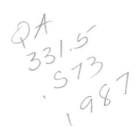

In memory of

Bruce Amert Luzader
Minnie Farson Luzader
Frances Wolff Stancl

Preface

This book is designed as a text for a first course in real analysis. It is specifically addressed to students who are unlikely to proceed to advanced degrees in mathematics and for whom their first course in real analysis will also be their last. It is our hope that the students who use this book will develop

- A solid understanding of the structure and properties of the real number system and real-valued functions
- A knowledge of the basic concepts and results of point-set topology
- An appreciation for the interplay between these two areas of mathematics
- An appreciation for the role of examples in suggesting generalizations
- An appreciation for the power of abstraction and its application to particular cases.

This book presents the standard material of a first course in real analysis: properties of the real numbers, functions on the reals, continuity, sequences and series, sequences of functions, integration, and differentiation are all covered. Mastery of this material will provide the student with a good understanding of the real numbers and real-valued functions. In addition we have also included material on the construction of the reals and cardinal arithmetic, topics which are seldom found in a text at this level, on the ground that these are so important and interesting that they should be made available to all students of mathematics, even those who will not proceed further in the discipline.

The major feature of this book lies in its presentation of the basic concepts of point-set topology in tandem with real analysis. Our reasons for including an introduction to point-set topology in a real analysis book are as follows:

- Point-set topology is a subject that the students to whom this book is addressed are unlikely to encounter elsewhere.
- Its ideas and results can easily be related to properties of the real numbers and real-valued functions, and often help in making these properties clearer.
- Its proofs tend to be rigorous but relatively nontechnical, and proving results in the general topological setting is seldom more difficult, and often easier, than proving them on the real line.
- The abstraction to the general topological setting helps students understand what a proof is and how proofs are constructed.
- The interplay between topology and analysis allows the students to observe and appreciate the interplay between abstraction and particularization that is so important to mathematics.
- Students find point-set topology an easily accessible and interesting subject.

We have attempted to interweave topology and analysis in such a way that each illuminates the other. Thus we continually proceed in two directions: we use knowledge of the real numbers and functions on the reals to suggest and illustrate topological generalizations, and we elucidate general topological concepts and results and immediately apply them to real analysis. This approach not only presents the basic ideas of point-set topology, but will also, we hope, make the power of generalization and abstraction clear to the student.

As a supplement to the standard definition-theorem-proof format, we have provided copious examples. There are also exercises at the end of each section; these present a spectrum of difficulty, ranging from simple applications of the results of the text to quite challenging problems. Some exercises are designed to supplement the text by introducing additional topics of interest. Hints are given for the more difficult exercises.

The organization of the text is straightforward. The sections concerning construction of the reals (Section 2.2), completion of metric spaces (Section 6.4), and function spaces and uniform approximation (Section 7.2) may be omitted without affecting later material. Section 5.3, on infinite series, may be omitted at the cost of eliminating some later exercises, while that on convergence of sequences of functions (Section 7.1) may be omitted at the cost of eliminating discussion of the integrability and differentiability of sequences of functions in Chapter 8. The material on set equivalence and cardinal numbers in the Appendix can be covered at any time after Chapter 1.

We would like to thank the reviewers for their criticisms and suggestions. Our special thanks to Dr. Nicholas N. Greenbaun of Trenton State College, who class-tested the manuscript and provided us with many helpful comments. Finally, our appreciation to the staff at Marcel Dekker, Inc. for their cooperation and assistance.

Donald L. Stancl
Mildred L. Stancl

Contents

Sets and Functions

In this book we shall study various sets and functions, paying particular attention to the set of real numbers and to functions whose values lie in the real numbers. This introductory chapter presents some basic definitions and results concerning sets and functions which will be used throughout the remainder of the text.

1.1 SETS

We begin our study of sets with a definition.

Definition 1.1.1 A *set* is a collection of objects, called the *elements* of the set. If object x is an element of set S, we say that x *belongs to S*, and write $x \in S$; if object x is not an element of set S, we say that x *does not belong to S*, and write $x \notin S$ (see Figure 1.1). The *empty set*, or *null set*, is the set which has no elements. The empty set is denoted by $\emptyset$.

We specify a set S either by listing its elements between braces or by writing a statement of the form

$$S = \{x \mid x \text{ has property } P\}$$

where P is some property which serves to define the elements of S unambiguously. The statement $S = \{x \mid x \text{ has property } P\}$ is read "S is the set of all elements x such that x has property P."

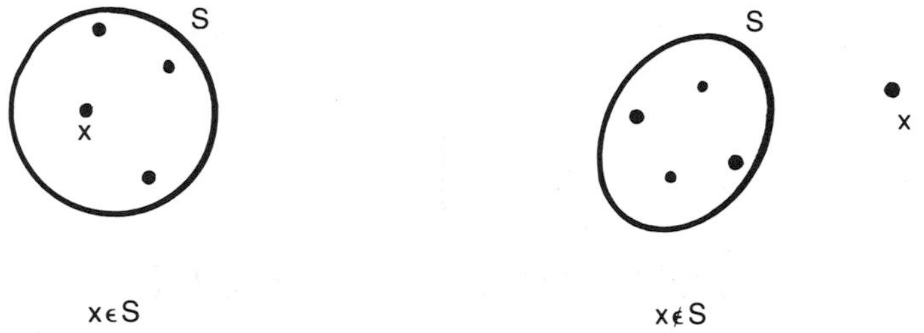

Figure 1.1 Element of a set.

Examples 1.1.2 1. The following notation will be standard throughout the text: the set of natural numbers (positive integers) will be denoted by $\boldsymbol{N}$, the set of integers by $\boldsymbol{Z}$, the set of rational numbers by $\boldsymbol{Q}$, and the set of real numbers by $\boldsymbol{R}$. Thus we may write

$$\boldsymbol{N} = \{1, 2, 3, 4, \ldots\}$$

$$\boldsymbol{Z} = \{0, 1, -1, 2, -2, \ldots\}$$

and

$$\boldsymbol{Q} = \left\{ \frac{m}{n} \,\middle|\, m \in Z,\ n \in Z,\ n \neq 0 \right\}$$

We assume that the reader is familiar with the arithmetic operations (addition, subtraction, multiplication, division, and exponentiation) and the order relationships ($<, \leqslant, >, \geqslant$) defined on these sets. The reader is undoubtedly aware that the set of real numbers $\boldsymbol{R}$ may be represented as a horizontal line called the *real line* (see Figure 1.2). Each point on the real line corresponds to a unique real number and each real number corresponds to a unique point on the line. We shall speak interchangeably of the set of real numbers $\boldsymbol{R}$ and the real line $\boldsymbol{R}$.

2. There may be several different ways of specifying the elements of a given set. For example, if

$$S = \{1, 2, 3\}$$

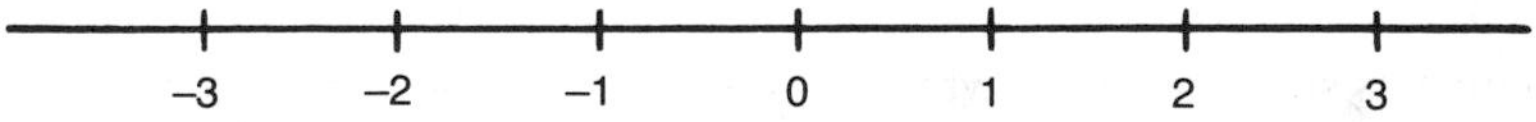

Figure 1.2 The real line.

we may also write

$$S = \{x \mid x \in N,\ x < 4\}$$

or

$$S = \{x \in Z \mid x^3 - 6x^2 + 11x - 6 = 0\}$$

3. Any statement of the form $\{x \mid x \text{ has property } P\}$ where P is a property which no object can satisfy serves to define the empty set $\emptyset$. For instance,

$$\emptyset = \{x \mid x \in R,\ x^2 < 0\}$$

since there is no real number whose square is negative.

4. Let $a \in R$ and $b \in R$, with $a < b$. The set

$$(a, b) = \{x \in R \mid a < x < b\}$$

is called an *open interval* in R (see Figure 1.3). The real numbers a and b are called the *endpoints* of (a, b). Note that the open interval (a, b) consists of all real numbers which are between a and b on the real line, but that the endpoints a and b do not belong to (a, b). The open interval $(0, 1)$ is referred to as the *open unit interval*.

a   b

Figure 1.3   The open interval (a, b).

5. Let $a \in R$ and $b \in R$, with $a \leqslant b$. The set

$$[a, b] = \{x \in R \mid a \leqslant x \leqslant b\}$$

is called the *closed interval* in R with *endpoints* a and b (see Figure 1.4). Note that the endpoints a and b belong to the closed interval $[a, b]$. Also, if $a = b$, then $[a, b] = [a, a] = \{a\}$, so a closed interval may consist of a single point. The interval $[0, 1]$ is the *closed unit interval*.

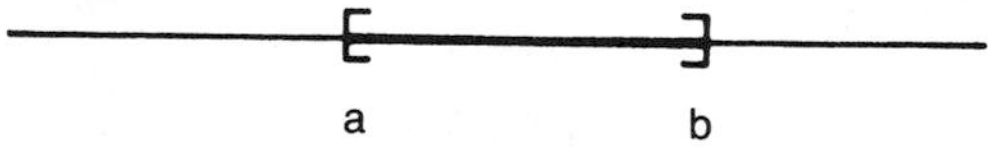

Figure 1.4 The closed interval $[a, b]$.

6. If $a \in \boldsymbol{R}$ and $b \in \boldsymbol{R}$ with $a < b$, then the sets

$$(a, b] = \{x \in \boldsymbol{R} \mid a < x \leqslant b\}$$

and

$$[a, b) = \{x \in \boldsymbol{R} \mid a \leqslant x < b\}$$

are *half-open intervals* in $\boldsymbol{R}$ with *endpoints* a and b (see Figure 1.5).

7. If $a \in \boldsymbol{R}$, then the sets

$$(a, +\infty) = \{x \in \boldsymbol{R} \mid a < x\}$$

and

$$(-\infty, a) = \{x \in \boldsymbol{R} \mid x < a\}$$

are called *open rays* in $\boldsymbol{R}$. The sets

$$[a, +\infty) = \{x \in \boldsymbol{R} \mid a \leqslant x\}$$

and

$$(-\infty, a] = \{x \in \boldsymbol{R} \mid x \leqslant a\}$$

are called *closed rays* in $\boldsymbol{R}$ (see Figure 1.6). The set $\boldsymbol{R}$ is both an open ray

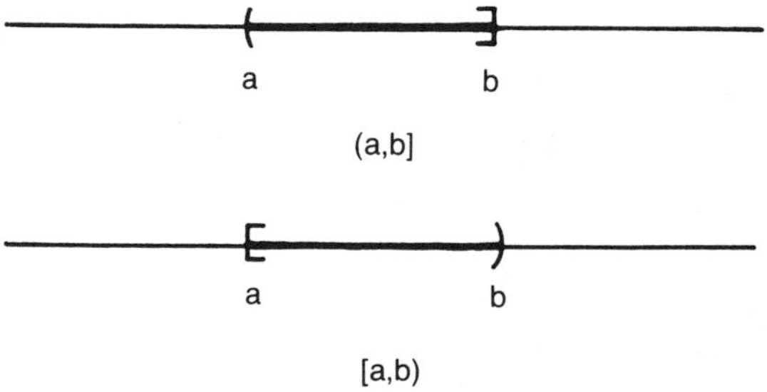

Figure 1.5 Half-open intervals.

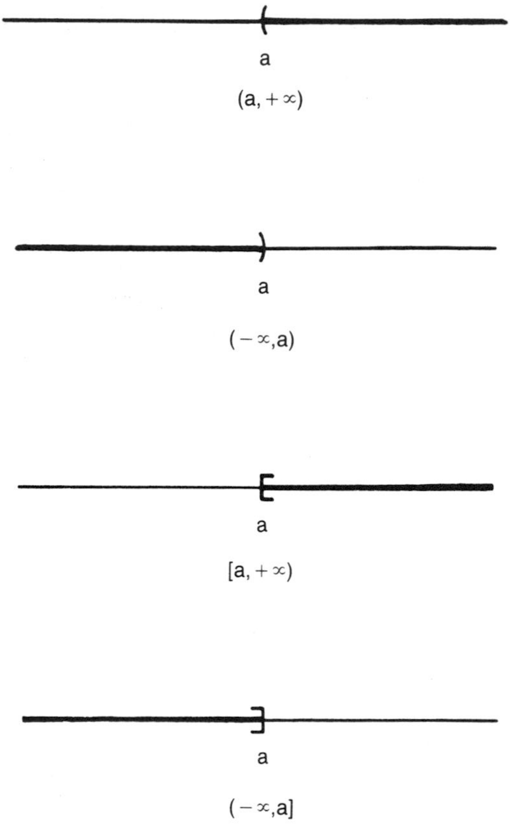

Figure 1.6 Rays.

and a closed ray and may be written as

$$\boldsymbol{R} = (-\infty, +\infty)$$

Next we define the notion of a subset of a given set.

Definition 1.1.3 A set S is a *subset* of a set T if and only if every element of S is also an element of T. If S is a subset of T, we write $S \subset T$ and say that S is *contained in* T. If S is not a subset of T, we write $S \not\subset T$ and say that S is *not contained in* T. If $S \subset T$ and $T \subset S$, then $S = T$. If $S \subset T$ but $S \neq T$, we say that S is a *proper subset* of T (see Figure 1.7).

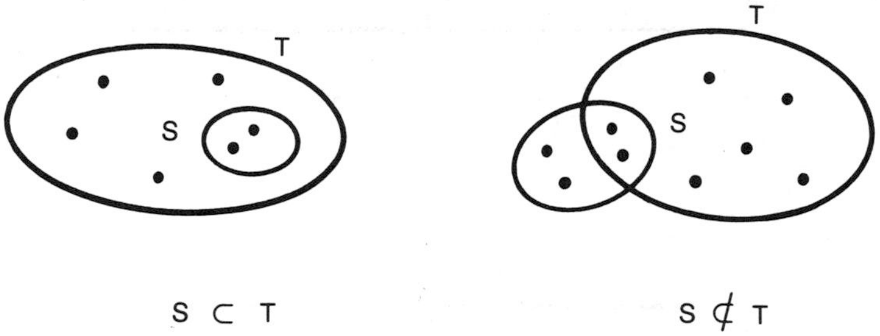

Figure 1.7 Set containment.

Examples 1.1.4 1. The following are true for any set S:

$$\emptyset \subset S \qquad S \subset S \qquad \text{and} \qquad S = S$$

2. For the sets $\boldsymbol{N}$, $\boldsymbol{Z}$, $\boldsymbol{Q}$, and $\boldsymbol{R}$ we have

$$\boldsymbol{N} \subset \boldsymbol{Z} \subset \boldsymbol{Q} \subset \boldsymbol{R}$$

Furthermore, $\boldsymbol{N}$ is a proper subset of $\boldsymbol{Z}$, $\boldsymbol{Z}$ is a proper subset of $\boldsymbol{Q}$, and $\boldsymbol{Q}$ is a proper subset of $\boldsymbol{R}$.

3. Let $a \in \boldsymbol{R}$ and $b \in \boldsymbol{R}$, with $a < b$; then

$$(a, b) \text{ is a proper subset of } [a, b]$$

$$(a, b) \text{ is a proper subset of } (a, b]$$

$$[a, b) \text{ is a proper subset of } [a, b]$$

and

$$(b, +\infty) \text{ is a proper subset of } (a, +\infty)$$

4. Every interval in $\boldsymbol{R}$, whether open, half-open, or closed, is a proper subset of $\boldsymbol{R}$. Every ray in $\boldsymbol{R}$, except for the ray $(-\infty, +\infty)$, is a proper subset of $\boldsymbol{R}$.

5. Let

$$S = \left\{ x \in \boldsymbol{R} \,\middle|\, \frac{1}{x} > 1 \right\} \qquad \text{and} \qquad T = (0, 1)$$

We claim that $S = T$. To prove this, we must show that $S \subset T$ and that

$T \subset S$. To show that $S \subset T$, let x be an arbitrary element of S. The condition $1/x > 1$ then implies that x is positive and less than 1; hence $x \in (0, 1) = T$. Therefore every element of S is also an element of T, and thus $S \subset T$. To show that $T \subset S$, let $y \in T$, so that $0 < y < 1$. But then $1/y > 1$ and hence $y \in S$. Therefore every element of T is also an element of S, and thus $T \subset S$.

The technique used in the previous example is the standard method by which set identities are proved. It may be summarized as follows: to prove that $S = T$, let x be an arbitrary element of S and show that $x \in T$, thus establishing the inclusion $S \subset T$; then let y be an arbitrary element of T and show that $y \in S$, thus establishing the inclusion $T \subset S$. The inclusions $S \subset T$ and $T \subset S$ then imply that $S = T$.

Now we are ready to consider how sets can be combined to form new sets. There are four common methods of doing this: by forming unions, intersections, complements, and Cartesian products of sets. We now define these set operations.

Definition 1.1.5 Let S and T be sets.

1. The *union* of S and T is the set $S \cup T$ defined by

$$S \cup T = \{x \mid x \in S \text{ or } x \in T\}$$

(see Figure 1.8a).

2. The *intersection* of S and T is the set $S \cap T$ defined by

$$S \cap T = \{x \mid x \in S \text{ and } x \in T\}$$

(see Figure 1.8b). If $S \cap T = \emptyset$, we say that S and T are *disjoint sets*.

3. The *complement* of T in S is the set $S - T$ defined by

$$S - T = \{x \mid x \in S \text{ and } x \notin T\}$$

(see Figure 1.8c).

4. An *ordered pair* with *first coordinate* x from S and *second coordinate* y from T is a pair (x, y) where $x \in S$ and $y \in T$. Two such ordered pairs (x, y) and (x', y') are equal if and only if $x = x'$ in S and $y = y'$ in T. The *Cartesian product* of S with T is the set $S \times T$ of all such ordered pairs, that is,

$$S \times T = \{(x, y) \mid x \in S, y \in T\}$$

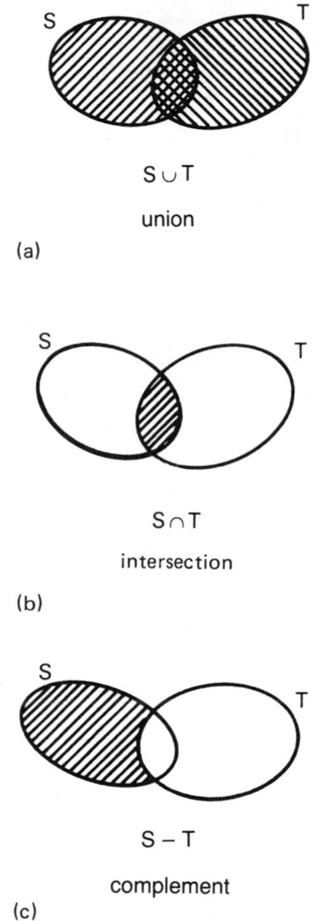

Figure 1.8 (a) Union, (b) intersection, (c) complement of sets.

Examples 1.1.6 1. Let $S = \{1, 2, 3\}$ and $T = \{3, 4\}$. We have

$$S \cup T = \{1, 2, 3, 4\} \qquad S \cap T = \{3\} \qquad S - T = \{1, 2\} \qquad T - S = \{4\}$$

$$S \times T = \{(1, 3), (1, 4), (2, 3), (2, 4), (3, 3), (3, 4)\}$$

and

$$T \times S = \{(3, 1), (3, 2), (3, 3), (4, 1), (4, 2), (4, 3)\}$$

2. Let

$$R^2 = R \times R = \{(x, y) \mid x \in R,\ y \in R\}$$

The set R^2 is called the *Euclidean plane*; it may be represented by the familiar Cartesian coordinate system as depicted in Figure 1.9.

3. Consider the intervals (0, 1), [1, 2], and [2, 3] in R. We have

$$(0, 1) \cup [1, 2] = (0, 2] \qquad (0, 1) \cap [1, 2] = \emptyset \qquad [1, 2] \cap [2, 3] = \{2\}$$

and

$$[1, 2] - [2, 3] = [1, 2) \qquad [1, 2] - (0, 1) = [1, 2]$$

Note that (0, 1) and [1, 2] are disjoint, but that [1, 2] and [2, 3] are not disjoint. The Cartesian product

$$[1, 2] \times [2, 3] = \{(x, y) \mid x \in [1, 2],\ y \in [2, 3]\}$$

is the subset of the Euclidean plane R^2 shown in Figure 1.10.

The algebraic rules which govern the set operations of union, intersection, and complementation are given in the following proposition.

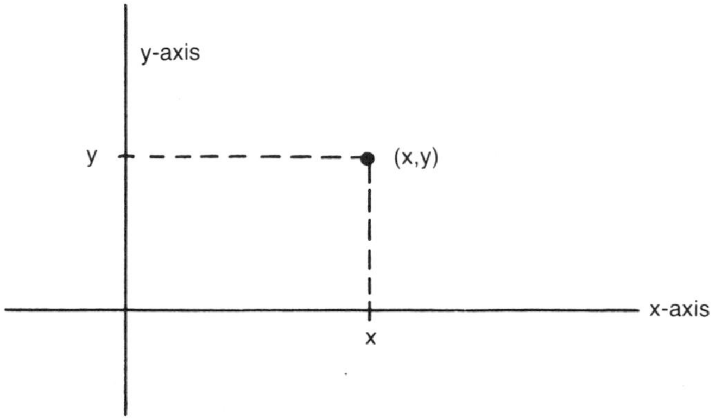

Figure 1.9 The Cartesian coordinate system for R^2.

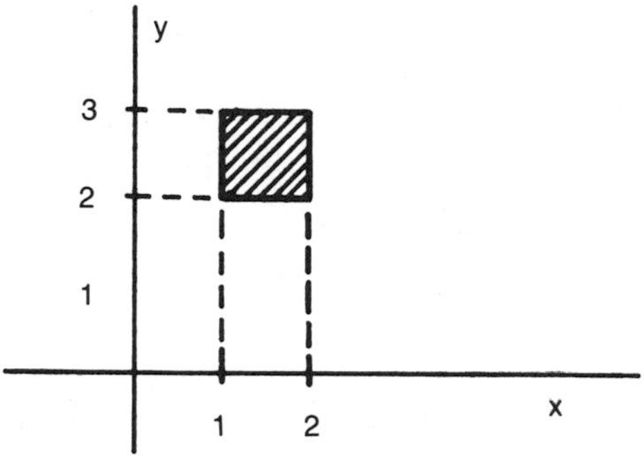

Figure 1.10 The Cartesian product $[1, 2] \times [2, 3]$.

Proposition 1.1.7 Let S, T, and V be sets. The following set identities hold:

1.	Idempotent laws	$S \cup S = S \qquad S \cap S = S$
2.	Commutative laws	$S \cup T = T \cup S \qquad S \cap T = T \cap S$
3.	Associative laws	$S \cup (T \cup V) = (S \cup T) \cup V$
		$S \cap (T \cap V) = (S \cap T) \cap V$
4.	Distributive laws	$S \cup (T \cap V) = (S \cup T) \cap (S \cup V)$
		$S \cap (T \cup V) = (S \cap T) \cup (S \cap V)$
5.	deMorgan's laws	$S - (T \cup V) = (S - T) \cap (S - V)$
		$S - (T \cap V) = (S - T) \cup (S - V)$

Proof: We shall prove the identity $S \cup (T \cap V) = (S \cup T) \cap (S \cup V)$ of (4). The remainder of the proof is similar and is left as an exercise.

Suppose that $S \cup (T \cap V) \neq \emptyset$ and let $x \in S \cup (T \cap V)$, so that either $x \in S$ or $x \in T \cap V$. If $x \in S$, then $x \in S \cup T$ and $x \in S \cup V$, so $x \in (S \cup T) \cap (S \cup V)$; if $x \in T \cap V$, then $x \in T$ and $x \in V$, so again $x \in S \cup T$ and $x \in S \cup V$, and hence $x \in (S \cup T) \cap (S \cup V)$. Therefore if $S \cup (T \cap V) \neq \emptyset$ then

$$[S \cup (T \cap V)] \subset [(S \cup T) \cap (S \cup V)]$$

But this inclusion is also true if $S \cup (T \cap V) = \emptyset$, because the empty set is a subset of every set; hence the inclusion always holds.

Now suppose that $(S \cup T) \cap (S \cup V) \neq \emptyset$ and let $y \in (S \cup T) \cap (S \cup V)$, so that $y \in S \cup T$ and $y \in S \cup V$; if $y \in S$, then surely $y \in S \cup (T \cap V)$, while if $y \notin S$ then $y \in T$ and $y \in V$, so that $y \in T \cap V$, and again $y \in S \cup (T \cap V)$. Therefore if $(S \cup T) \cap (S \cup V) \neq \emptyset$ then

$$[(S \cup T) \cap (S \cup V)] \subset [S \cup (T \cap V)]$$

But, as above, this inclusion is also true if $(S \cup T) \cap (S \cup V) = \emptyset$, and hence it always holds. Thus we have shown that $S \cup (T \cap V) = (S \cup T) \cap (S \cup V)$. □

It will sometimes be necessary for us to distinguish between finite and infinite sets. We will postpone the formal definitions of these terms until Section 1.3, but for now we will consider a set to be finite if its elements can be counted and the counting process terminates; if the elements of a set cannot be counted, or if they can but the counting process does not terminate, then we consider the set to be infinite. Thus the set $S = \{a, b, c\}$ is finite; the set $S = \{1, \ldots, n\}$ is finite for any $n \in \boldsymbol{N}$; and the empty set is finite. On the other hand, the set $\boldsymbol{N}$ is infinite because if we count its elements the counting process will not terminate. For the same reason, the sets $\boldsymbol{Z}$ and $\boldsymbol{Q}$ are infinite. The set of real numbers $\boldsymbol{R}$ is infinite because its elements cannot be counted. (We shall prove this fact in Chapter 2.)

We may form unions and intersections of any finite number of sets by taking them two at a time. For instance, suppose we have sets S_1, S_2, S_3, and S_4. Then we may form

$$S_2 \cup S_4 \qquad (S_2 \cup S_4) \cup S_3 \qquad \text{and} \qquad ((S_2 \cup S_4) \cup S_3) \cup S_1$$

But the commutative and associative laws of Proposition 1.1.7 (together with mathematical induction) imply that in such a union neither the order of the sets nor the manner of their grouping matters. Thus we may write

$$((S_2 \cup S_4) \cup S_3) \cup S_1 = S_1 \cup S_2 \cup S_3 \cup S_4$$

Similar remarks apply to intersections of finitely many sets.

In addition to thus being able to form unions and intersections of finitely many sets, we would like to be able to form unions and intersections of infinitely many sets. An efficient way to do this is to use the concept of an index set.

Definition 1.1.8 Let A be a nonempty set. (A may be finite or infinite.) If to each element $\alpha \in A$ there corresponds a set X_α, then the elements of A are said to *index* the family of sets $\{X_\alpha \mid \alpha \in A\}$, and A is called an *index* *set*. We define the union and intersection of the family of sets $\{X_\alpha \mid \alpha \in A\}$ as follows:

$$\bigcup_{\alpha \in A} X_\alpha = \{x \mid x \in X_\alpha \text{ for some } \alpha \in A\}$$

$$\bigcap_{\alpha \in A} X_\alpha = \{x \mid x \in X_\alpha \text{ for all } \alpha \in A\}$$

The sets $\{X_\alpha \mid \alpha \in A\}$ are *mutually disjoint* if and only if $X_\alpha \cap X_\beta = \emptyset$ whenever $\alpha \in A$, $\beta \in A$, and $\alpha \neq \beta$.

Examples 1.1.9 1. Let $A = \{1, 2, 3\}$, and for each $\alpha \in A$ let X_α denote the closed interval $[\alpha, \alpha + 3]$ in $\boldsymbol{R}$; then

$$\bigcup_{\alpha \in A} X_\alpha = X_1 \cup X_2 \cup X_3 = [1, 4] \cup [2, 5] \cup [3, 6] = [1, 6]$$

and

$$\bigcap_{\alpha \in A} X_\alpha = X_1 \cap X_2 \cap X_3 = [1, 4] \cap [2, 5] \cap [3, 6] = [3, 4]$$

Here the index set A is finite and we have taken the union and intersection of finitely many closed intervals in $\boldsymbol{R}$.

2. Let $A = \boldsymbol{N}$, and for each $n \in \boldsymbol{N}$ let $X_n = \{1, \ldots, n\}$; then

$$\bigcup_{\alpha \in A} X_\alpha = \bigcup_{n \in N} X_n = N \qquad \text{and} \qquad \bigcap_{\alpha \in A} X_\alpha = \bigcap_{n \in N} X_n = \{1\}$$

Here the index set is infinite, and we have taken the union and intersection of infinitely many sets, each of which was a subset of $\boldsymbol{N}$.

3. For each $n \in \boldsymbol{N}$, let X_n be the half-open interval $[n - 1, n)$ in $\boldsymbol{R}$; then

$$\bigcup_{n \in N} X_n = [0, +\infty) \qquad \text{and} \qquad \bigcap_{n \in N} X_n = \emptyset$$

(see Exercise 15 of Section 1.1). Note that the sets $\{X_n\}$ are mutually disjoint and that $\bigcap_{n \in N} X_n = \emptyset$.

4. The previous example suggests that if $\{X_\alpha \mid \alpha \in A\}$ is a family of mutually disjoint sets, then

$$\bigcap_{\alpha \in A} X_\alpha = \emptyset$$

This is indeed true (see Exercise 16 in Section 1.1). However, the intersection may be empty even if no two of the sets $\{X_\alpha \mid \alpha \in A\}$ are mutually disjoint. To see this, let

$$X_n = \left(0, \frac{1}{2^n}\right)$$

for each $n \in N$. No two of the sets $\{X_n \mid n \in N\}$ are disjoint, for if $m \geqslant n$ then $X_m \subset X_n$ and hence $X_n \cap X_m \neq \emptyset$. However, it is easy to check that

$$\bigcap_{n \in N} X_n = \emptyset$$

(see Exercise 15 in Section 1.1).

5. For each $r \in \boldsymbol{R}$, let X_r be the ordered pair $(r, r+1)$ in $\boldsymbol{R}^2$; then

$$\bigcup_{r \in R} X_r = \{(r, r+1) \mid r \in \boldsymbol{R}\}$$

If we represent $\boldsymbol{R}^2$ by means of the Cartesian coordinate system of Example 2 in Examples 1.1.6, the above union is the line with slope 1 and y-intercept 1 (see Figure 1.11).

Now that we can form unions and intersections of arbitrarily many sets, we must generalize the distributive laws and deMorgan's laws of Proposition 1.1.7.

Proposition 1.1.10 Let Y be a set. If A is a nonempty set and

$$\{X_\alpha \mid \alpha \in A\}$$

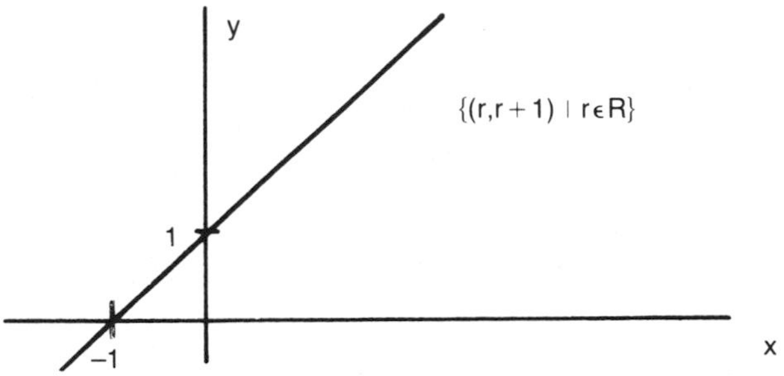

Figure 1.11 A union of ordered pairs in $\boldsymbol{R}^2$.

is a family of sets indexed by A, then

1. $$Y \cup \left(\bigcap_{\alpha \in A} X_\alpha \right) = \bigcap_{\alpha \in A} (Y \cup X_\alpha)$$

2. $$Y \cap \left(\bigcup_{\alpha \in A} X_\alpha \right) = \bigcup_{\alpha \in A} (Y \cap X_\alpha)$$

3. $$Y - \bigcup_{\alpha \in A} X_\alpha = \bigcap_{\alpha \in A} (Y - X_\alpha)$$

4. $$Y - \bigcap_{\alpha \in A} X_\alpha = \bigcup_{\alpha \in A} (Y - X_\alpha)$$

Proof: We will prove (1), leaving the remainder of the proof as an exercise. For ease of notation, we write $\bigcup_\alpha$ for $\bigcup_{\alpha \in A}$ and $\bigcap_\alpha$ for $\bigcap_{\alpha \in A}$.

We wish to show that $Y \cup (\bigcap_\alpha X_\alpha) = \bigcap_\alpha (Y \cup X_\alpha)$. This is the generalization to arbitrarily many sets of the identity we proved in Proposition 1.1.7. The reader should compare the proof we give here with that of Proposition 1.1.7.

Suppose $Y \cup (\bigcap_\alpha X_\alpha) \neq \emptyset$. If $x \in Y \cup (\bigcap_\alpha X_\alpha)$, then either $x \in Y$ or $x \in \bigcap_\alpha X_\alpha$. If $x \in Y$, then $x \in Y \cup X_\alpha$ for all $\alpha \in A$; if $x \in \bigcap_\alpha X_\alpha$, then $x \in X_\alpha$ for all $\alpha \in A$ and hence $x \in Y \cup X_\alpha$ for all $\alpha \in A$. Thus, in either case, $x \in Y \cup X_\alpha$ for all $\alpha \in A$, so $x \in \bigcap_\alpha (Y \cup X_\alpha)$. Therefore we have shown that if $Y \cup (\bigcap_\alpha X_\alpha) \neq \emptyset$, then

$$\left[Y \cup \left(\bigcap_\alpha X_\alpha \right) \right] \subset \left[\bigcap_\alpha (Y \cup X_\alpha) \right]$$

and the inclusion also holds if $Y \cup (\bigcap_\alpha X_\alpha) = \emptyset$.

Now suppose $\bigcap_\alpha (Y \cup X_\alpha) \neq \emptyset$. If $y \in \bigcap_\alpha (Y \cup X_\alpha)$, then $y \in Y \cup X_\alpha$ for all $\alpha \in A$. If $y \in Y$, then $y \in Y \cup (\bigcap_\alpha X_\alpha)$; if $y \notin Y$, then $y \in X_\alpha$ for all $\alpha \in A$, and hence $y \in \bigcap_\alpha X_\alpha$, which implies that $y \in Y \cup (\bigcap_\alpha X_\alpha)$. Therefore

$$\left[\bigcap_\alpha (Y \cup X_\alpha) \right] \subset \left[Y \cup \left(\bigcap_\alpha X_\alpha \right) \right]$$

if $\bigcap_\alpha (Y \cup X_\alpha) \neq \emptyset$, and since the inclusion also holds when $\bigcap_\alpha (Y \cup X_\alpha) = \emptyset$, we are done. □

To conclude this section, we offer a few remarks concerning "if and only if" proofs. (Some of the exercises which follow request "if and only if" proofs.) Suppose A and B are mathematical statements. Then the phrase "A if and only if B" means that A implies B and also that B implies A; that is, it means that if A is true, then B must be true and also that if B is true, then A must be true. Therefore in order to prove that a statement of the form "A if and only if B" holds, it is necessary to do two things: (a) assume that A is true and show that this implies the truth of B, and (b) assume that B is true and show that this implies the truth of A. Thus every "if and only if" proof comes in two parts, which are called the *converses* of one another.

EXERCISES

1. Let $S = \{a, b, c, d\}$, $T = \{a, c\}$, and $V = \{c, d, e\}$. Find

$$S \cup T \qquad S \cup T \cup V \qquad S \cap T \qquad S \cap T \cap V \qquad S \cup (T \cap V)$$

$$S \cap (T \cup V) \qquad S - T \qquad T - S \qquad S - (T \cap V) \qquad V - (S - T)$$

$$T \times V \qquad \text{and} \qquad V \times T$$

2. Let $S = \{n \in \boldsymbol{N} \mid n = 3k \text{ for some } k \in \boldsymbol{N}\}$, $T = \{n \in \boldsymbol{N} \mid n = 4k \text{ for some } k \in \boldsymbol{N}\}$, and $V = \{1, 2, 3, 5, 7, 11\}$. Find

$$S \cup T \qquad S \cap T \qquad S \cup T \cup V \qquad S \cap T \cap V$$

$$S - T \qquad T - S \qquad \text{and} \qquad V - (S \cup T)$$

3. Prove that any interval in $\boldsymbol{R}$ (open, closed, or half-open) can be written as the intersection of two properly chosen rays in $\boldsymbol{R}$.
4. How many elements are there in the set 0? In the set $\{0\}$? In the set $\{0, \{0\}\}$?
5. Let $S = \{a, b, c\}$. Let $V = \{T \mid T \text{ is a subset of } S\}$. Write out the elements of the set V.
6. Let $n \in \boldsymbol{N}$ and let $S = \{a_1, \ldots, a_n\}$. Prove that S has 2^n distinct subsets.
7. Let S and T be sets. Prove that $S - T = \emptyset$ if and only if $S \subset T$.
8. Let S and T be sets. Prove that $S = (S \cap T) \cup (S - T)$ and that $S \cap T$ and $S - T$ are disjoint.

9. Let S, T, T_1, and T_2 be sets, with $T = T_1 \cup T_2$. Prove that $S \times T = (S \times T_1) \cup (S \times T_2)$.
10. Let S, T, and V be sets.
    (a) Prove that if $V \subset T$, then $(S - T) \subset (S - V)$.
    (b) Suppose that both T and V are subsets of S. Prove that $V \subset T$ if and only if $(S - T) \subset (S - V)$.
    (c) Suppose that both T and V are subsets of S. Prove that $T = S - V$ if and only if $V = S - T$.
11. Prove (1) and (2) of Proposition 1.1.7.
12. Prove (3) of Proposition 1.1.7.
13. Prove the second distributive law in (4) of Proposition 1.1.7.
14. Prove (5) of Proposition 1.1.7.
15. (a) For each $n \in \boldsymbol{N}$, let X_n be the interval $[n - 1, n)$ in $\boldsymbol{R}$. Show that $\bigcup_{n \in N} X_n = [0, +\infty)$ and that $\bigcap_{n \in N} X_n = \emptyset$.
    (b) For each $n \in \boldsymbol{N}$, let X_n be the interval $(0, 1/2^n)$ in $\boldsymbol{R}$. Show that $\bigcup_{n \in N} X_n = (0, 1/2)$ and that $\bigcap_{n \in N} X_n = \emptyset$.
16. Let A be an index set. Prove that if the sets $\{X_\alpha \mid \alpha \in A\}$ are mutually disjoint, then $\bigcap_{\alpha \in A} X_\alpha = \emptyset$.
17. Prove Proposition 1.1.10.
18. Let A be an index set, and for each $\alpha \in A$, let X_α be an open interval in $\boldsymbol{R}$. Prove that if $\bigcap_{\alpha \in A} X_\alpha \neq \emptyset$, then $\bigcup_{\alpha \in A} X_\alpha$ is either an open interval or an open ray.

1.2 FUNCTIONS

In order to study sets more thoroughly, we need to be able to compare one set with another. This means that we must have some method of "pairing" their elements and leads us to the concept of a function from one set to another.

Definition 1.2.1 Let S and T be sets. A *function from S to T* is a subset f of $S \times T$ such that

1. For each $x \in S$ there is some $y \in T$ such that the ordered pair $(x, y) \in f$.
2. If (x, y) and (x, y') are elements of f, then $y = y'$ in T.

The set S is called the *domain* of the function and the set T the *range* of the function. If $S = T$, we refer to a function from S to S as a *function on S*.

Note that according to this definition, a function from S to T is a subset of the Cartesian product $S \times T$ and hence is a set of ordered pairs, with the first coordinate of each ordered pair being an element of the domain S and the second an element of the range T. Condition (1) of the definition says that every element of S must appear as the first coordinate of some ordered pair which belongs to the function; in other words, every element of the domain must get "used" by the function. Condition (2) says that no element of S can appear as the first coordinate in more than one ordered pair which belongs to the function; in other words, no element of the domain can get "used" more than once by the function. There are no restrictions on the use of the elements of the range T: the function may "use" some or all of the elements of T, and it may "use" them more than once.

Examples 1.2.2 1. Let $S = \{1, 2, 3\}$ and $T = \{\alpha, \beta, \gamma\}$, and consider the subsets of $S \times T$:

$$f = \{(1, \alpha), (2, \beta), (3, \beta)\} \qquad g = \{(1, \gamma), (2, \alpha), (3, \beta)\}$$

$$h = \{(2, \beta), (3, \gamma)\} \qquad \text{and} \qquad k = \{(1, \alpha), (1, \beta), (2, \gamma), (3, \gamma)\}$$

Since each element of S appears as the first coordinate of some ordered pair belonging to f, and no element of S appears as the first coordinate of more than one ordered pair belonging to f, f is a function from S to T. Note that not every element of the range T appears as the second coordinate of an ordered pair in f; note also that the element $\beta \in T$ appears as a second coordinate more than once.

The set g is also a function from S to T; in this case, every element of T does appear as the second coordinate of some ordered pair belonging to the function, and no element of T appears more than once.

The set h is *not* a function from S to T because it violates condition (1) of the definition: $1 \in S$ but there is no $y \in T$ such that $(1, y) \in h$. (However, h *is* a function from the set $\{2, 3\}$ to T.)

The set k is *not* a function from S to T because the element 1 of S appears as the first coordinate of more than one ordered pair belonging to k.

2. Consider the subset f of $\boldsymbol{N} \times \boldsymbol{Q}$ given by $f = \{(n, 1/n) \mid n \in \boldsymbol{N}\}$. We claim that f is a function from $\boldsymbol{N}$ to $\boldsymbol{Q}$. This follows because if $n \in \boldsymbol{N}$, then $(n, 1/n) \in f$, and if (n, y) and (n, y') both belong to f, then $y = 1/n = y'$.

3. Let

$$f = \{(x, x^2) \mid x \in \boldsymbol{R}\} \qquad \text{and} \qquad g = \{(x^2, x) \mid x \in \boldsymbol{R}\}$$

The subset f is a function on $\boldsymbol{R}$ (i.e., a function from $\boldsymbol{R}$ to $\boldsymbol{R}$), because if $x \in \boldsymbol{R}$, then $(x, x^2) \in f$ and if $(x, y) \in f$ and $(x, y') \in f$, then $y = x^2 = y'$. However, g is not a function on $\boldsymbol{R}$, because, for instance, $(1, 1)$ and $(1, -1)$ both belong to g.

4. Let

$$f = \{(x, x) \mid x \in \boldsymbol{R}, x \geqslant 0\} \cup \{(x, -x) \mid x \in \boldsymbol{R}, x < 0\}$$

then f is a function on $\boldsymbol{R}$, called the *absolute value function*. Figure 1.12 depicts f as a subset of $\boldsymbol{R} \times \boldsymbol{R}$: such a depiction is called the *graph* of the function. Every function whose domain and range are subsets of $\boldsymbol{R}$ has a graph.

We have defined a function from S to T as a subset of $S \times T$ having certain properties. It is often convenient to think of a function as a "rule of correspondence" or "mapping" which assigns to *each* element of its domain a *unique* element of its range. See Figure 1.13 for a depiction of this viewpoint. In line with this approach, we adopt the following notation: we will write $f: S \to T$ to signify that f is a function from S to T and write $f(x) = y$ if and only if the ordered pair $(x, y) \in f$.

Examples 1.2.3 1. Let $S = \{1, 2, 3\}$ and $T = \{\alpha, \beta, \gamma\}$, and let $f: S \to T$ be the function defined by

$$f(1) = \alpha \qquad \text{and} \qquad f(2) = f(3) = \beta$$

This is the function f of the first example in Examples 1.2.2. The function

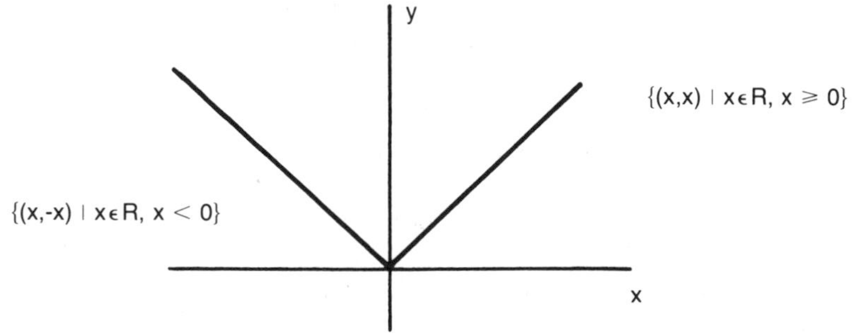

Figure 1.12 The graph of the absolute value function.

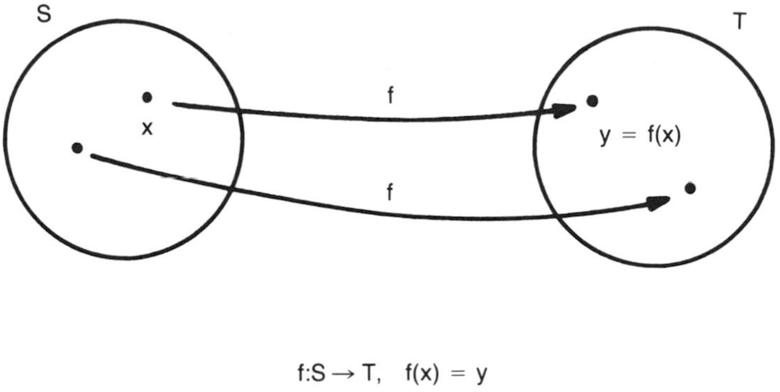

Figure 1.13 A function as a mapping.

g of the same example may be written as $g\colon S \to T$, where

$$g(1) = \gamma \qquad g(2) = \alpha \qquad g(3) = \beta$$

2. Let $f\colon \boldsymbol{R} \to \boldsymbol{R}$ be given by $f(x) = x^2$ for all $x \in \boldsymbol{R}$. This is the function f of the third example in Examples 1.2.2. Its graph is depicted in Figure 1.14.

3. Let $f\colon \boldsymbol{R} \to \boldsymbol{R}$ be given by $f(x) = |x|$ for all $x \in \boldsymbol{R}$, where by definition,

$$|x| = \begin{cases} x & \text{if } x \geqslant 0 \\ -x & \text{if } x < 0 \end{cases}$$

This is the absolute value function introduced in Example 4 of Examples 1.2.2.

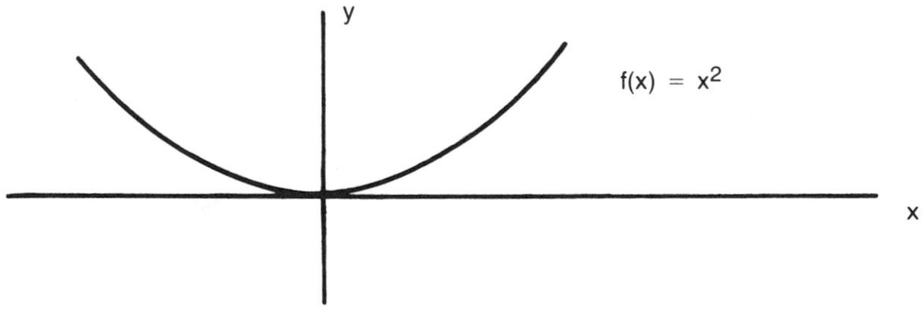

Figure 1.14 The graph of $f(x) = x^2$.

4. Let us show that the correspondence from $S=[0,+\infty)$ to $T=[0,+\infty)$ given by $f(x)=\sqrt{x}$ for all $x\in[0,+\infty)$ is a function. Here $\sqrt{x}$ denotes the nonnegative square root of x. We must show that

a. For each x in the domain, $f(x)$ is defined and belongs to the range.
b. If $f(x)=y$ and $f(x)=y'$, then $y=y'$.

But if $x\in[0,+\infty)$, then $x\geqslant 0$, so $f(x)=\sqrt{x}\geqslant 0$, and thus $f(x)$ is defined and belongs to $[0,+\infty)$. Hence (a) is satisfied. Furthermore, if $f(x)=y$ and $f(x)=y'$, then $y=\sqrt{x}=y'$, so (b) is satisfied. Therefore $f(x)=\sqrt{x}$ is a function from $[0,+\infty)$ to $[0,+\infty)$.

Now we are ready to consider two special types of functions: onto functions and one-to-one functions.

Definition 1.2.4 Let S and T be sets and let $f\colon S\to T$.

1. If for every $y\in T$ there is some $x\in S$ such that $f(x)=y$, then f is said to be a function from S *onto* T.
2. If $f(x)=f(z)$ implies that $x=z$, then f is said to be a *one-to-one function.*

Thus $f\colon S\to T$ is onto if every element of its range gets "used," and it is one-to-one if no two elements of its domain are assigned to the same element of its range (see Figures 1.15 and 1.16).

Examples 1.2.5 1. The function f of Example 1 in Examples 1.2.3 is not a function of S onto T because $\gamma\in T$ and there is no $x\in S$ such that $f(x)=\gamma$. Furthermore, f is not one-to-one because $f(2)=\beta$ and $f(3)=\beta$. However, the function g of the same example is both onto and one-to-one.

2. The function of Example 2 in Examples 1.2.3 is not a function of $\boldsymbol{R}$ onto $\boldsymbol{R}$ because, for instance, $-1\in\boldsymbol{R}$ but there is no $x\in\boldsymbol{R}$ such that $f(x)=-1$. It is not one-to-one because, for instance, $f(-1)=f(1)=1$.

3. The absolute value function of Example 3 in Examples 1.2.3 is neither onto $\boldsymbol{R}$ nor one-to-one. (Why?) The square root function of Example 4 in Examples 1.2.3 is onto $[0,+\infty)$ because if $y\in[0,+\infty)$, then

$$y^2\in[0,+\infty) \qquad\text{and}\qquad f(y^2)=\sqrt{y^2}=y$$

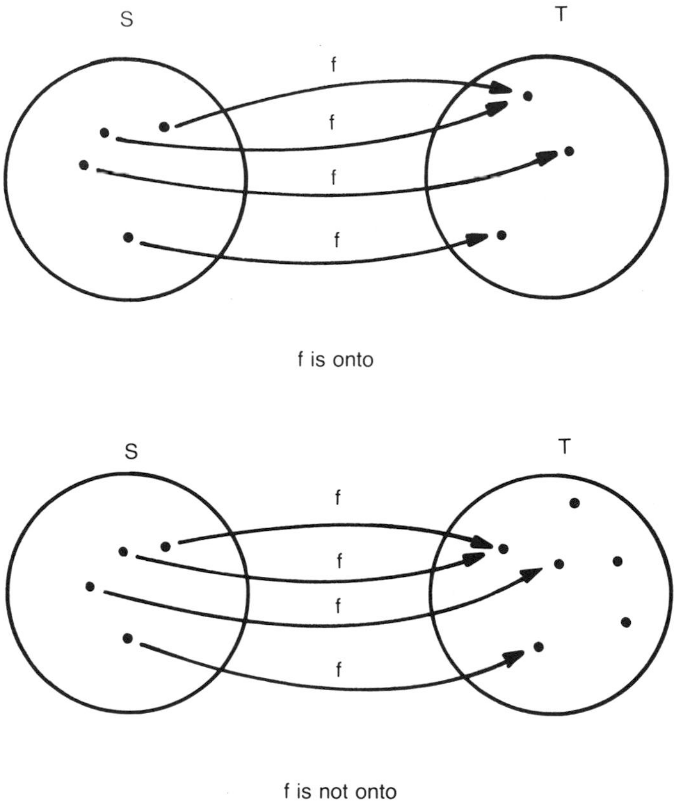

Figure 1.15 The concept of an onto function.

It is one-to-one because if $x \in [0, +\infty)$, $z \in [0, +\infty)$ and $\sqrt{x} = \sqrt{z}$, then $x = z$.

4. Let S and T be any nonempty sets and let $c \in T$. Let $f: S \to T$ be given by $f(x) = c$ for all $x \in S$. Such a function is called a *constant function*. A constant function cannot be onto if its range contains more than one element; it cannot be one-to-one if its domain contains more than one element.

If $f: S \to T$ and U is a subset of S, we can consider the subset of T which consists of all elements $y \in T$ such that $y = f(x)$ for some $x \in U$. Similarly, if V is a subset of T, we can consider the subset of S which consists of all elements x of S such that $f(x) \in V$.

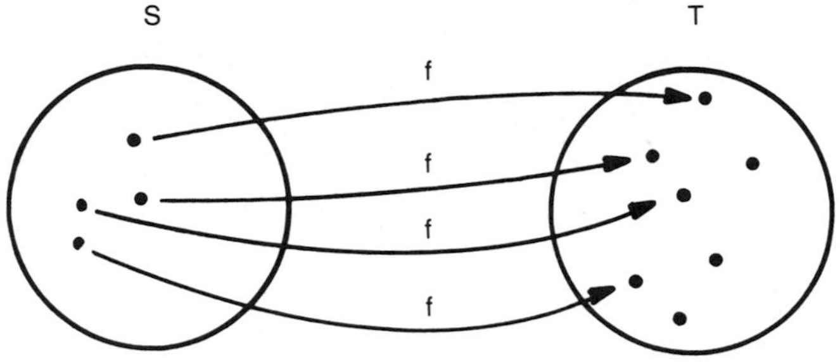

f is one-to-one

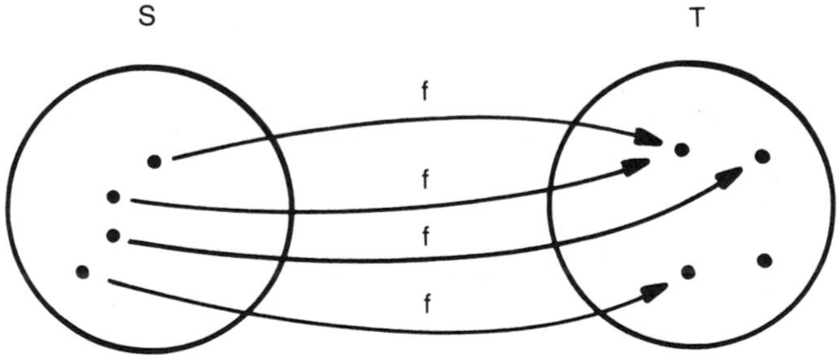

f is not one-to-one

Figure 1.16 The concept of a one-to-one function

Definition 1.2.6 Let $f\colon S \to T$, and let $U \subset S$ and $V \subset T$.

1. The *image of U under f*, denoted by $f(U)$, is the subset of T defined by

$$f(U) = \{f(x) \in T \mid x \in U\}$$

2. The *preimage of V under f*, denoted by $f^{-1}(V)$, is the subset of S defined by

$$f^{-1}(V) = \{x \in S \mid f(x) \in V\}$$

(see Figure 1.17).

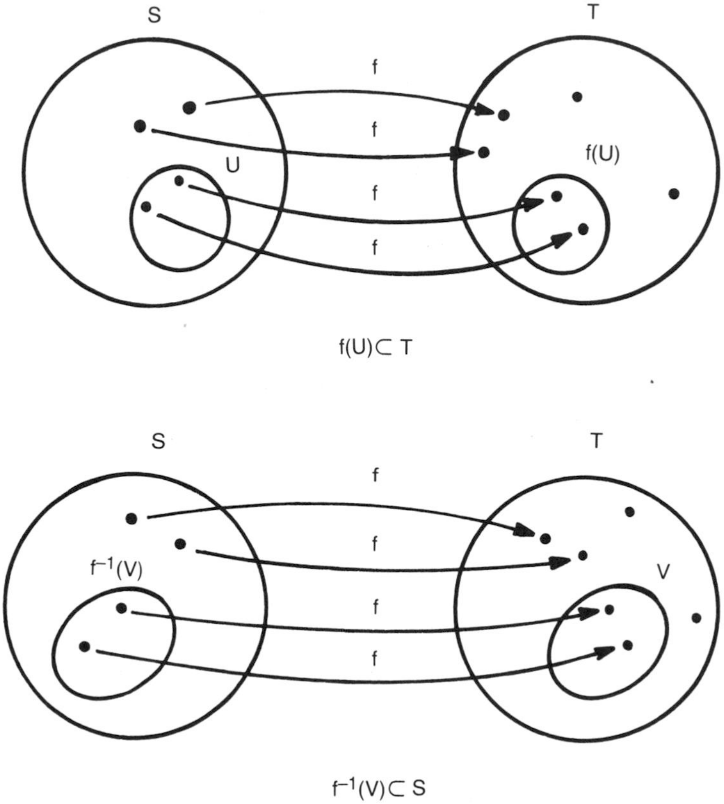

Figure 1.17 Image and preimage.

Examples 1.2.7 1. Let $S = \{a, b, c\}$, $T = \{1, 2, 3, 4\}$, and $f\colon S \to T$ be given by

$$f(a) = 3 \qquad f(b) = 4 \qquad f(c) = 3$$

If $U = \{a, b\}$, then $f(U) = \{3, 4\}$. If $V = \{1, 2, 4\}$, then $f^{-1}(V) = \{b\}$.

2. Let S and T be any sets and $f\colon S \to T$ any function from S to T. We have $f(\emptyset) = \emptyset$, $f^{-1}(\emptyset) = \emptyset$, $f^{-1}(T) = S$, and $f(U) \subset f(S)$ for all $U \subset S$. (Check these facts.)

3. Let $f\colon \boldsymbol{R} \to \boldsymbol{R}$ be given by $f(x) = x^2$ for all $x \in \boldsymbol{R}$. We will show that the image of $[0, 2]$ under f is $[0, 4]$ and that the preimage of $[0, 4]$ under f is $[-2, 2]$.

By definition, $f([0,2]) = \{x^2 \in \boldsymbol{R} \mid x \in [0,2]\}$. If $x \in [0,2]$, then certainly $f(x) = x^2 \in [0,4]$; hence $f([0,2]) \subset [0,4]$. On the other hand, if $x \in [0,4]$, then $\sqrt{x} \in [0,2]$ and $f(\sqrt{x}) = x \in [0,4]$; hence $[0,4] \subset f([0,2])$. Therefore $f([0,2]) = [0,4]$.

Now consider $f^{-1}([0,4]) = \{x \in \boldsymbol{R} \mid x^2 \in [0,4]\}$. If $x \in [-2,2]$, then clearly $f(x) = x^2 \in [0,4]$; hence $[-2,2] \subset f^{-1}([0,4])$. On the other hand, if $f(x) = x^2 \in [0,4]$, then $x = \pm\sqrt{x^2} \in [-2,2]$, so $f^{-1}([0,4]) \subset [-2,2]$. Therefore $f^{-1}([0,4]) = [-2,2]$.

Our next proposition tells us how set operations are affected by functions. It states that taking preimages preserves containments, unions, intersections, and complements, while taking images preserves containments and unions. Intersections and complements are not in general preserved by taking images.

Proposition 1.2.8 Let S and T be sets and $f: S \to T$. Let U_1 and U_2 be subsets of S and V_1 and V_2 be subsets of T.

1. If $U_1 \subset U_2$, then $f(U_1) \subset f(U_2)$.
   If $V_1 \subset V_2$, then $f^{-1}(V_1) \subset f^{-1}(V_2)$.
2. $f(U_1 \cup U_2) = f(U_1) \cup f(U_2)$
   $f^{-1}(V_1 \cup V_2) = f^{-1}(V_1) \cup f^{-1}(V_2)$
3. $f(U_1 \cap U_2) \subset [f(U_1) \cap f(U_2)]$
   $f^{-1}(V_1 \cap V_2) = f^{-1}(V_1) \cap f^{-1}(V_2)$
4. $f(U_1 - U_2) \subset f(U_1)$
   $f^{-1}(V_1 - V_2) = f^{-1}(V_1) - f^{-1}(V_2)$

Proof: We will prove (3), leaving the remainder of the proof as an exercise.

Suppose $U_1 \cap U_2 \neq \emptyset$ and let $x \in U_1 \cap U_2$; then $x \in U_1$ and $x \in U_2$, so $f(x) \in f(U_1)$ and $f(x) \in f(U_2)$. Thus $f(x) \in f(U_1) \cap f(U_2)$. Hence $f(U_1 \cap U_2) \subset [f(U_1) \cap f(U_2)]$ when $U_1 \cap U_2 \neq \emptyset$, and this inclusion also holds when $U_1 \cap U_2 = \emptyset$. (Why?)

Now assume that $f^{-1}(V_1 \cap V_2) \neq \emptyset$ and let $x \in f^{-1}(V_1 \cap V_2)$. Then $f(x) \in V_1 \cap V_2$, and therefore $x \in f^{-1}(V_1)$ and also $x \in f^{-1}(V_2)$. Hence $x \in f^{-1}(V_1) \cap f^{-1}(V_2)$, and thus when $f^{-1}(V_1 \cap V_2) \neq \emptyset$, then $f^{-1}(V_1 \cap V_2) \subset [f^{-1}(V_1) \cap f^{-1}(V_2)]$. This inclusion also holds when $f^{-1}(V_1 \cap V_2) = \emptyset$. (Why?)

Finally, if $f^{-1}(V_1) \cap f^{-1}(V_2) \neq \emptyset$ and $x \in f^{-1}(V_1) \cap f^{-1}(V_2)$, then $x \in f^{-1}(V_1)$ and $x \in f^{-1}(V_2)$, so $f(x) \in V_1$ and $f(x) \in V_2$. Therefore $f(x) \in V_1 \cap V_2$, which implies that $x \in f^{-1}(V_1 \cap V_2)$. Hence when

$f^{-1}(V_1) \cap f^{-1}(V_2) \neq \emptyset$, we have

$$[f^{-1}(V_1) \cap f^{-1}(V_2)] \subset f^{-1}(V_1 \cap V_2)$$

and this inclusion also holds when $f^{-1}(V_1) \cap f^{-1}(V_2) = \emptyset$. Therefore $f^{-1}(V_1 \cap V_2) = f^{-1}(V_1) \cap f^{-1}(V_2)$. □

Example 1.2.9 Let $f\colon \mathbf{R} \to \mathbf{R}$ be given by $f(x) = x^2$ for all $x \in \mathbf{R}$. Let $U_1 = [-2, 0]$ and $U_2 = [0, 2]$. We have $U_1 \cap U_2 = \{0\}$ and thus

$$f(U_1 \cap U_2) = f(0) = 0$$

whereas

$$f(U_1) \cap f(U_2) = [0, 4] \cap [0, 4] = [0, 4]$$

This shows that in general $f(U_1 \cap U_2) \neq f(U_1) \cap f(U_2)$. Of course Proposition 1.2.8 says that $f(U_1 \cap U_2) \subset [f(U_1) \cap f(U_2)]$, and this is certainly the case here.

Now consider $f(U_1 - U_2)$ and $f(U_1) - f(U_2)$. We have

$$f(U_1 - U_2) = f([-2, 0)) = (0, 4]$$

whereas

$$f(U_1) - f(U_2) = [0, 4] - [0, 4] = \emptyset$$

Thus in general $f(U_1 - U_2) \neq f(U_1) - f(U_2)$. Note that Proposition 1.2.8 states that $f(U_1 - U_2) \subset f(U_1)$, and that this is so here.

We now turn to the question of how to use known functions to create new ones. Our first result in this regard says that we can create a new function from an old one simply by restricting the domain of the old one.

Proposition 1.2.10 Let $f\colon S \to T$ be a function from a set S to a set T and let U be a subset of S. Define g by setting $g(x) = f(x)$ for all $x \in U$; then $g\colon U \to T$ is a function called the *restriction of f to U*. Furthermore, if f is one-to-one, so is g.

The proof of Proposition 1.2.10 will be left as an exercise.

Example 1.2.11 Let $f\colon \mathbf{R} \to \mathbf{R}$ be the absolute value function. If $U \subset [0, +\infty)$, the restriction of f to U is the function $g\colon U \to \mathbf{R}$ given by $g(x) = x$ for all $x \in U$. If $U \subset (-\infty, 0)$, the restriction of f to U is the function $g\colon U \to \mathbf{R}$ given by $g(x) = -x$ for all $x \in U$.

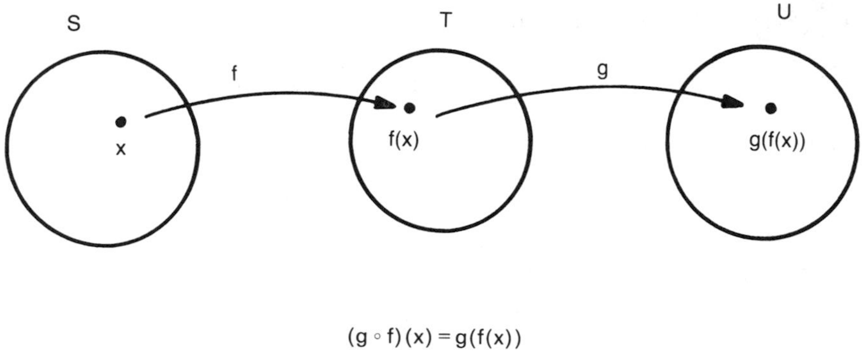

Figure 1.18 Composition of functions.

Next we show how to create new functions from old by means of a process called *composition of functions*. The idea of composition of functions is this: if $f: S \to T$ and $g: T \to U$, then for each $x \in S$, the element $f(x) \in T$, and we can apply g to $f(x)$ to obtain a unique element of U. In this manner we can assign to each element of S a unique element of U, thus constructing a function from S to U. Figure 1.18 depicts the process of composition of functions.

Proposition 1.2.12 Let S, T, and U be sets, let $f: S \to T$ and $g: T \to U$, and define $g \circ f$ as follows:

$$(g \circ f)(x) = g(f(x)) \qquad \text{for all } x \in S$$

Then $(g \circ f): S \to U$ is a function called the *composite of g with f*. In both f and g are onto, so is $g \circ f$; if both f and g are one-to-one, so is $g \circ f$.

Proof: To show that $g \circ f$ is a function from S to U, we must show that for each $x \in S$, $(g \circ f)(x)$ is a unique element of U. But since f is a function from S to T, $f(x)$ is a unique element of T for each $x \in S$ and since g is a function from T to U, $g(f(x))$ is a unique element of U. This establishes that $g \circ f$ is a function from S to U.

Now we show that if f and g are both onto, so is $g \circ f$. If $z \in U$, then because g is onto, there is some $y \in T$ such that $g(y) = z$. But similarly, since f is onto, there is some $x \in S$ such that $f(x) = y$. Therefore

$$(g \circ f)(x) = g(f(x)) = g(y) = z$$

Finally we prove that if f and g are both one-to-one, so is $g \circ f$. If $(g \circ f)(x) = (g \circ f)(x')$, for some $x \in S$, $x' \in S$, then $g(f(x)) = g(f(x'))$. But single g is one-to-one, we must have $f(x) = f(x')$, and then since f is one-to-one, it follows that $x = x'$. □

Examples 1.2.13 1. Let $S = \{1, 2, 3\}$, $T = \{\alpha, \beta, \gamma\}$, and $U = \{u, v, w\}$. Let $f: S \to T$ and $g: T \to U$ be given by

$$f(1) = \beta \qquad f(2) = \alpha \qquad f(3) = \beta \qquad \text{and} \qquad g(\alpha) = g(\beta) = u \qquad g(\gamma) = w$$

We then have $g \circ f: S \to U$ given by

$$(g \circ f)(1) = g(f(1)) = g(\beta) = u$$

$$(g \circ f)(2) = g(f(2)) = g(\alpha) = u$$

and

$$(g \circ f)(3) = g(f(3)) = g(\beta) = u$$

Hence $g \circ f$ is a constant function from S to U.

2. Let $f: \boldsymbol{R} \to \boldsymbol{R}$ and $g: \boldsymbol{R} \to \boldsymbol{R}$ be given by $f(x) = 2x$ and $g(x) = x + 1$, for all $x \in \boldsymbol{R}$. We then have $(g \circ f): \boldsymbol{R} \to \boldsymbol{R}$ given by

$$(g \circ f)(x) = g(f(x)) = g(2x) = 2x + 1$$

Since f and g are both one-to-one and onto, $g \circ f$ is one-to-one and onto. In this case we can also form the composite function $f \circ g$. Thus we have $f \circ g$ given by

$$(f \circ g)(x) = f(g(x)) = f(x + 1) = 2(x + 1)$$

and $f \circ g$ is one-to-one and onto. Note that although both composite functions $g \circ f$ and $f \circ g$ are defined, they are not equal.

The final method that we will consider for creating new functions from old ones is that of forming the *inverse function* of a given one-to-one function. We do this as follows: if $f: S \to T$ is one-to-one, then each $y \in f(S)$ corresponds to a unique $x \in S$, namely, the unique element x such that $f(x) = y$. The inverse of f is a function from $f(S)$ to S, which assigns the element x to the element y. See Figure 1.19.

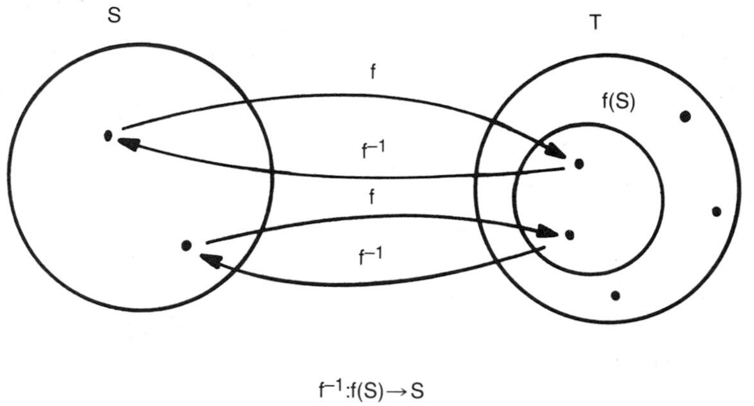

Figure 1.19 Inverse function.

Proposition 1.2.14 Let S and T be sets, let $f\colon S \to T$ be a one-to-one function, and for each $y \in f(S)$, define $f^{-1}(y) = x$ if and only if $x \in S$ and $f(x) = y$; then $f^{-1}\colon f(S) \to S$ is a one-to-one onto function called the *inverse function of f*. Furthermore,

$$(f^{-1} \circ f)(x) = x \qquad \text{for all } x \in S$$

and

$$(f \circ f^{-1})(y) = y \qquad \text{for all } y \in f(S)$$

Proof: To show that f^{-1} is a function from $f(S)$ to S, we must show that if $y \in f(S)$, then $f^{-1}(y)$ is a unique element of S. But by definition, $f^{-1}(y) = x$, where $x \in S$ and $f(x) = y$. Since $y \in f(S)$, such an element x exists, and if x were not unique, that is, if there were an element $z \in S$, $z \neq x$, such that $f(z) = y$, then f would not be one-to-one.

To show that f^{-1} is onto, note that if $x \in S$, then $f(x) = y \in f(S)$, and thus $f^{-1}(y) = x$. To show that f^{-1} is one-to-one, suppose that $y \in f(S)$ and also $y' \in f(S)$, with $f^{-1}(y) = f^{-1}(y')$; then there is some $x \in S$ such that $f(x) = y$ and $f(x) = y'$. But since f is a function, we must have $y = y'$.

To conclude the proof, let $x \in S$ and set $y = f(x)$; then

$$(f^{-1} \circ f)(x) = f^{-1}(f(x)) = f^{-1}(y) = x$$

Similarly, if $y \in f(S)$ and $x = f^{-1}(y)$, then $f(x) = y$ and

$$(f \circ f^{-1})(y) = f(f^{-1}(y)) = f(x) = y \quad \square$$

We emphasize here that the inverse function f^{-1} exists if and only if the function f is one-to-one; if f is not one-to-one, we cannot form f^{-1}. Also, the inverse function f^{-1} should not be confused with the preimage $f^{-1}(V)$ of a subset V of the range of f: this preimage is a subset of the domain, and it always exists, whether or not f is one-to-one.

Examples 1.2.15 1. Let $S = \{1, 2\}$, $T = \{\alpha, \beta, \gamma\}$, and let $f\colon S \to T$ be given by $f(1) = \alpha$, $f(2) = \beta$. Since f is one-to-one, f^{-1} exists, and $f^{-1}\colon \{\alpha, \beta\} \to S$ is given by $f^{-1}(\alpha) = 1, f^{-1}(\beta) = 2$.

2. Let $f\colon \boldsymbol{R} \to \boldsymbol{R}$ be given by $f(x) = x + 1$ for all $x \in \boldsymbol{R}$. Since f is one-to-one, f^{-1} exists. Furthermore, since f is onto (why?), $f(\boldsymbol{R}) = \boldsymbol{R}$ and hence $f^{-1}\colon \boldsymbol{R} \to \boldsymbol{R}$. Since $(f \circ f^{-1})(x) = x$ for all $x \in \boldsymbol{R}$, we have

$$x = (f \circ f^{-1})(x) = f(f^{-1}(x)) = f^{-1}(x) + 1$$

for all $x \in \boldsymbol{R}$. Therefore $f^{-1}(x) = x - 1$ for all $x \in \boldsymbol{R}$. Note also that

$$(f^{-1} \circ f)(x) = f^{-1}(f(x)) = f^{-1}(x + 1) = (x + 1) - 1 = x$$

as required.

3. Let $f\colon [0, +\infty) \to [0, +\infty)$ be given by $f(x) = \sqrt{x}$ for all $x \in [0, +\infty)$. Since f is one-to-one and onto, $f^{-1}\colon [0, +\infty) \to [0, +\infty)$. Thus for all $x \in [0, +\infty)$,

$$x = (f \circ f^{-1})(x) = f(f^{-1}(x)) = \sqrt{f^{-1}(x)}$$

which implies that $f^{-1}(x) = x^2$ for all $x \in [0, +\infty)$.

EXERCISES

1. For each of the following, show that f either is or is not a function from S to T. If it is a function and its domain and range are subsets of $\boldsymbol{R}$, draw its graph.
   (a) $S = \{1, 2, 3, 4\}$, $T = \{\alpha, \beta, \gamma\}$,
   $f = \{(1, \alpha), (2, \beta), (3, \beta), (4, \alpha), (4, \gamma)\}$.
   (b) $S = \{1, 2, 3, 4\}$, $T = \{\alpha, \beta, \gamma\}$,
   $f = \{(1, \alpha), (2, \alpha), (3, \beta), (4, \gamma)\}$.
   (c) $S = \boldsymbol{Q}$, $T = \boldsymbol{Z}$, $f = \{(n/m, n) \mid n \in \boldsymbol{Z}, m \in \boldsymbol{Z}, m \neq 0\}$.
   (d) $S = \boldsymbol{Q}$, $T = \boldsymbol{Z}$, $f = \{(n/m, n + m) \mid n \in \boldsymbol{Z}, m \in \boldsymbol{Z}, m \neq 0\}$.
   (e) $S = T = \boldsymbol{R}$, $f = \{(x, y) \mid x \in \boldsymbol{R}, y \in \boldsymbol{R}, x^2 + y^2 = 1\}$.

(f) $S = T = \boldsymbol{R}, f = \{(x, ax + b) \mid x \in \boldsymbol{R}\}$. Here a and b are fixed real numbers.
(g) $S = T = \boldsymbol{R}, f = \{(x, \sqrt{x^2 + 1}) \mid x \in \boldsymbol{R}\}$.
(h) $S = T = \boldsymbol{R}$, f given by $f(x) = x^2 + x + 1$ for all $x \in \boldsymbol{R}$.
(i) $S = T = \boldsymbol{R}$, f given by

$$f(x) = \begin{cases} x & \text{if } x \neq 0 \\ 1 & \text{if } x = 0 \end{cases}$$

(j) $S = T = \boldsymbol{R}$, f given by

$$f(x) = \begin{cases} x & \text{if } x \geqslant 0 \\ x + 1 & \text{if } x < 0 \end{cases}$$

2. For each of the functions of Exercise 1, state whether or not the function is onto and whether or not it is one-to-one. Justify your statements.
3. Let $n \in \boldsymbol{N}$, $m \in \boldsymbol{N}$, and let S be a set which has n distinct elements and T a set which has m distinct elements. Let $f: S \to T$. Prove each of the following:
(a) If f is onto, then $n \geqslant m$.
(b) If f is one-to-one, then $n \leqslant m$.
4. Let $S = \{1, 2, 3\}$, $T = \{\alpha, \beta, \gamma, \delta\}$, and let $f: S \to T$, $g: S \to T$ be given by

$$f(1) = \gamma \qquad f(2) = \beta \qquad f(3) = \alpha$$

and

$$g(1) = g(2) = \gamma \qquad g(3) = \delta$$

(a) State whether f and g are one-to-one and onto.
(b) If $U = \{1, 3\}$ and $V = \{\alpha, \gamma, \delta\}$, find $f(U)$, $g(U)$, $f^{-1}(V)$, and $g^{-1}(V)$.
5. Let $f: \boldsymbol{R} \to \boldsymbol{R}$, $g: \boldsymbol{R} \to \boldsymbol{R}$, and $h: \boldsymbol{R} \to \boldsymbol{R}$ be given by

$$f(x) = 2x + 1 \quad \text{for all } x \in \boldsymbol{R} \qquad g(x) = x^2 + 1 \quad \text{for all } x \in \boldsymbol{R}$$

and

$$h(x) = \begin{cases} 2x & \text{if } x \geqslant \frac{1}{2} \\ x - 1 & \text{if } x < \frac{1}{2} \end{cases}$$

(a) Find $f([0, 1])$, $g((0, 1])$, and $h((0, 1))$.
(b) Find $f^{-1}([0, 1])$, $g^{-1}((0, 1])$, and $h^{-1}((0, 1))$.

6. Prove (1) of Proposition 1.2.8.
7. Prove (2) of Proposition 1.2.8.
8. Prove (4) of Proposition 1.2.8.
9. Prove Proposition 1.2.10.
10. Let $S = \{1, 2, 3, 4\}$, $T = \{\alpha, \beta, \gamma\}$, and $U = \{u, v, w, z\}$. Let $f: S \to T$ and $g: T \to U$ be given by

$$f(1) = \beta \qquad f(2) = \gamma \qquad f(3) = \alpha \qquad f(4) = \beta$$

and

$$g(\alpha) = v \qquad g(\beta) = z \qquad g(\gamma) = u$$

Write out the defining equations for $f \circ g$.

11. Let $f: \boldsymbol{R} \to \boldsymbol{R}$ be given by $f(x) = x^2$ for all $x \in \boldsymbol{R}$. Let $g: \boldsymbol{R} \to \boldsymbol{R}$ be given by $g(x) = 2x + 1$ for all $x \in \boldsymbol{R}$.
(a) Write out the defining equations for $f \circ g$ and $g \circ f$. Is $f \circ g = g \circ f$?
(b) Write out the defining equations for $f^2 = f \circ f$, $g^2 = g \circ g$, $f^3 = f \circ f^2$, and $g^3 = g \circ g^2$.
12. Give an example of functions $f: \boldsymbol{R} \to \boldsymbol{R}$ and $g: \boldsymbol{R} \to \boldsymbol{R}$ such that $f \neq g$ but $f \circ g = g \circ f$.
13. For each of the functions of Exercise 1 above, state whether or not the inverse function f^{-1} exists. If f^{-1} does exist, find a description or defining equation for it. Check your answer by computing $(f \circ f^{-1})(y)$ and $(f^{-1} \circ f)(x)$.
14. Let S, T, and V be sets, with $V \subset T$, and let $f: S \to T$. Prove that $f^{-1}(V) = \emptyset$ if and only if $V \cap f(S) = \emptyset$.

1.3 FINITE AND INFINITE SETS

Suppose we have two sets and we wish to compare the number of elements which they contain. One way to do this is to pair the elements of one set with those of the other in a one-to-one manner: if every element of the first set is paired with an element of the second set, and if each element of the second set is paired with some element of the first one, then the sets must have the same number of elements. We make this notion precise with the following definition.

Definition 1.3.1 A set S is *equivalent* to a set T if there exists a function $f: S \to T$ which is one-to-one and onto.

Examples 1.3.2 1. Any set S is equivalent to itself, since $f: S \to S$ defined by $f(x) = x$ for all $x \in S$ is a one-to-one onto function.

2. Let $\boldsymbol{E} = \{2k \mid k \in \boldsymbol{N}\} = \{2, 4, 6, \ldots\}$, and let $f: \boldsymbol{N} \to \boldsymbol{E}$ be given by $f(n) = 2n$ for all $n \in \boldsymbol{N}$. It is easy to check that f is one-to-one and onto. Hence the set $\boldsymbol{N}$ of natural numbers is equivalent to the set $\boldsymbol{E}$ of even natural numbers.

3. For all $z \in \boldsymbol{Z}$, let $f: \boldsymbol{Z} \to \boldsymbol{N}$ be given by

$$f(z) = \begin{cases} 2z & \text{if } z \in \boldsymbol{N} \\ -2z + 1 & \text{if } z \in \boldsymbol{Z} - \boldsymbol{N} \end{cases}$$

Since f is a one-to-one, onto function (check this), $\boldsymbol{Z}$ is equivalent to $\boldsymbol{N}$.

4. Let $[a, b]$ and $[c, d]$ be closed intervals in $\boldsymbol{R}$, with $a < b$ and $c < d$. Let $f: [a, b] \to [c, d]$ be given by

$$f(x) = \frac{d - c}{b - a}(x - a) + c$$

for all $x \in [a, b]$. The function f is one-to-one and onto (check this), and thus any two closed intervals in $\boldsymbol{R}$ which contain more than one point are equivalent. The same function may be used to show that any two open intervals in $\boldsymbol{R}$ are equivalent.

Now we are ready to give a rigorous definition of finite and infinite sets.

Definition 1.3.3 A set is *infinite* if it is equivalent to a proper subset of itself. A set is *finite* if it is not infinite.

Examples 1.3.4 1. The set $\boldsymbol{N}$ of natural numbers is infinite because it is equivalent to the set $\boldsymbol{E}$ of even natural numbers, and $\boldsymbol{E}$ is a proper subset of $\boldsymbol{N}$.

2. The empty set is finite, because if it were not, it would be equivalent to a proper subset of itself, and this is impossible since the empty set has no proper subsets.

3. For any $n \in \boldsymbol{N}$, the set $\{1, \ldots, n\}$ is finite. (See Exercise 5 at the end of this section.)

4. Let $[a, b]$ be a closed interval in $\boldsymbol{R}$, with $a < b$. Choose real numbers c and d such that $a < c < d < b$. Since the interval $[c, d]$ is a proper subset of $[a, b]$ and since any two closed intervals having more than one point are equivalent by Example 4 of Examples 1.3.2, it follows that $[a, b]$ is infinite. Thus any closed interval having more than one point is infinite. Similar arguments show that open intervals, half-open intervals, and rays in $\boldsymbol{R}$ are infinite.

Intuitively, we may think of equivalent sets as being of the same "size." Thus if two sets are equivalent, then either both of them must be finite or both must be infinite. Furthermore, if $S \subset T$ and T is finite, then S must be at least as "small" as T, and hence S must itself be finite. On the other hand, if $S \subset T$ and S is infinite, then T must be at least as "large" as S and hence must be infinite. The following proposition and its corollary prove these facts.

Proposition 1.3.5

1. Any set equivalent to an infinite set is infinite.
2. Any set which contains an infinite set is infinite.

Proof: 1. If the set S is equivalent to the infinite set T, then there exists a one-to-one function f from S onto T and also a one-to-one function g from T onto some proper subset V of T. Consider

$$(f^{-1} \circ g) \circ f: S \to f^{-1}(V)$$

See Figure 1.20. The function $(f^{-1} \circ g) \circ f$ is one-to-one and onto (why?),

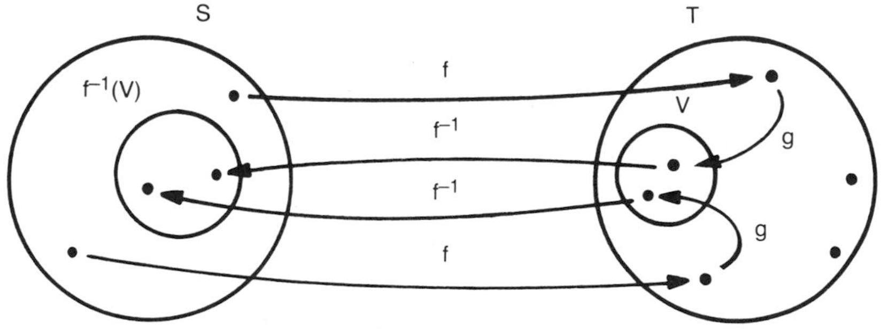

Figure 1.20 $(f^{-1} \circ g) \circ f: S \to f^{-1}(V)$.

and since $T - V \neq \emptyset$, we have

$$S - f^{-1}(V) = f^{-1}(T) - f^{-1}(V) = f^{-1}(T - V) \neq \emptyset$$

Thus $(f^{-1} \circ g) \circ f$ is a one-to-one function from S onto a proper subset of S, so S is infinite.

2. Let T contain the infinite set S. Since S is infinite, there exists a one-to-one function f from S onto some proper subset U of S. Define $g: T \to U$ as follows:

$$g(x) = \begin{cases} f(x) & \text{if } x \in S \\ x & \text{if } x \in T - S \end{cases}$$

It is easy to check that g is a one-to-one function from T onto its proper subset $(T - S) \cup U$. Hence T is infinite. □

Corollary 1.3.6

1. Any set equivalent to a finite set is finite.
2. Any set which is contained in a finite set is finite.

The proof of the corollary is left as an exercise.

Example 1.3.7 The set $\boldsymbol{Z}$ of integers, the set $\boldsymbol{Q}$ of rational numbers, and the set $\boldsymbol{R}$ of real numbers are all infinite sets because each of them contains the infinite set $\boldsymbol{N}$.

There is more than one kind of infinity; that is, infinite sets differ in the type or "intensity" of their infiniteness. As we shall see, the sets $\boldsymbol{N}$, $\boldsymbol{Z}$, and $\boldsymbol{Q}$ all possess the same type of infinity, that which is the least "intense." The set $\boldsymbol{R}$ of real numbers, on the other hand, has a different and more "intense" type of infiniteness. Our next result is the first step toward a classification of the different types of infinity.

Proposition 1.3.8 If S is a nonempty set, then either there is some $n \in \boldsymbol{N}$ such that S is equivalent to $\{1, \ldots, n\}$, or else S contains a subset equivalent to $\boldsymbol{N}$.

Proof: Suppose that S is not equivalent to $\{1, \ldots, n\}$, for any $n \in \boldsymbol{N}$. Since S is nonempty, we may choose an element $x_1 \in S$. If $S - \{x_1\} = \emptyset$, then $S = \{x_1\}$; but since the set $\{x_1\}$ is obviously equivalent to $\{1\}$, this cannot occur. Hence $S - \{x_1\} \neq \emptyset$.

Now suppose $n \geqslant 1$ and assume that we have selected n distinct elements $x_1, \ldots, x_n$ of S. If $S - \{x_1, \ldots, x_n\} = \emptyset$, then S would be equivalent to $\{1, \ldots, n\}$. Therefore $S - \{x_1, \ldots, x_n\} \neq \emptyset$, and we may select an element $x_{n+1} \in S$ such that $x_{n+1} \notin \{x_1, \ldots, x_n\}$. Thus by induction we produce a subset $\{x_1, \ldots, x_n, x_{n+1}, \ldots\}$ of distinct elements of S, and this subset is clearly equivalent to $\boldsymbol{N}$ by means of the function f defined by $f(x_n) = n$ for all $n \in \boldsymbol{N}$. □

Corollary 1.3.9 Every finite set is equivalent to $\{1, \ldots, n\}$, for some $n \in \boldsymbol{N}$. Every infinite set contains a subset equivalent to $\boldsymbol{N}$.

Definition 1.3.10 An infinite set is *denumerable* if it is equivalent to the set $\boldsymbol{N}$ of natural numbers. A set is *countable* if it is finite or denumerable. A set which is not countable is said to be *uncountable*.

Examples 1.3.11 1. The sets $\boldsymbol{N}$ of natural numbers and $\boldsymbol{Z}$ of integers are denumerable, hence countable.

2. For any $n \in \boldsymbol{N}$, the set $\{1, \ldots, n\}$ is finite, hence countable.

3. By Corollary 1.3.9, every infinite set contains a denumerable subset. Therefore the denumerable sets are the "smallest" infinite sets.

A countable set is just that: a set whose elements can be counted. For if the elements of a set can be counted, then the counting process establishes a correspondence between the set and $\boldsymbol{N}$ or between the set and a subset $\{1, \ldots, n\}$ of $\boldsymbol{N}$, and such a correspondence shows that the set is either denumerable or finite, as the case may be. Of course, in order to count the elements of a set, we must be able to list them in such a way that each element appears somewhere in the list, and when we count them we must make sure that every element gets counted. Summarizing these notions, we have the *countability criterion*:

> A nonempty set is countable if and only if its elements may be listed according to some well-defined scheme and then counted using some well-defined procedure.

Let us use the countability criterion to show that the set $\boldsymbol{Q}$ of rational numbers is countable. We list the rationals in order of increasing denominators, as follows:

$$
\begin{array}{cccccccc}
\frac{0}{1}, & \frac{1}{1}, & \frac{-1}{1}, & \frac{2}{1}, & \frac{-2}{1}, & \frac{3}{1}, & \frac{-3}{1}, & \ldots \\
\frac{0}{2}, & \frac{1}{2}, & \frac{-1}{2}, & \frac{2}{2}, & \frac{-2}{2}, & \frac{3}{2}, & \frac{-3}{2}, & \ldots \\
\frac{0}{3}, & \frac{1}{3}, & \frac{-1}{3}, & \frac{2}{3}, & \frac{-2}{3}, & \frac{3}{3}, & \frac{-3}{3}, & \ldots \\
\frac{0}{4}, & \frac{1}{4}, & \frac{-1}{4}, & \frac{2}{4}, & \frac{-2}{4}, & \frac{3}{4}, & \frac{-3}{4}, & \ldots \\
\vdots & \vdots & \vdots & \vdots & \vdots & \vdots & \vdots & \vdots
\end{array}
$$

Clearly, every element of $\boldsymbol{Q}$ appears in this list. (Some elements appear more than once, but in fact this does not matter: counting some elements of the set more than once will not affect its countability. This is a consequence of the Schroeder–Bernstein Theorem, which is proved in the appendix. See also Proposition A.3 of the appendix.)

Now we count the elements of $\boldsymbol{Q}$ by counting along diagonals as indicated by the arrows:

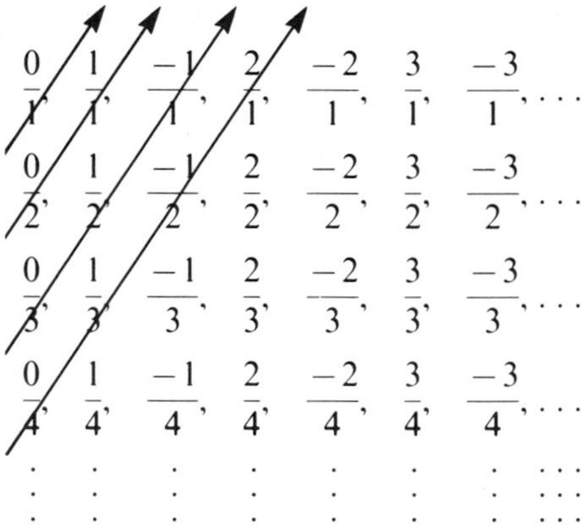

Thus we have demonstrated that the set $\boldsymbol{Q}$ is countable.

So far we have not produced any examples of uncountable sets. It is a major result about the set $\boldsymbol{R}$ of real numbers that $\boldsymbol{R}$ is uncountable. We intend to prove this result, but in order to do so we will need the following proposition.

Proposition 1.3.12 Every subset of a countable set is countable.

A formal proof of this proposition is given in the appendix. Here we will outline an argument based on the countability criterion. Thus, suppose that S is a countable set, so that there exists a list of the elements of S and a procedure whereby the elements of S can be counted. But if T is contained in S, the elements of T will appear in the list for S and will be counted as S is counted. Therefore T must be countable.

Now we are ready to prove that the reals are uncountable. The proof proceeds by showing that the closed unit interval $[0, 1]$ is uncountable, and this is accomplished by associating with each $x \in [0, 1]$ a unique decimal expansion and then showing that no matter how we attempt to list all such expansions, we can always find an expansion which is not in the list. The argument is called "Cantor's Diagonalization Procedure."

Theorem 1.3.13 (Cantor) The set $\boldsymbol{R}$ of real numbers is uncountable.

Proof: If $\boldsymbol{R}$ is countable, then so is $[0, 1]$, by Proposition 1.3.12. Therefore it suffices to show that $[0, 1]$ is uncountable.

Each $x \in [0, 1]$ has a decimal expansion of the form

$$x = 0.a_1a_2a_3 \ldots$$

where $a_i \in \{0, 1, \ldots, 9\}$ for each i. Some numbers have two such expansions: for instance

$$\frac{1}{2} = 0.5000 \ldots = 0.4999 \ldots$$

However, if a number does have two such expansions, all the digits from some point on in one of them will be zeros, while all the digits from some point on in the other will be nines. Let us agree that if $x \in [0, 1]$ has two decimal expansions, we will always use the one which has a tail of nines. With this agreement in force, we may consider that every $x \in [0, 1]$ has associated with it a unique decimal expansion.

Now assume that $[0, 1]$ is countable. Since $[0, 1]$ is infinite, it must be denumerable, and hence we must be able to index the elements of $[0, 1]$ by the natural numbers, thus:

$$[0, 1] = \{x_1, x_2, \ldots\}$$

Suppose that for each $n \in N$ we let $0.a_{n1}a_{n2}a_{n3} \ldots$ be the unique decimal expansion of x_n, so that

$$\begin{aligned} x_1 &= 0.a_{11}a_{12}a_{13} \ldots \\ x_2 &= 0.a_{21}a_{22}a_{23} \ldots \\ &\vdots \\ x_n &= 0.a_{n1}a_{n2}a_{n3} \ldots \\ &\vdots \end{aligned}$$

Define $y \in [0, 1]$ as follows:

$$y = 0.b_1b_2b_3 \ldots$$

where for each $n \in N$,

$$b_n = \begin{cases} 1 & \text{if } a_{nn} \neq 1 \\ 2 & \text{if } a_{nn} = 1 \end{cases}$$

Obviously $y \in [0, 1]$ and hence $y = x_n$, for some $n \in N$, and therefore the unique decimal expansions for y and x_n must be the same. But this is impossible, because for every $n \in N$, the expansion for y differs in its nth digit from the expansion for x_n. Thus we have arrived at a contradiction, and therefore [0, 1] must be uncountable. □

Corollary 1.3.14 Every subset of $\boldsymbol{R}$ which contains an open interval is uncountable.

The proof of the corollary will be left as an exercise.

There is much more to be said about set equivalence, finite and infinite sets, and the various types of infinity than we have been able to present here. The reader interested in pursuing these topics may begin by studying the appendix.

EXERCISES

1. Prove:
   (a) If set S is equivalent to set T, then set T is equivalent to set S.

(b) If set S is equivalent to set T and set T is equivalent to set U, then set S is equivalent to set U.

2. (a) Prove that $\boldsymbol{N}$ is equivalent to $\{1, 3, 5, \ldots\}$ (the set of odd natural numbers).
   (b) Let $k \in \boldsymbol{Z}$. Prove that $\boldsymbol{N}$ is equivalent to $\{nk \mid n \in \boldsymbol{N}\}$.
3. Prove that the function of Example 4 of Examples 1.3.2 is one-to-one and onto, and then show that
   (a) Any two open intervals (a, b) and (c, d) are equivalent.
   (b) Any two half-open intervals $[a, b)$ and $[c, d)$ are equivalent.
   (c) Any two closed rays $[a, +\infty)$ and $[b, +\infty)$ are equivalent.
4. Prove that $\boldsymbol{R}$ is equivalent to the open interval $(-1, 1)$ by showing that the function f defined by

$$f(x) = \frac{x}{1 + |x|} \qquad \text{for all } x \in \boldsymbol{R}$$

   is one-to-one and onto.
5. Let $n \in \boldsymbol{N}$. Show that $\{1, \ldots, n\}$ is finite.
6. Prove Corollary 1.3.6.
7. (a) Is the intersection of finite sets always a finite set? Justify your answer.
   (b) Is the union of finite sets always a finite set? Justify your answer.
8. Let S and T be countable sets. Prove that
   (a) $S \cap T$ is countable.
   (b) $S \cup T$ is countable.
   (c) $S \times T$ is countable.
9. Let $I = \{x \in \boldsymbol{R} \mid x \notin \boldsymbol{Q}\}$. The set I is the set of *irrational* numbers. Is I countable or uncountable? Justify your answer.
10. Prove Corollary 1.3.14.

2
The Real Numbers

The set $\boldsymbol{R}$ of real numbers will be of fundamental importance to us. Throughout this text we will always apply our general results concerning sets and functions to the set $\boldsymbol{R}$ and in turn use our knowledge of the properties of $\boldsymbol{R}$ as a guide to further generalization. In this chapter we discuss some of the basic properties of the real number system and also present a method for constructing the system of real numbers from the system of rational numbers.

2.1 PROPERTIES OF REAL NUMBERS

As stated in Chapter 1, we assume that the reader is familiar with the properties of the arithmetic operations (addition, subtraction, multiplication, division, and exponentiation) and the order relations ($<$, $\leqslant$, $>$, $\geqslant$) on $\boldsymbol{R}$. We shall not develop these here.

The first property of the real numbers which we wish to discuss is that of distance between numbers. Between any two real numbers there is a well-defined distance, the definition of which depends on the concept of the absolute value of a real number.

Definition 2.1.1 Let $x \in \boldsymbol{R}$. The *absolute value of* x, denoted by $|x|$, is defined by

$$|x| = \begin{cases} x & \text{if } x \geqslant 0 \\ -x & \text{if } x < 0 \end{cases}$$

If $x \in \boldsymbol{R}$ and $y \in \boldsymbol{R}$, we define the *distance between x and y* to be $|x - y|$. See Figure 2.1.

Examples 2.1.2 We have

$$|0| = 0 \qquad |4| = 4 \qquad |-4| = 4 \qquad |5\sqrt{2}| = 5\sqrt{2} \qquad |4-1| = 3 \qquad |1-4| = 3$$

Our first proposition lists some of the properties of absolute value and distance in $\boldsymbol{R}$.

Proposition 2.1.3 If $x \in \boldsymbol{R}$ and $y \in \boldsymbol{R}$, then

1. $|x| \geqslant 0$.
2. $|x| = 0$ if and only if $x = 0$.
3. If a is a positive real number, then $|x| \leqslant a$ if and only if $-a \leqslant x \leqslant a$.
4. $|xy| = |x||y|$.
5. (triangle inequality) $|x + y| \leqslant |x| + |y|$.

Proof: We prove only (5), the triangle inequality, leaving the remainder of the proof as an exercise.

To prove the triangle inequality, we proceed by cases.

Case 1: Both x and y are nonnegative: then $x \geqslant 0$, $y \geqslant 0$, and $x + y \geqslant 0$, so

$$|x + y| = x + y = |x| + |y|$$

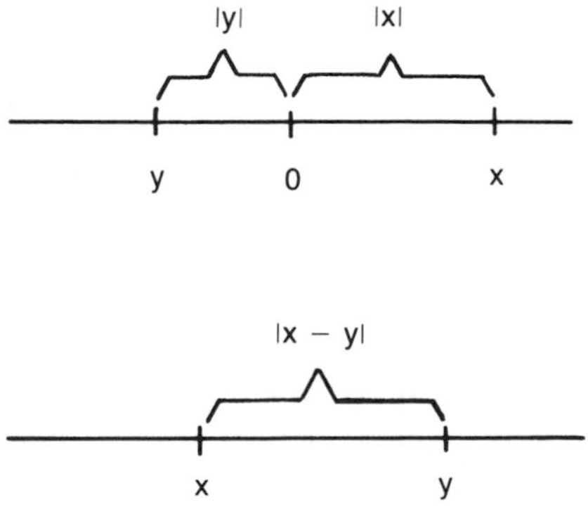

Figure 2.1 Absolute value and distance in $\boldsymbol{R}$.

Case 2: Both x and y are negative: then $x<0$, $y<0$, and $x+y<0$, so

$$|x+y| = -(x+y) = -x-y = |x| + |y|$$

Case 3: One of x or y negative and the other is nonnegative: suppose that $y<0\leqslant x$, so that either $x+y\geqslant 0$ or $x+y<0$. If $x+y\geqslant 0$, then $|x+y| = x+y$. But $y<0$ implies that $y<-y$, so

$$|x+y| = x+y < x-y = |x| + |y|$$

If $x+y<0$, then $|x+y| = -(x+y) = -x-y$. But $x\geqslant 0$ implies that $-x\leqslant x$, so

$$|x+y| = -x-y \leqslant x-y = |x| + |y|$$

This completes the proof of the triangle inequality. □

Corollary 2.1.4 If x, y, and z are real numbers, then

1. $|x-y| \geqslant 0$;
2. $|x-y| = 0$ if and only if $x=y$;
3. If a is a positive real number, then $|x-y| \leqslant a$ if and only if $y-a \leqslant x \leqslant y+a$;
4. $|x-y| = |y-x|$;
5. (triangle inequality) $|x-y| \leqslant |x-z| + |y-z|$.

The proof of the corollary is left as an exercise.

We now turn to another property of the real number system, the property known as boundedness. This is a property possessed by certain subsets of $\boldsymbol{R}$ which depends on the order relationships among real numbers.

Definition 2.1.5 Let S be a nonempty subset of $\boldsymbol{R}$. If there exists a real number α such that $x\leqslant\alpha$ for all $x\in S$, then S is *bounded above by* α, and α is an *upper bound* for S. Similarly, if there exists a real number β such that $\beta\leqslant x$ for all $x\in S$, then S is *bounded below by* β, and β is a *lower bound for* S. The set S is *bounded* if it is both bounded above and bounded below; otherwise S is *unbounded*.

Examples 2.1.6 1. Let S be the open unit interval $(0, 1)$ in $\boldsymbol{R}$. Any real number $\alpha \geqslant 1$ is an upper bound for $(0, 1)$, and any real number $\beta \leqslant 0$ is a lower bound for $(0, 1)$. Thus $(0, 1)$ is a bounded subset of $\boldsymbol{R}$. Note that the upper and lower bounds are not unique: clearly, if α is an upper bound for a set S, then so is any $\alpha' > \alpha$. Similarly, if β is a lower bound for S, then so is β' for any $\beta' < \beta$.

2. Let S be the open ray $(-\infty, 1)$ in $\boldsymbol{R}$. Any real number $\alpha \geqslant 1$ is an upper bound for $(-\infty, 1)$, but this ray has no lower bound. Thus $(-\infty, 1)$ is bounded above and unbounded below, so is an unbounded subset of $\boldsymbol{R}$.

3. The set $\boldsymbol{N}$ of natural numbers is bounded below, but unbounded above. The set $\boldsymbol{Z}$ of integers and the set $\boldsymbol{Q}$ of rational numbers are unbounded above and unbounded below.

Consider again the open unit interval $(0, 1)$ in $\boldsymbol{R}$. Although any real number $\alpha \geqslant 1$ is an upper bound for $(0, 1)$, it is clear that $\alpha = 1$ is the smallest possible upper bound for the set. When we need to work with an upper bound of a subset of $\boldsymbol{R}$, it will be convenient to select the smallest one (if there is a smallest one). Similar remarks apply to lower bounds. This leads us to our next definition.

Definition 2.1.7 Let S be a nonempty subset of $\boldsymbol{R}$. A real number α is a *least upper bound*, or *supremum*, for S if

1. α is an upper bound for S and
2. For every α' which is an upper bound for S, $\alpha \leqslant \alpha'$.

Similarly, a real number β is a *greatest lower bound*, or *infimum*, for S if

3. β is a lower bound for S and
4. For every β' which is a lower bound for S, $\beta \geqslant \beta'$.

We shall prove immediately that if a subset of $\boldsymbol{R}$ has a least upper bound, then that least upper bound must be unique, and similarly for greatest lower bounds.

Proposition 2.1.8 Let S be a nonempty subset of $\boldsymbol{R}$. If S has a least upper bound, then that least upper bound is unique. If S has a greatest lower bound, then that greatest lower bound is unique.

Proof: We prove the proposition only for the case of least upper bounds.

Suppose α_1 and α_2 are both least upper bounds for the set S. Then by (2) of Definition 2.1.7, α_1 an upper bound and α_2 a least upper bound implies that $\alpha_2 \leqslant \alpha_1$, while α_2 an upper bound and α_1 a least upper bound implies that $\alpha_1 \leqslant \alpha_2$. Therefore $\alpha_1 = \alpha_2$. □

Proposition 2.1.8 allows us to define *the* least upper bound and *the* greatest lower bound of a set.

Definition 2.1.9 Let S be a nonempty subset of $\boldsymbol{R}$. If they exist, the unique least upper bound of S is denoted by $\sup S$ and the unique greatest lower bound of S by $\inf S$.

Examples 2.1.10 1. We claim that $\sup (0, 1) = 1$ and $\inf (0, 1) = 0$. Since we know that 1 is an upper bound for $(0, 1)$, to show that 1 is the least upper bound it suffices to show that if α is any upper bound for $(0, 1)$ then $1 \leqslant \alpha$. Suppose the contrary; that is, suppose that α is an upper bound for $(0, 1)$ such that $\alpha < 1$. Since α is an upper bound for $(0, 1)$, we must have $0 < \alpha$. But then $x = \alpha + (1 - \alpha)/2$ is in $(0, 1)$ and $\alpha < x$, thus contradicting the fact that α is an upper bound for $(0, 1)$. Therefore $\alpha < 1$ is impossible, so $\sup (0, 1) = 1$. A similar argument shows that $\inf (0, 1) = 0$. Note that in this example $\sup S$ and $\inf S$ are not elements of the set S.

2. The argument of the previous example shows that $\sup [0, 1] = 1$ and also that $\inf [0, 1] = 1$. Note that here both $\sup S$ and $\inf S$ are elements of S.

3. Let $S = \{1/n \mid n \in \boldsymbol{N}\}$. We claim that $\sup S = 1$: for 1 is certainly an upper bound for S, and no real number $\alpha < 1$ can be an upper bound for S because $1 \in S$. Note that $\sup S \in S$.

4. If S is not bounded above, then of course $\sup S$ cannot exist; if S is not bounded below, then $\inf S$ cannot exist. Hence $\sup \boldsymbol{N}$, $\sup \boldsymbol{Z}$, and $\sup \boldsymbol{Q}$ do not exist, nor do $\inf \boldsymbol{Z}$ and $\inf \boldsymbol{Q}$. However, $\inf \boldsymbol{N} = 1$.

The preceding examples demonstrate that $\sup S$, if it exists, may or may not be an element of S, and similarly for $\inf S$. In none of the examples, however, was S bounded above but without a least upper bound, or bounded below and without a greatest lower bound; this is because such situations cannot occur in $\boldsymbol{R}$. This is a very important fact about the real numbers; it is known as the Completeness Property for $\boldsymbol{R}$.

Completeness Property for *R* Every nonempty subset of ***R*** which is bounded above has a least upper bound. Every nonempty subset of ***R*** which is bounded below has a greatest lower bound.

The Completeness Property for ***R*** is really a theorem about the real number system. We will not prove it here, for it is not possible to prove that ***R*** has the Completeness Property while working entirely within the real number system: if we attempt to construct a proof of the Completeness Property using only facts about ***R***, we will be forced to rely on some other property of ***R*** which is equivalent to the Completeness Property, and hence our proof will be circular. The proper way to show that ***R*** has the Completeness Property is to *construct* the system ***R*** in such a way that the Completeness Property becomes true in ***R*** *by construction*. We will do this in Section 2.2 of this chapter, but for now let us accept the Completeness Property as a fact and use it to obtain more properties of the real number system. The first of these is the well-known Archimedean Property for ***R***.

Theorem 2.1.11 (Archimedean Property for ***R***) Given any two positive real numbers x and y, there exists $m \in \boldsymbol{N}$ such that $mx > y$.

Proof: Let $S = \{nx \mid n \in \boldsymbol{N}\}$; then S is a nonempty subset of ***R***. If there is no $m \in \boldsymbol{N}$ such that $mx > y$, then y is an upper bound for S, and hence by the Completeness Property, S has a least upper bound. Let $\sup S = \alpha$.

Since $(n+1)x = nx + x$ is an element of S for all $n \in \boldsymbol{N}$, we must have $nx + x \leqslant \alpha$ for all $n \in \boldsymbol{N}$, and thus $nx \leqslant \alpha - x$ for all $n \in \boldsymbol{N}$. But this implies that $\alpha - x$ is an upper bound for S which is less than the least upper bound α, a contradiction. □

Theorem 2.1.12 Between any two distinct real numbers there is a rational number.

Proof: We must show that if x and y are real numbers with $x < y$, then there is some rational number r such that $x < r < y$. If $x < 0$ and $y > 0$, we may take $r = 0$, so that it suffices to prove the theorem for the cases $0 \leqslant x < y$ and $x < y \leqslant 0$.

Assume first that $0 < x < y$. By the Archimedean Property for ***R***, there exist natural numbers m and n such that $mx < 1$ and $n(y - x) > 1$. Let k be the larger of m and n; k is a natural number, and since $kx > 1$ and $k(y - x) > 1$, we have $1/k < x$ and $1/k < (y - x)$. Again by the

Archimedean Property for $\boldsymbol{R}$, there exists a natural number p such that $p(1/k) > x$. Now consider the rational numbers

$$\frac{1}{k}, \frac{2}{k}, \ldots, \frac{p}{k}$$

Since

$$\frac{1}{k} < x < \frac{p}{k}$$

there exists a rational number q/k, where $1 < q < p - 1$, such that

$$\frac{q}{k} < x < \frac{q+1}{k}$$

But then

$$x < \frac{q+1}{k} = \frac{q}{k} + \frac{1}{k} < x + \frac{1}{k} < x + y - x = y$$

Letting $r = (q+1)/k$, we have shown the existence of a rational number between x and y when $0 < x < y$.

Now suppose $0 \leqslant x < y$. Since $0 \leqslant x < (x+y)/2 < y$, by what we have just shown there is a rational number between $(x+y)/2$ and y, and hence between x and y. Therefore the theorem is proved in the case that $0 \leqslant x < y$.

Finally, if $x < y \leqslant 0$, then $0 \leqslant -y < -x$, and by what we have just proved there is a rational number r such that $-y < r < -x$. But then $x < -r < y$, and since r rational implies $-r$ rational, we are done. □

Corollary 2.1.13 If x is any positive real number, then there is some $n \in \boldsymbol{N}$ such that $0 < 1/n < x$.

Proof: There is some rational number m/n, where $m \in \boldsymbol{N}$ and $n \in \boldsymbol{N}$, such that $0 < m/n < x$. Hence $0 < 1/n < m/n < x$. □

Example 2.1.14 1. Consider the set $S = \{1/n \mid n \in \boldsymbol{N}\}$. Previously we showed that $\sup S = 1$. Now we claim that $\inf S = 0$: clearly 0 is a lower bound for S, and if $\beta > 0$, then by the corollary there is some element of S between 0 and β, so β cannot be a lower bound for S.

2. Let $S = \{x \in \boldsymbol{R} \mid x \in \boldsymbol{Q} \text{ and } x^2 < 2\}$. The set S is nonempty and bounded above, so $\sup S$ exists. Obviously, the real number $\sqrt{2}$ is an upper bound for S and $\sup S$ is positive. If $\sup S < \sqrt{2}$, then there is a

rational number r such that

$$0 < \sup S < r < \sqrt{2}$$

But then $r \in \boldsymbol{Q}$ and $r^2 < 2$, so $r \in S$, and thus $\sup S < r$ contradicts the fact that $\sup S$ is an upper bound for S. Therefore $\sup S = \sqrt{2}$.

The preceding example implies that the system $\boldsymbol{Q}$ of rational numbers does *not* have the Completeness Property. For consider the subset S of $\boldsymbol{Q}$ defined by

$$S = \{r \in \boldsymbol{Q} \mid r^2 < 2\}$$

Then S is bounded above in $\boldsymbol{Q}$ (by the rational number 2, for instance), but S has no least upper bound in $\boldsymbol{Q}$; if it did, the least upper bound would have to be $\sqrt{2}$, and $\sqrt{2}$ is not a rational number. (Exercise 12 at the end of this section asks you to fill in the details of this argument.) Hence the Completeness Property fails to hold in $\boldsymbol{Q}$. This is the major defect of $\boldsymbol{Q}$ as a number system.

We conclude this section by constructing a subset of the closed unit interval [0, 1] known as the *Cantor ternary set*. The Cantor ternary set is more subtle in its form and properties than the intervals and rays which we have studied so far, and it is a fruitful source of examples and counterexamples in analysis.

Example 2.1.15 (Cantor Ternary Set) We begin with a closed unit interval [0, 1]. We remove the open middle third (1/3, 2/3) from [0, 1], thus obtaining the set

$$C_1 = \left[0, \frac{1}{3}\right] \cup \left[\frac{2}{3}, 1\right]$$

Next we remove from C_1 its open middle thirds (1/9, 2/9) and (7/9, 8/9) to obtain

$$C_2 = \left[0, \frac{1}{9}\right] \cup \left[\frac{2}{9}, \frac{1}{3}\right] \cup \left[\frac{2}{3}, \frac{7}{9}\right] \cup \left[\frac{8}{9}, 1\right]$$

Now we remove the open middle thirds from C_2 to obtain

$$C_3 = \left[0, \frac{1}{27}\right] \cup \left[\frac{2}{27}, \frac{1}{9}\right] \cup \left[\frac{2}{9}, \frac{7}{27}\right] \cup \left[\frac{8}{27}, \frac{1}{3}\right] \cup \left[\frac{2}{3}, \frac{19}{27}\right] \cup \left[\frac{20}{27}, \frac{7}{9}\right]$$
$$\cup \left[\frac{8}{9}, \frac{25}{27}\right] \cup \left[\frac{26}{27}, 1\right]$$

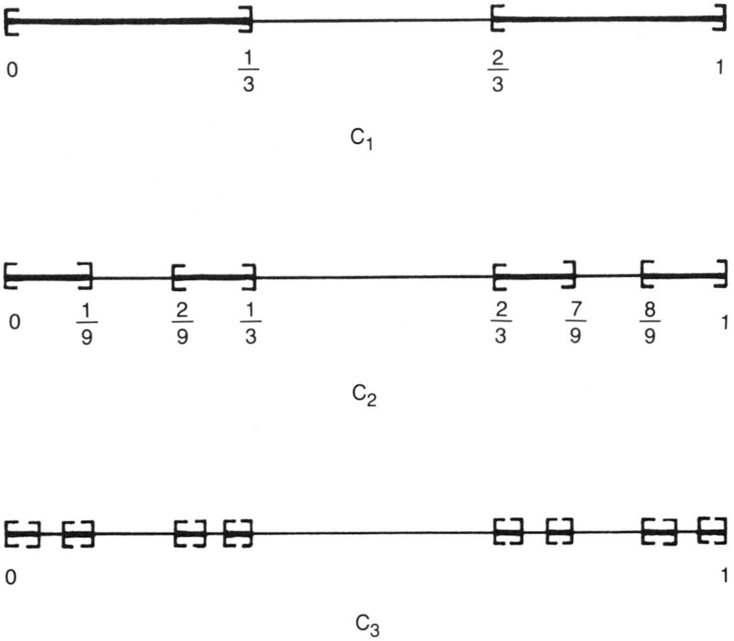

Figure 2.2 Construction of the Cantor set.

(See Figure 2.2.) Continuing in this manner, for each $n \in N$ we define a set C_n. The Cantor ternary set C is then given by

$$C = \bigcap_{n \in N} C_n$$

Thus the Cantor ternary set C consists of exactly those elements of $[0, 1]$ which remain after all open middle thirds have been removed. Another way of writing the Cantor set is to let

$$C = [0, 1] - \bigcup_{n \in N} S_n$$

where for each $n \in N$,

$$S_n = \left(\frac{1}{3^n}, \frac{2}{3^n}\right) \cup \left(\frac{7}{3^n}, \frac{8}{3^n}\right) \cup \cdots \cup \left(\frac{3^n - 2}{3^n}, \frac{3^n - 1}{3^n}\right)$$

(See Exercise 13 at the end of this section.)

The Cantor set is nonempty, since $0 \in C$ and $1 \in C$. In fact, every real number of the form $1/3^n$, $n \in \boldsymbol{N}$ is an element of C, so C is an infinite set. Since we obtain C from $[0, 1]$ by removing most of the elements of $[0, 1]$, we feel intuitively that C must be small compared to $[0, 1]$. In some ways this is so: for instance, C does not contain any open interval. (See Exercise 16 at the end of this section.) Also, the length of C is zero, because for each $n \in \boldsymbol{N}$, $C \subset C_n$ and C_n has length $(2/3)^n$. Therefore in these senses C is indeed a smaller set than $[0, 1]$. On the other hand, C is uncountable (see Exercise 15 at the end of this section), and hence in this sense it is just as large as $[0, 1]$.

EXERCISES

1. Find all values of x which satisfy the given equality or inequality.
   (a) $|x - 2| = 3$
   (b) $|x - 2| < 3$
   (c) $|3x + 5| \leqslant 6$
   (d) $|2 - x| \geqslant 4$
   (e) $|5 + 2x| \geqslant 6$
   (f) $|x + 1| < |x - 3|$
2. (a) Prove 1, 2, 3, and 4 of Proposition 2.1.3.
   (b) Prove Corollary 2.1.4.
3. Let $x \in \boldsymbol{R}$, $y \in \boldsymbol{R}$. Prove:
   (a) $|-x| = |x|$
   (b) $|x/y| = |x|/|y|$ $(y \neq 0)$
   (c) $|x - y| \leqslant |x| + |y|$
   (d) $||x| - |y|| \leqslant |x - y|$
4. Let S and T be bounded subsets of $\boldsymbol{R}$. Prove that $S \cup T$ and $S \cap T$ are bounded.
5. Prove or give a counterexample: if X_α is a bounded subset of $\boldsymbol{R}$ for each $\alpha \in A$, then $\bigcup_{\alpha \in A} X_\alpha$ is bounded.
6. Let $[a, b]$ be a closed interval in $\boldsymbol{R}$. Prove that $\inf [a, b] = a$ and $\sup [a, b] = b$.
7. Let I be a nonempty subset of $\boldsymbol{R}$ having the following property: whenever $x \in I$, $y \in I$, and $x \leqslant y$, then $[x, y] \subset I$. Prove that I is an interval or a ray.
8. Let $x \in \boldsymbol{R}$ and let $S = \{q \in \boldsymbol{Q} \mid q < x\}$. Show that $\sup S = x$.
9. Let S be a nonempty subset of $\boldsymbol{R}$ which is bounded above. Let

$$T = \{x \in \boldsymbol{R} \mid x \text{ is an upper bound for } S\}.$$

   Prove that $\inf T = \sup S$.

10. Let S be a nonempty subset of $\boldsymbol{R}$. Prove that if x is an upper bound for S such that $x \in S$, then $x = \sup S$.
11. Let S and T be nonempty subsets of $\boldsymbol{R}$, with T bounded and $S \subset T$.
    (a) Prove that $\sup S \leqslant \sup T$ and that $\inf T \leqslant \inf S$.
    (b) Give an example to show that even if S is a proper subset of T, it may be the case that $\sup S = \sup T$ and $\inf S = \inf T$.
12. Fill in the details of the argument following Example 2.1.14 to show that the system $\boldsymbol{Q}$ of rational numbers does not have the Completeness Property.
13. Show that the two expressions for the Cantor ternary set C given in Example 2.1.15 are equivalent.
14. Just as every $x \in [0, 1]$ has a decimal expansion, so does every $x \in [0, 1]$ have a ternary expansion

    $$x = 0.a_1a_2a_3 \ldots \qquad \text{where } a_i \in \{0, 1, 2\}$$

    (Such a ternary expansion for x may be obtained by writing $x = a_1/3 + a_2/3^2 + a_3/3^3 + \cdots$.) Prove that the Cantor set C consists of exactly those elements of $[0, 1]$ whose ternary expansions have every digit $a_i \in \{0, 2\}$.
15. If $x \in [0, 1]$ has two ternary expansions, one of them ends in a tail of zeros while the other ends in a tail of twos. Use this fact and the result of Exercise 14 to prove that the Cantor set C is uncountable.
16. Prove that the Cantor set C does not contain an open interval.

2.2 CONSTRUCTION OF THE REAL NUMBERS*

In this section we will construct the set $\boldsymbol{R}$ of real numbers from the set $\boldsymbol{Q}$of rational numbers by what is known as the Dedekind cut method. Once this has been done, it will be a relatively simple matter to prove that $\boldsymbol{R}$ has the Completeness Property.

We assume that the operations of arithmetic and the usual order relations have been defined on $\boldsymbol{Q}$. As noted in the discussion following Example 2.1.14 of this chapter, $\boldsymbol{Q}$ does not have the Completeness Property. The reason this is so is that, if we think of $\boldsymbol{Q}$ as consisting of the rational points on a line, there are gaps between the points which represent elements of $\boldsymbol{Q}$. (For instance, the real number $\sqrt{2}$ represents such a gap. See Figure 2.3.) The Dedekind cut method constructs $\boldsymbol{R}$ from $\boldsymbol{Q}$

*This section is optional.

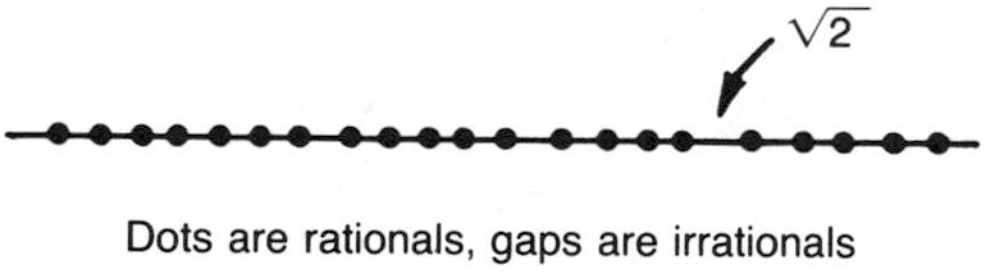

Figure 2.3 $\sqrt{2}$ as a gap in the rationals.

by filling in these gaps. We begin by defining the concept of a Dedekind cut.

Definition 2.2.1 A nonempty proper subset S of $\boldsymbol{Q}$ is a *Dedekind cut* if

1. $s \in S$, $q \in \boldsymbol{Q}$, $q < s$ imply that $q \in S$ and
2. $s \in S$ implies that there is some $t \in S$ such that $s < t$.

Note that a Dedekind cut is a subset of $\boldsymbol{Q}$. Condition (1) of the definition says that if a particular rational number s belongs to a Dedekind cut, then so do all rationals less than s; condition (2) says that a Dedekind cut contains no largest element. Thus we see intuitively that a Dedekind cut consists either of all the rationals less than (but not equal to) some specific rational, or of all the rationals less than some particular gap in $\boldsymbol{Q}$ (see Figure 2.4).

Proposition 2.2.2 If $r \in \boldsymbol{Q}$ and $S_r = \{p \in \boldsymbol{Q} \mid p < r\}$, then S_r is a Dedekind cut.

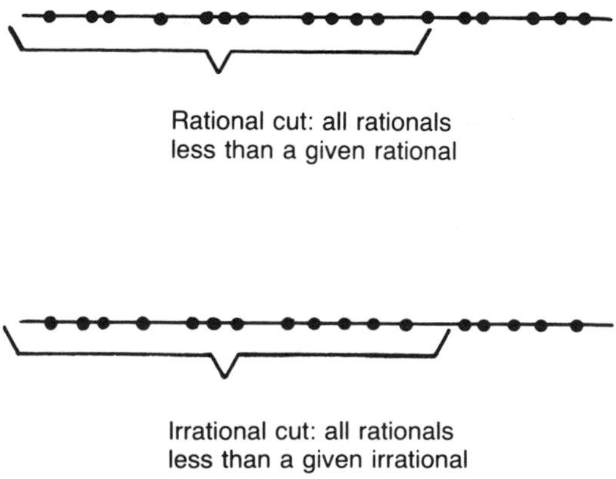

Figure 2.4 Dedekind cuts.

Proof: The set S_r is obviously a nonempty proper subset of $\boldsymbol{Q}$, and if $s \in S_r$ and $q \in \boldsymbol{Q}$ with $q < s$, then $q < s < r$, so $q \in S_r$. Furthermore, if $s \in S_r$ and we let $t = (s + r)/2$, then $s < t < r$ and $t \in S_r$. Therefore (1) and (2) of Definition 2.2.1 are satisfied, and hence S_r is a Dedekind cut. □

Definition 2.2.3 A Dedekind cut which is of the form $S_r = \{p \in \boldsymbol{Q} \mid p < r\}$ for some $r \in \boldsymbol{Q}$ is called a *rational cut*. A Dedekind cut which is not a rational cut is called an *irrational cut*.

Now we are ready to define the set of real numbers.

Definition 2.2.4 A *real number* is a Dedekind cut. The *set* $\boldsymbol{R}$ *of real numbers* is the set of all Dedekind cuts. A real number is *rational* if it is a rational Dedekind cut; otherwise it is *irrational*.

Note that there is a one-to-one correspondence between rational numbers and rational Dedekind cuts. (See Exercise 1 at the end of this section.) Therefore if we identify each rational cut with the rational number which determines it, we may consider $\boldsymbol{Q}$ to be a subset of $\boldsymbol{R}$.

We have now constructed $\boldsymbol{R}$ as the set of all Dedekind cuts. All the usual properties of $\boldsymbol{R}$ may be proved using Definitions 2.2.1 and 2.2.4 and the known properties of $\boldsymbol{Q}$. For instance, if x and y are real numbers (Dedekind cuts), we may define their sum as follows:

$$x + y = \{p + q \mid p \in x \text{ and } q \in y\}$$

It is easy to check that $x + y$ is a Dedekind cut, and hence a real number, and thus addition is defined in $\boldsymbol{R}$. Since our immediate objective here is to show that $\boldsymbol{R}$ has the Completeness Property, we shall not pursue this further. (But see the exercises at the end of this section).

Before we can prove that $\boldsymbol{R}$ has the Completeness Property, we must define the order relation $<$ on $\boldsymbol{R}$.

Definition 2.2.5 Let x and y be real numbers (Dedekind cuts), and define $x < y$ if x is a proper subset of y. Define $x \leqslant y$ if $x < y$ or $x = y$.

Now we are ready to prove that $\boldsymbol{R}$ has the Completeness Property. Actually, this is so by construction. The reason that $\boldsymbol{Q}$ does not have the Completeness Property is that there are gaps between elements of $\boldsymbol{Q}$; but we have constructed $\boldsymbol{R}$ by filling in these gaps with Dedekind cuts. Hence $\boldsymbol{R}$ must have the Completeness Property.

Theorem 2.2.6 The system $\boldsymbol{R}$ of real numbers has the Completeness Property.

Proof: Let X be a nonempty subset of $\boldsymbol{R}$ which is bounded above. We will show that X has a least upper bound.

Every real number $x \in X$ is a Dedekind cut and hence is a nonempty proper subset of $\boldsymbol{Q}$ which satisfies properties (1) and (2) of Definition 2.2.1. Let

$$S = (q \in \boldsymbol{Q} \mid q \in x \text{ for some } x \in X\}$$

Thus

$$S = \bigcup_{x \in X} x$$

Since X is nonempty, S is a nonempty subset of $\boldsymbol{Q}$. Furthermore, since X is bounded above, there is some Dedekind cut which does not belong to X, and thus there is some rational number which does not belong to S. Therefore S is a proper subset of $\boldsymbol{Q}$.

Let us show that S is a Dedekind cut. If $s \in S$, then $s \in x$ for some $x \in X$. Because x is a Dedekind cut, we must have $q \in x$ for all rational q such that $q < s$. But $q \in x$ implies that $q \in S$. This shows that (1) of Definition 2.2.1 is satisfied. To prove that (2) is satisfied, we note that if $s < S$, then, since x is a Dedekind cut, there is some $t \in x$ such that $s < t$. But $t \in x$ implies that $t \in S$. Therefore S is a Dedekind cut, and hence a real number.

If $x \in X$ then $x \in S$ and therefore by definition $x \leqslant S$. Thus S is an upper bound for X. On the other hand, if T is any upper bound for X, then for all $x \in X$ we must have $x \leqslant T$, or $x \subset T$, so that

$$S = \bigcup_{x \in X} x \subset T$$

and hence $S \leqslant T$. Thus S is a least upper bound for X, and we have shown that every nonempty subset of $\boldsymbol{R}$ which is bounded above has a least upper bound.

To complete the proof we must show that every nonempty subset Y of $\boldsymbol{R}$ which is bounded below has a greatest lower bound. We outline an argument which establishes this fact, leaving the details to the reader.

(See Exercise 7 at the end of this section.) Let

$$X = \{x \in \boldsymbol{R} \mid x \leqslant y \text{ for all } y \in Y\}$$

then X is a nonempty subset of $\boldsymbol{R}$ which is bounded above, and hence by what we have already proved, X has a least upper bound; but a least upper bound for X is a greatest lower bound for Y. □

EXERCISES

1. Prove that two rational Dedekind cuts S_r and S_q are equal if and only if $r = q$.
2. If x and y are real numbers (Dedekind cuts), define

$$x + y = \{p + q \mid p \in x \text{ and } q \in y\}$$

   Prove that $x + y$ is a Dedekind cut and hence a real number.
3. In the notation of Proposition 2.2.2, S_0 is the rational Dedekind cut determined by the rational number 0. Define the real number 0 by setting $0 = S_0$ and then use the definition of addition given in Exercise 2 to show that for every real number x, $x + 0 = x$.
4. Let x, y be real numbers. If x is irrational, define

$$-x = \{q \in Q \mid -q \notin x\}$$

   If x is rational, say $x = S_r$, where $r \in \boldsymbol{Q}$, then define

$$-x = \{q \in \boldsymbol{Q} \mid -q \notin x \text{ and } -q \neq r\}$$

   (a) Prove that $-x$ is a real number (Dedekind cut).
   (b) Show that $x + y = 0$ (see Exercise 3) if and only if $y = -x$. This allows us to define subtraction in $\boldsymbol{R}$. (How?)
   (c) Prove that $0 < x$ if and only if $-x < 0$.
5. Let x, y be real numbers. Define multiplication for nonnegative real numbers as follows: if $0 \leqslant x$ and $0 \leqslant y$, define

$$xy = 0 \cup \{qp \mid q \in x, p \in y, 0 \leqslant q, 0 \leqslant p\}$$

   (a) Prove that xy is a Dedekind cut.
   (b) Prove that $x0 = 0$.

(c) In the notation of Proposition 2.2.2, S_1 is the rational Dedekind cut determined by the rational number 1. Define the real number 1 by setting $1 = S_1$ and the then use the definition of multiplication to show that $x1 = x$.

6. If x is a real number, $x > 0$, prove that there is a unique real number y such that $xy = 1$. This allows us to define division in $\boldsymbol{R}$. (How?)
7. Fill in the details for the argument outlined in the last paragraph of the proof of Theorem 2.2.6.

3
Topology

Thus far we have been concerned mainly with combining sets by means of set operations and comparing them using functions. In so doing we have been looking at sets primarily from an external point of view, and have not examined their internal structures in any detail. It is important to consider the internal structures of sets, however. For instance, the set $\boldsymbol{R}$ of real numbers is more than just an uncountable collection of elements: certain subsets of $\boldsymbol{R}$ (such as the open intervals) have important properties of their own, and we cannot understand the real numbers thoroughly unless we examine these subsets and their properties. In this chapter we will study the internal structures of sets by considering certain collections of their subsets known as topologies. We begin by introducing the concept of a topological space and then explore such spaces with the aid of continuous functions. All our general results about topology can be applied to the set $\boldsymbol{R}$ of real numbers, and beginning in this chapter and throughout the rest of the text, the reader should always give thought to what our general definitions and results say about $\boldsymbol{R}$.

3.1 TOPOLOGICAL SPACES

We commence with the definition of a topological space.

Definition 3.1.1 A *topological space* is a pair (X, τ), where X is a set and τ is a collection of subsets of X such that

1. $X \in \tau$ and $\emptyset \in \tau$.
2. If A is an index set and $G_\alpha \in \tau$ for each $\alpha \in A$, then $\bigcup_{\alpha \in A} G_\alpha \in \tau$.
3. If A is a finite index set and $G_\alpha \in \tau$ for each $\alpha \in A$, then $\bigcap_{\alpha \in A} G_\alpha \in \tau$.

Such a collection τ of subsets of X is called a *topology on* X, and X is then referred to as the *space*, or *underlying point set*, of τ. Also, the elements of X are called the *points* of the topological space. A subset G of X is an *open set in the topology* τ if $G \in \tau$; a subset F of X is a *closed set in the topology* τ if its complement $X - F \in \tau$.

Note that a collection τ of subsets of a set X is a topology on X if and only if both the empty set and X itself belong to τ; *any* union of sets belonging to τ is a set belonging to τ; and any intersection of *finitely many* sets belonging to τ is a set belonging to τ.

Examples 3.1.2 1. Let $X = \{x, y, z\}$ and let $\tau = \{\emptyset, \{x\}, \{y\}, \{x, y\}, X\}$. The collection τ satisfies the conditions of Definition 3.1, so (X, τ) is a topological space, that is, τ is a topology on X. The open sets in τ are the sets

$$\emptyset, \{x\}, \{y\}, \{x, y\}, \text{ and } X$$

The closed sets in τ are the complements of the open sets, namely

$$X, \{y, z\}, \{x, z\}, \{z\}, \text{ and } \emptyset$$

Notice that $\emptyset$ and X are subsets of X which are both open and closed in the topology. Also, in this example every subset of the topological space is either open, closed, or both open and closed.

2. Let $X = \{x, y, z\}$ and let

$$\tau_1 = \{\emptyset, \{x\}, X\}$$

$$\tau_2 = \{\emptyset, \{x\}, \{x, y\}, \{x, z\}, X\}$$

$$\tau_3 = \{\emptyset, \{x\}, \{y\}, X\}$$

$$\tau_4 = \{\emptyset, \{x, y\}, \{x, z\}, X\}$$

The collections τ_1 and τ_2 are topologies on X (check this), but τ_3 and τ_4 are not: τ_3 is not a topology on X because

$$\{x\} \cup \{y\} = \{x, y\} \notin \tau_3$$

while τ_4 is not a topology on X because

$$\{x, y\} \cap \{x, z\} = \{x\} \notin \tau_4$$

Notice that there are subsets of X which are neither open nor closed in the topology τ_1.

3. Let X be any set and let $\tau = \{S \mid S \text{ is a subset of } X\}$; then τ is a topology on X, called the *discrete topology* on X. Every subset of the space is both open and closed in the discrete topology.

4. Let X be any set and let $\tau = \{\emptyset, X\}$; then τ is a topology on X, called the *indiscrete topology* on X. In the indiscrete topology, the empty set and the entire space are the only open sets and also the only closed sets.

Assigning a topology to a set is called *topologizing* the set. There are always at least two ways to topologize a set, for we may always assign it either the discrete topology or the indiscrete topology.

Now consider the set $\boldsymbol{R}$ of real numbers. We can of course topologize $\boldsymbol{R}$ by assigning it either the discrete or the indiscrete topology, but these are not very interesting. Instead, we will use the open intervals as the basis for a topology on $\boldsymbol{R}$. This topology, called the natural topology on $\boldsymbol{R}$, will be of fundamental importance to us throughout the remainder of this text.

Proposition 3.1.3 Let τ consist of all subsets S of $\boldsymbol{R}$ which have the property that if $x \in S$ then there is some open interval $I_x \subset S$ such that $x \in I_x$. The collection τ is a topology on $\boldsymbol{R}$, called the *natural topology* on $\boldsymbol{R}$.

Proof: We must verify that the collection τ satisfies the conditions of Definition 3.1.1. It is certainly the case that if $x \in \boldsymbol{R}$, then there is some open interval $I_x \subset \boldsymbol{R}$ such that $x \in I_x$. Therefore $\boldsymbol{R} \in \tau$.

To show that the empty set belongs to τ, we must demonstrate that if $x \in \emptyset$ then there is some open interval $I_x \subset \emptyset$ such that $x \in I_x$. But there is no x which is an element of $\emptyset$, so there is nothing to prove. (When a situation like this occurs, we say that the condition is vacuously satisfied.) Hence $\emptyset \in \tau$.

Now let A be an index set and suppose that $G_\alpha \in \tau$ for all $\alpha \in A$. If $\bigcup_{\alpha \in A} G_\alpha = \emptyset$, then $\bigcup_{\alpha \in A} G_\alpha \in \tau$ by what we have just proved, so let us assume that $\bigcup_{\alpha \in A} G_\alpha \neq \emptyset$. If $x \in \bigcup_{\alpha \in A} G_\alpha$, then there is some $\beta \in A$ such that $x \in G_\beta$. But since $G_\beta \in \tau$, there is some open interval $I_x \subset G_\beta \subset \bigcup_{\alpha \in A} G_\alpha$ such that $x \in I_x$, and therefore the set $\bigcup_{\alpha \in A} G_\alpha \in \tau$.

Finally, let A be a finite index set and suppose that $G_\alpha \in \tau$ for all $\alpha \in A$. Again, if $\bigcap_{\alpha \in A} G_\alpha = \emptyset$, we are done, so assume that $\bigcap_{\alpha \in A} G_\alpha \neq \emptyset$ and let $x \in \bigcap_{\alpha \in A} G_\alpha$. Then $x \in G_\alpha$ for each $\alpha \in A$, and since $G_\alpha \in \tau$, for each $\alpha \in \tau$ there is some open interval $(a_\alpha, b_\alpha) \subset G_\alpha$ such that $x \in (a_\alpha, b_\alpha)$. Because A is a finite set, the set $\{a_\alpha \mid \alpha \in A\}$ is finite and therefore has a largest element: let $a = \max \{a_\alpha \mid \alpha \in A\}$ be this largest element. Similarly, let $b = \min \{b_\alpha \mid \alpha \subset A\}$. Then $x \in (a, b) \subset (a_\alpha, b_\alpha)$ for all $\alpha \in A$, and hence (a, b) is an open interval contained in $\bigcap_{\alpha \in A} G_\alpha$ such that $x \in (a, b)$. Therefore $\bigcap_{\alpha \in A} G_\alpha \in \tau$, and we are done. □

Note that a set is open in the natural topology on $\boldsymbol{R}$ if and only if we can surround each of its points with an open interval which is itself contained in the set. Thus open intervals are open sets in the natural topology, as are open rays. On the other hand, closed intervals and closed rays are not open sets: for example, if $[a, b]$ is a closed interval, it is not possible to surround the endpoint a with an open interval which is contained within $[a, b]$. Closed intervals and closed rays are closed sets in the natural topology, however, because they are complements of open sets. We collect these and a few other facts in a proposition, the proof of which is left to the reader.

Proposition 3.1.4 In the natural topology on $\boldsymbol{R}$,

1. Every open interval is an open set.
2. Every open ray is an open set.
3. Every closed ray is a closed set.
4. Every closed interval is a closed set.
5. Every one-element subset of $\boldsymbol{R}$ is a closed set.
6. Every finite subset of $\boldsymbol{R}$ is a closed set.

We have seen that a given set may be topologized in several different ways. But since a topology is itself a set, we can compare topologies which have the same underlying point set by means of set containment.

Definition 3.1.5 Let τ_1 and τ_2 be topologies on the set X. If τ_1 is a proper subset of τ_2, we say that τ_1 is *smaller than* τ_2, or that τ_2 is *larger than* τ_1, and write $\tau_1 \subset \tau_2$. If $\tau_1 \subset \tau_2$ and $\tau_2 \subset \tau_1$ then $\tau_1 = \tau_2$. If $\tau_1 \not\subset \tau_2$ and $\tau_2 \not\subset \tau_1$, we say that τ_1 and τ_2 are *not comparable*.

Note that if τ_1 and τ_2 are topologies on X then $\tau_1 \subset \tau_2$ if and only if every set which is open in τ_1 is also open in τ_2. Thus if τ_1 is smaller than τ_2, then τ_1 has fewer open sets than τ_2, and if $\tau_1 = \tau_2$, then τ_1 and τ_2 have exactly the same open sets.

Examples 3.1.6 1. Let $X = \{x, y, z\}$, $\tau_1 = \{\emptyset, \{x\}, X\}$, and $\tau_2 = \{\emptyset, \{x\}, \{x, y\}, X\}$. Both τ_1 and τ_2 are topologies on X, and τ_1 is a proper subset of τ_2. Hence τ_1 is smaller than τ_2.

2. Let $X = \{x, y, z\}$, $\tau_1 = \{\emptyset, \{x\}, X\}$, and $\tau_2 = \{\emptyset, \{y\}, X\}$. Then τ_1 and τ_2 are not comparable.

3. Let τ_1 be the natural topology on $\boldsymbol{R}$ and let τ_2 be defined as follows:

$$\tau_2 = \{G \subset \boldsymbol{R} \mid G = \emptyset \text{ or } G \text{ is a union of open intervals}\}$$

We claim that $\tau_1 = \tau_2$. Thus τ_2 is merely another way of describing the natural topology on $\boldsymbol{R}$. Note that it is not necessary to prove directly that τ_2 is a topology on $\boldsymbol{R}$: if we can show that τ_1 and τ_2 have the same elements, then τ_2 must satisfy the conditions of Definition 3.1.1 because we have already shown that τ_1 does. Therefore in order to prove the claim, it will be sufficient to show $\tau_1 \subset \tau_2$ and $\tau_2 \subset \tau_1$.

Let $G \in \tau_2$. If $G = \emptyset$, then $G \in \tau_1$. If $G \neq \emptyset$, then G is a union of open intervals, which are open sets in the natural topology, and therefore $G \in \tau_1$.

Now let $G \in \tau_1$. If $G = \emptyset$, then $G \in \tau_2$. If $G \neq \emptyset$, then for each $x \in G$ there is some open interval I_x such that $x \in I_x \subset G$. But this implies that $G = \bigcup_{x \in G} I_x$. Therefore $G \in \tau_2$, and the claim is proved.

By virtue of the previous example, we can think of natural topology on $\boldsymbol{R}$ as being generated by the open intervals, in the sense that every nonempty open set in the natural topology is a union of open intervals and conversely every such union is an open set. The following useful result further characterizes these open sets as countable unions of mutually disjoint subsets of $\boldsymbol{R}$.

Proposition 3.1.7 If G is a nonempty open set in the natural topology on $\boldsymbol{R}$, then G is a union of countably many mutually disjoint subsets of $\boldsymbol{R}$, each of which is either an open interval or an open ray.

Proof: Let $x \in G$. Since G is open in the natural topology, there is some open interval I_x such that $x \in I_x \subset G$. Let J_x denote the union of all subsets of G which contain x and are either open intervals or open rays. It is easy to check that J_x must itself be either an open interval or an open ray. (See Exercise 18 of Section 1.1, Chapter 1.)

We claim that if $x \in G$ and $y \in G$, then either $J_x \cap J_y = \emptyset$ or $J_x = J_y$. For if $J_x \cap J_y \neq \emptyset$ then $J_x \cup J_y$ must be either an open interval or an open ray, and since $x \in J_x \cup J_y$ and $(J_x \cup J_y) \subset G$, the definition of J_x tells us that $(J_x \cup J_y) \subset J_x$. Similarly, $(J_x \cup J_y) \subset J_y$. Therefore $J_x = J_y$ whenever $J_x \cap J_y \neq \emptyset$. Clearly $G = \bigcup_{x \in G} J_x$, and thus we have established that G is a union of sets each of which is either an open interval or an open ray, and that the distinct sets in the union are mutually disjoint. But any union of mutually disjoint subsets of $\boldsymbol{R}$ is countable, because each set in the union must contain a rational number which does not belong to any of the other sets in the union, and hence there is a one-to-one correspondence between the sets in the union and a subset of the rational numbers. □

We have been examining the natural topology on $\boldsymbol{R}$ in some detail and shall continue to do so throughout this text, for the natural topology is, as its name implies, the topology on $\boldsymbol{R}$ that is most useful to us. It is not the only nontrivial topology on $\boldsymbol{R}$, however. For example, suppose we let

$$\tau = \{\emptyset\} \cup \{\boldsymbol{R}\} \cup \{(a, +\infty) \mid a \in \boldsymbol{R}\}$$

Then τ is a topology on $\boldsymbol{R}$ called the *open ray topology*. It is neither the discrete nor the indiscrete topology, and it is not the natural topology because it is smaller than the natural topology. We leave the proof of these facts to the reader. (See Exercise 10 at the end of this section.)

We conclude this section with a method for constructing new topological spaces from old ones. The next proposition says that, given a topological space (X, τ) we can topologize any subset Y of X by intersecting Y with the open sets of τ.

Proposition 3.1.8 Let (X, τ) be a topological space and let Y be a subset of X. If

$$\tau' = \{G' \subset Y \mid G' = Y \cap G \text{ for some } G \in \tau\}$$

then τ' is a topology on Y called the *subspace topology induced on Y by τ*.

Proof: Since $X \in \tau$ and $\emptyset \in \tau$, it follows that $Y = Y \cap X \in \tau'$ and $\emptyset = Y \cap \emptyset \in \tau'$.

Let A be an index set and suppose that $G'_\alpha \in \tau'$ for all $\alpha \in A$; then for each $\alpha \in A$ we have $G'_\alpha = Y \cap G_\alpha$ for some $G_\alpha \in \tau$. Thus

$$\bigcup_{\alpha \in A} G'_\alpha = \bigcup_{\alpha \in A} (Y \cap G_\alpha) = Y \cap \left(\bigcup_{\alpha \in A} G_\alpha \right)$$

where $G_\alpha \in \tau$ for each $\alpha \in A$. Since τ is a topology, $\bigcup_{\alpha \in A} G_\alpha \in \tau$, and therefore $\bigcup_{\alpha \in A} G'_\alpha = Y \cap (\bigcup_{\alpha \in A} G_\alpha) \in \tau'$.

Now let A be a finite index set and suppose that $G'_\alpha \in \tau'$ for all $\alpha \in A$; then for each $\alpha \in A$ we have $G'_\alpha = Y \cap G_\alpha$ for some $G_\alpha \in \tau$, and hence

$$\bigcap_{\alpha \in A} G'_\alpha = \bigcap_{\alpha \in A} (Y \cap G_\alpha) = Y \cap \left(\bigcap_{\alpha \in A} G_\alpha \right)$$

where $G_\alpha \in \tau$ for each $\alpha \in A$. Since τ is a topology, $\bigcap_{\alpha \in A} G_\alpha \in \tau$, and therefore $\bigcap_{\alpha \in A} G'_\alpha \in \tau'$. □

Examples 3.1.9 1. Let $X = \{x, y, z\}$ and $\tau = \{\emptyset, \{x, y\}, X\}$. Let $Y = \{y, z\}$. The subspace topology induced on Y by τ is the topology $\tau' = \{\emptyset, \{y\}, Y\}$.

2. Consider the closed unit interval $[0, 1]$ in $\boldsymbol{R}$. Let us describe the subspace topology induced on $[0, 1]$ by the natural topology on $\boldsymbol{R}$. A subset G of $[0, 1]$ is open in the subspace topology if and only if it is the intersection of $[0, 1]$ with a set which is open in the natural topology on $\boldsymbol{R}$; that is, if and only if it is the intersection of $[0, 1]$ with a union of open intervals. It is easy to check that the intersection $[0, 1]$ with any open interval is either $\emptyset$, $[0, 1]$, an open interval (a, b), or a half-open interval of the form $(a, 1]$ or $[0, b)$. Therefore a nonempty proper subset of $[0, 1]$ is open in the subspace topology if and only if it is a union of intervals of the form (a, b), $(a, 1]$, or $[0, b)$, where $0 \leqslant a < b \leqslant 1$.

EXERCISES

1. Let $X = \{x, y\}$. Find all topologies on X. For each topology, find all the open sets, all the closed sets, all the sets which are both open and closed, and all the subsets of X which are neither open nor closed in the topology.

2. Repeat Exercise 1 for the set $X = \{x, y, z\}$. (There are 14 topologies on this set.)
3. Let $\tau = \{S \subset \boldsymbol{N} \mid S = \boldsymbol{N} \text{ or } S \text{ is a finite subset of } \boldsymbol{N}\}$. Is τ a topology on $\boldsymbol{N}$? Justify your answer.
4. Let $\tau = \{S \subset \boldsymbol{N} \mid S = \emptyset \text{ or } \boldsymbol{N} - S \text{ is a finite subset of } \boldsymbol{N}\}$. Is τ a topology on $\boldsymbol{N}$? Justify your answer.
5. Prove Proposition 3.1.4.
6. Prove that in the natural topology on $\boldsymbol{R}$ half-open intervals are neither open nor closed.
7. For each topology τ of Exercise 1 above, list all the topologies which are smaller than τ, all those which are larger than τ, and all those which are not comparable to τ.
8. For each topology τ of Exercise 2 above, list all the other topologies of Exercise 2 which are smaller than τ, those which are larger than τ, and those which are not comparable to τ.
9. Prove that the indiscrete topology is the smallest topology on any set and that the discrete topology is the largest topology on any set.
10. Let τ be the open ray topology on $\boldsymbol{R}$ defined by

$$\tau = \{\emptyset\} \cup \{\boldsymbol{R}\} \cup \{(a, +\infty) \mid a \in \boldsymbol{R}\}$$

Prove that τ is a topology on $\boldsymbol{R}$, that it is neither the discrete nor the indiscrete topology, and that it is smaller than the natural topology.
11. Considering the natural numbers $\boldsymbol{N}$ and the rational numbers $\boldsymbol{Q}$ as subsets of $\boldsymbol{R}$,
    (a) Show that the natural topology on $\boldsymbol{R}$ induces the discrete topology on $\boldsymbol{N}$;
    (b) Describe the open sets in the subspace topology induced on $\boldsymbol{Q}$ by the natural topology on $\boldsymbol{R}$.
12. Let $\boldsymbol{R}$ have the open ray topology of Exercise 10 above. Describe the open sets in the subspace topologies induced on $\boldsymbol{N}$ and on $\boldsymbol{Q}$ by the open ray topology.
13. Let X be a set, Y a nonempty subset of X, and τ_1 and τ_2 topologies on X with $\tau_1 \neq \tau_2$. Is it possible for τ_1 and τ_2 to induce the same subspace topology on Y? Justify your answer.
14. Let τ_1 and τ_2 be topologies on a set X and define

$$\tau_1 \cap \tau_2 = \{G \subset X \mid G \in \tau_1 \text{ and } G \in \tau_2\}$$

and

$$\tau_1 \cup \tau_2 = \{G \subset X \mid G \in \tau_1 \text{ or } G \in \tau_2\}$$

(a) Prove that $\tau_1 \cap \tau_2$ is a topology on X which is smaller than or equal to τ_1 and also smaller than or equal to τ_2.

(b) Give an example to show that $\tau_1 \cup \tau_2$ need not be a topology.

15. Let X be a set and $\mathscr{F}$ be a collection of subsets of X such that

(a) $\emptyset \in \mathscr{F}$ and $X \in \mathscr{F}$;

(b) If A is an index set and $F_\alpha \in \mathscr{F}$ for all $\alpha \in A$, then $\bigcap_{\alpha \in A} F_\alpha \in \mathscr{F}$;

(c) If A is a finite index set and $F_\alpha \in \mathscr{F}$ for all $\alpha \in A$, then $\bigcup_{\alpha \in A} F_\alpha \in \mathscr{F}$.

Let $\tau = \{G \subset X \mid X - G \in \mathscr{F}\}$. Prove that τ is a topology on X and that $\mathscr{F}$ is the collection of closed sets for the topology τ.

16. Consider the Euclidean plane $\boldsymbol{R}^2 = \boldsymbol{R} \times \boldsymbol{R}$. A subset S of $\boldsymbol{R}^2$ is an *open rectangle* if it is of the form

$$S = \{(x, y) \mid a < x < b, c < y < d\}$$

for some real numbers a, b, c, and d. A subset D of $\boldsymbol{R}^2$ is an *open disc with center* (h, k) *and radius* r if it is of the form

$$D = \{(x, y) \mid (x - h)^2 + (y - k)^2 < r^2\}$$

for some real numbers h and k and some positive real number r.

(a) Let $\tau_1 = \{G \subset \boldsymbol{R}^2 \mid G = \emptyset$ or G is a union of open rectangles$\}$. Prove that τ_1 is a topology on $\boldsymbol{R}^2$. This topology is called the *product topology* on $\boldsymbol{R}^2$.

(b) Let $\tau_2 = \{G \subset \boldsymbol{R}^2 \mid G = \emptyset$ or G is a union of open discs$\}$. Prove that τ_2 is the product topology of part (a).

17. Let X be a set, τ be a topology on X, and $x \in X$. A subset V of X is a *neighborhood of* x if there is some open set $G \in \tau$ such that $G \subset V$ and $x \in G$.

(a) Let $X = \{x, y, z\}$ and $\tau = \{\emptyset, \{x\}, \{x, y\}, X\}$. Find all neighborhoods of each element of X.

(b) Suppose $X = \boldsymbol{R}$ and τ is the natural topology. Show that V is a neighborhood of x if and only if there is some open interval I such that $x \in I$ and $I \subset V$.

(c) Suppose τ is the discrete topology. For any $x \in X$, describe all neighborhoods of x. Do the same if τ is the indiscrete topology.

(d) Prove that a subset G of X is open in τ if and only if G is a neighborhood of each of its points.

3.2 OPEN SETS AND CLOSED SETS

An understanding of open sets and closed sets is fundamental to an understanding of topological spaces. In this section we characterize open and closed sets in a general topological space. We remark here that if (X, τ) is a topological space and τ is the only topology on X which is under consideration, we will speak of "the topological space X" rather than "the topological space (X, τ)." We will also refer to "open sets in X" rather than use the more precise language "sets open in the topology on X," and similarly for closed sets. This should cause no confusion.

Our first proposition collects some elementary properties of open sets and closed sets. Its proof follows directly from the definitions of these concepts and will be left to the reader.

Proposition 3.2.1 Let X be a topological space and A be an index set. If for all $\alpha \in A$, G_α is an open set in X and F_α a closed set in X, then

1. $\bigcup_{\alpha \in A} G_\alpha$ is an open set in X and $\bigcap_{\alpha \in A} F_\alpha$ is a closed set in X.
2. If A is finite, $\bigcap_{\alpha \in A} G_\alpha$ is an open set in X and $\bigcup_{\alpha \in A} F_\alpha$ a closed set in X.

Suppose that (X, τ) is a topological space. According to the definition of open set, a subset G of X is open in X if and only if $G \in \tau$. This is what is known as a *global characterization* of open sets: if we wish to use this definition to test whether or not a given set is open, we must check to see whether or not the set as a whole satisfies the required condition. It would be convenient for us to have a *local characterization* of open sets, that is, one which would enable us to test whether or not a given set is open by checking to see if an arbitrary element of the set, rather than the set as a whole, satisfies some condition. The next proposition affords us just such a local characterization of open sets. Note that the characterization is similar to the property we used to define the natural topology on $\boldsymbol{R}$.

Proposition 3.2.2 Let X be a topological space. A subset G of X is open in X if and only if for each $x \in G$ there is some open set H_x such that $x \in H_x$ and $H_x \subset G$.

Proof: Let G be open in X. If $G = \emptyset$ then it is vacuously true that for each $x \in G$ there is some open set H_x such that $x \in H_x \subset G$. (Why?) If $G \neq \emptyset$ and $x \in G$, choose $H_x = G$. Thus, in either case, if G is open there is an open set H_x such that $x \in H_x \subset G$.

To prove the converse, let G be a subset of X such that for each $x \in G$ there is some open set H_x such that $x \in H_x \subset G$. But then $G = \bigcup_{x \in G} H_x$ is a union of open sets and hence is itself open. □

We now have two characterizations for open sets: a global one which says that a set G is open if and only if it belongs to the topology, and a local one which says that a set G is open if and only if we can surround each of its points with an open set which is itself contained in G.

Now we turn our attention to closed sets. We have a global characterization of closed sets: a subset F of the topological space X is closed in X if and only if its complement $X - F$ is open in X. We would like to obtain a local characterization of closed sets. In order to do so, we must introduce the concept of a *limit point* of a set. This concept is important not only for closed sets, but for all sets in a topological space.

Definition 3.2.3 Let X be a topological space and let S be a subset of X. A point $x \in X$ is a *limit point* of S if whenever G is open in X and $x \in G$, then

$$(G \cap S) - \{x\} \neq \emptyset$$

The set of all limit points of S, called the *derived set* of S, is denoted by $\dot{S}$.

Note that x is a limit point of S if every open set containing x also contains at least one point of S different from x. Note also that a limit point of S need not be an element of S.

Examples 3.2.4 1. Let $X = \{x, y, z\}$ and $\tau = \{\emptyset, \{x\}, \{x, y\}, X\}$, and consider the set $S = \{x\}$. The element x is *not* a limit point of S because $\{x\}$ is an open set containing x which does not contain any point of S different from x. However, y is a limit point of S because the only open sets which contain y are $\{x, y\}$ and X, and each of these contains a point of S different from y. Similarly, z is a limit point of S. Therefore

$$\dot{S} = \{y, z\}$$

Note that S is not a closed set and that $\dot{S} \not\subset S$.

Now consider the subset $T = \{y, z\}$ of X. The element x is not a limit point of T because $\{x\}$ is an open set containing x which does not contain a point of T different from x. Furthermore, y is not a limit point of T because $\{x, y\}$ is an open set containing y which does not contain a point of T different from y. However, z is a limit point of T since the only open set containing z is X, and X contains a point of T different from z. Therefore

$$\dot{T} = \{z\}$$

Note that T is closed and that $\dot{T} \subset T$.

2. Let $\boldsymbol{R}$ have the natural topology and consider the set $\boldsymbol{Z}$ of integers as a subset of $\boldsymbol{R}$. For any $x \in \boldsymbol{R}$ we can always find an open interval which contains x and does not contain any element of $\boldsymbol{Z}$, except possibly x if $x \in \boldsymbol{Z}$. Therefore no element of $\boldsymbol{R}$ can be a limit point of $\boldsymbol{Z}$, and thus

$$\dot{\boldsymbol{Z}} = \emptyset$$

Now consider the set $\boldsymbol{Q}$ of rational numbers as a subset of $\boldsymbol{R}$. If $x \in \boldsymbol{R}$ and G is any open set containing x, then there is some open interval (a, b) such that $x \in (a, b) \subset G$. But then there is a rational number q such that $a < q < x$, and hence $q \in (\boldsymbol{Q} \cap G) - \{x\}$. Therefore every $x \in \boldsymbol{R}$ is a limit point of $\boldsymbol{Q}$, and thus

$$\dot{\boldsymbol{Q}} = \boldsymbol{R}$$

3. Let $\boldsymbol{R}$ have the natural topology and let S be a nonempty subset of $\boldsymbol{R}$ which is bounded above. If $\sup S \notin S$ and (a, b) is an open interval which contains $\sup S$, then (a, b) must contain a point of S, for otherwise the endpoint b would be a least upper bound for S which is less than $\sup S$. Hence if $\sup S \notin S$ then $\sup S$ is a limit point of S. A similar result holds for greatest lower bounds.

4. If $\boldsymbol{R}$ has the natural topology and $S = (a, b)$ is an open interval and $T = [a, b]$ a closed interval in $\boldsymbol{R}$, then

$$\dot{S} = [a, b] \qquad \text{and} \qquad \dot{T} = [a, b]$$

We leave the proof to the reader. (See Exercise 7 at the end of this section.) Note that S is not closed and does not contain all its limit points, while T is closed and does contain all its limit points.

Now we use the concept of limit point to obtain a local characterization of closed sets. The characterization says that a set is closed if and only if it contains all its limit points; it will appear as a corollary to the next proposition.

Proposition 3.2.5 If X is a topological space and S is a subset of X, then $S \cup \dot{S}$ is closed in X.

Proof: We prove that $S \cup \dot{S}$ is closed by showing that its complement is open. Since this is certainly the case if the complement is empty, we suppose that the complement of $S \cup \dot{S}$ is nonempty.

Let $x \in X - (S \cup \dot{S})$. Since $x \notin \dot{S}$, x is not a limit point of S and hence there must exist an open set G such that $x \in G$ and $(G \cap S) - \{x\} = \emptyset$. But since also $x \notin S$, it follows that in fact $G \cap S = \emptyset$. Furthermore, $G \cap \dot{S} = \emptyset$, for if this were not so, G would contain a limit point of S, and therefore a point of S different from the limit point, thus contradicting the fact that $G \cap S = \emptyset$. We have now shown that $G \cap S = \emptyset$ and that $G \cap \dot{S} = \emptyset$, and thus that G is contained in $X - (S \cup \dot{S})$. Therefore every element x in the complement of $S \cup \dot{S}$ is contained in an open set G, which is itself contained in the complement. Hence the complement is open, and we are done. □

Corollary 3.2.6 Let X be a topological space. A subset F of X is closed in X if and only if F contains all its limit points.

Proof: If F contains all its limit points, then $F = F \cup \dot{F}$, which is closed by the proposition. Conversely, suppose F is closed but does not contain all its limit points. In this case there is some $x \in X$ such that $x \notin F$ and x is a limit point of F. But then $x \in X - F$, which is an open set containing no point of F, and this contradicts the fact that x is a limit point of F. □

Now that we have considered open sets and closed sets, what about sets which are neither open nor closed? For example, consider the half-open interval $(a, b]$ in $\boldsymbol{R}$. This set, which is neither open nor closed in the natural topology, contains a "largest open set," namely (a, b), and is contained in a "smallest closed set," namely $[a, b]$. Thus if we wished we could study $(a, b]$ by analyzing the open set (a, b), the closed set $[a, b]$, and the "boundary set" $\{a, b\}$. This sort of dissection of a set can be done in any topological space, provided we define "largest open set," "smallest closed set," and "boundary set" properly.

Definition 3.2.7 Let X be a topological space and let S be a subset of X.

1. An open set G in X is the *largest open set contained in* S if G is contained in S and whenever H is an open set contained in S then $H \subset G$.

2. A closed set F in X is the *smallest closed set containing* S if F contains S and whenever E is a closed set containing S then $F \subset E$.

Note that if a largest open set contained in S exists, it must be unique: for if G and G' are both largest open sets contained in S then necessarily $G \subset G'$ and $G' \subset G$, so that $G = G'$. A similar remark applies to smallest closed sets containing S. Therefore use of the phrases "*the* largest open set" and "*the* smallest closed set" in the definition is justified. Furthermore, the largest open set G contained in S always exists: let $G = \bigcup H$, where H ranges over all open sets contained in S. Of course, if the only open set contained in S is the empty set, then $G = \emptyset$. Similarly, the smallest closed set F containing S always exists: let $F = \bigcap E$, where E ranges over all closed sets containing S. If the only closed set containing S is the space X, then $F = X$.

Definition 3.2.8 Let X be a topological space and let S be a subset of X.

1. The *interior* of S, denoted by S°, is the largest open set contained in S.
2. The *closure* of S, denoted by $\bar{S}$, is the smallest closed set containing S.
3. The *boundary* of S, denoted by bd S, is defined by bd $S = \bar{S} - S^\circ$.

The remarks preceding the definition show that for any subset S of a topological space X, the interior, closure, and boundary of S exist and are unique.

Examples 3.2.9 1. Let $X = \{x, y, z\}$ and $\tau = \{\emptyset, \{x\}, \{x, y\}, X\}$. Let

$$S = \{x\} \qquad T = \{y\} \qquad \text{and} \qquad V = \{x, z\}$$

We have

$$S^\circ = S \qquad \bar{S} = X \qquad \text{and} \qquad \text{bd } S = \{y, z\}$$

$$T^\circ = \emptyset \qquad \bar{T} = \{y, z\} \qquad \text{and} \qquad \text{bd } T = \{y, z\}$$

$$V^\circ = \{x\} \qquad \bar{V} = X \qquad \text{and} \qquad \text{bd } V = \{y, z\}$$

2. Let $\boldsymbol{R}$ have the natural topology and let S be any of the intervals (a, b), $[a, b]$, $(a, b]$, or $[a, b)$; then

$$S^\circ = (a, b) \qquad \bar{S} = [a, b] \qquad \text{and} \qquad \text{bd } S = \{a, b\}$$

Also

$$\boldsymbol{Z}^\circ = \emptyset \qquad \bar{\boldsymbol{Z}} = \boldsymbol{Z} \qquad \text{and} \qquad \text{bd } \boldsymbol{Z} = \boldsymbol{Z}$$

and

$$\boldsymbol{Q}^\circ = \emptyset \qquad \bar{\boldsymbol{Q}} = \boldsymbol{R} \qquad \text{and} \qquad \text{bd } \boldsymbol{Q} = \boldsymbol{R}$$

The interior, closure, and boundary can be used to characterize open and closed sets.

Proposition 3.2.10 If X is a topological space and S is a subset of X, then

1. $S^\circ \subset S \subset \bar{S}$.
2. S is open in X if and only if $S = S^\circ$.
3. S is closed in X if and only if $S = \bar{S}$.

Proof: Part (1) is true by definition. To prove (2), note that if S is open, then it is the largest open set contained in S; conversely, the interior S° of S is open by definition. The proof of (3) is similar and is left to the reader. (See Exercise 11 at the end of this section.) □

Corollary 3.2.11 If X is a topological space and S is a subset of X, then

1. S is open in X if and only if $S \cap \text{bd } S = \emptyset$.
2. S is closed in X if and only if $\text{bd } S \subset S$.
3. S is both open and closed in X if and only if $\text{bd } S = \emptyset$.

Proof: We prove (1), leaving the rest as an exercise. (See Exercise 17 at the end of this section.)

If S is open, then $S = S^\circ$, and this implies that $\text{bd } S = \bar{S} - S$. But then

$$S \cap \text{bd } S = S \cap (\bar{S} - S) = \emptyset$$

Conversely, if $S \cap \text{bd } S = \emptyset$, then $S \cap (\bar{S} - S^\circ) = \emptyset$, and since $S - S^\circ$ is contained in $\bar{S} - S^\circ$, it follows that $S - S^\circ = \emptyset$. But this implies that $S = S^\circ$ and thus that S is open. □

Definition 3.2.12 Let X be a topological space and let S be a subset of X. If $\bar{S} = X$, we say that *S is dense in X*.

Examples 3.2.13 1. Let $X = \{x, y, z\}$ and $\tau = \{\emptyset, \{x\}, \{x, y\}, X\}$. Then $\{x\}$ is dense in X but $\{y\}$ is not dense in X.

2. If [0, 1] has the subspace topology induced by the natural topology on $\boldsymbol{R}$, then (0, 1) is dense in [0, 1].

3. Let $\boldsymbol{R}$ have the natural topology. The set $\boldsymbol{Q}$ of rational numbers is dense in $\boldsymbol{R}$.

We conclude this section with another characterization of the interior, closure, and boundary of a set.

Proposition 3.2.14 If X is a topological space and S is a subset of X, then

1. $S^{\circ} = \{x \in S \mid x \in G \text{ for some open set } G \subset S\}$.
2. $\bar{S} = S \cup \dot{S}$.
3. $\text{bd } S = \{x \in X \mid \text{if } x \in G,\ G \text{ open, then } G \cap S \neq \emptyset \text{ and } G \cap (X - S) \neq \emptyset\}$.

Proof: We prove (1) and (2), leaving (3) as an exercise. (See Exercise 18 at the end of this section.)

Let us put $T = \{x \in S \mid x \in G \text{ for some open set } G \subset S\}$: then in order prove (1) we must show that $S^{\circ} = T$. If $S^{\circ} = \emptyset$, then the only open set contained in S is $\emptyset$, and T must be empty also. On the other hand, if $T = \emptyset$ then S cannot contain a nonempty open set and hence $S^{\circ} = \emptyset$. Therefore (1) holds if either S° or T is empty, so we assume that these sets are not empty.

If $x \in S^{\circ}$ then $x \in T$, since S° is an open set contained in S. Therefore $S^{\circ} \subset T$. Conversely, if $x \in T$, then there is some open set G such that $x \in G$ and $G \subset S$. But then $G \subset S^{\circ}$, because S° is the largest open set contained in S. Therefore $T \subset S^{\circ}$, and we have proved (1).

Now we prove (2). Since $S \subset \bar{S}$, every limit point of S is also a limit point of $\bar{S}$. (See Exercise 8 at the end of this section.) Since $\bar{S}$ is closed, it contains all its limit points. Therefore $\bar{S}$ contains all the limit points of S, and it follows that $(S \cup \dot{S}) \subset \bar{S}$. On the other hand, since $S \cup \dot{S}$ is a closed set containing S, we must have $\bar{S} \subset (S \cup \dot{S})$, and we are done. □

Corollary 3.2.15 Let X be a topological space. A subset S of X is dense in X if and only if every point of X is either a point of S or a limit point of S.

EXERCISES

1. Prove Proposition 3.2.1.
2. Give an example of a topological space X and a collection $\{G_{\alpha} \mid \alpha \in A\}$ of open sets in X such that $\bigcap_{\alpha \in A} G_{\alpha}$ is not an open set in X.

3. Give an example of a topological space X and a collection $\{F_\alpha \mid \alpha \in A\}$ of closed sets in X such that $\bigcup_{\alpha \in A} F_\alpha$ is not a closed set in X.
4. Let $[0, 1]$ have the subspace topology induced by the natural topology on $\boldsymbol{R}$. Prove that the Cantor ternary set is closed but not open in the subspace topology.
5. Let X be a discrete topological space and let S be any subset of X. Show that $\dot{S} = \emptyset$.
6. Let X be an indiscrete topological space and let S be any nonempty subset of X. Show that either $\dot{S} = X$ or $\dot{S} = X - S$.
7. Let $\boldsymbol{R}$ have the natural topology and let $S = (a, b)$ and $T = [a, b]$. Prove that $\dot{S} = \dot{T} = [a, b]$.
8. Let X be a topological space, and let S and T be subsets of X such that $S \subset T$. Prove that $\dot{S} \subset \dot{T}$.
9. Let $\boldsymbol{R}$ have the open ray topology, and let $S = [0, 1]$. Find $\dot{\boldsymbol{Q}}$, $\dot{\boldsymbol{Z}}$, and $\dot{S}$ in this topology.
10. Let $[0, 1]$ have the subspace topology induced by the natural topology on $\boldsymbol{R}$. Let C denote the Cantor ternary set. Show that $\dot{C} = C$.
11. Prove (3) of Proposition 3.2.10.
12. Let $X = \{x, y, z, w\}$ and $\tau = \{\emptyset, \{x\}, \{x, y\}, \{x, y, z\}, X\}$. Let $S = \{x, y\}$ and $T = \{x, z\}$. Find $\dot{S}$, S°, $\bar{S}$, bd S, $\dot{T}$, T°, $\bar{T}$, and bd T.
13. Let $\boldsymbol{R}$ have the open ray topology. Find $\boldsymbol{Q}^\circ$, $\bar{\boldsymbol{Q}}$, bd $\boldsymbol{Q}$, $\boldsymbol{Z}^\circ$, $\bar{\boldsymbol{Z}}$, and bd $\boldsymbol{Z}$. If $S = [0, 1]$, find S°, $\bar{S}$ and bd S.
14. Let the Euclidean plane $\boldsymbol{R}^2$ have the product topology. (See Exercise 16 of Section 3.1.) Let D be an open disc in $\boldsymbol{R}^2$ and L a line in $\boldsymbol{R}^2$. Find $\dot{D}$, D°, $\bar{D}$, bd D, $\dot{L}$, L°, $\bar{L}$, and bd L. Is L open in $\boldsymbol{R}^2$? closed in $\boldsymbol{R}^2$?
15. Let X be a topological space and let S be a subset of X. Let $\bar{S}$ have the subspace topology induced by the topology on X. Prove that S is dense in $\bar{S}$.
16. Give an example of topological space X which is
    (a) Infinite and such that every nonempty subset of X is dense in X.
    (b) Infinite and such that no proper subset of X is dense in X.
    (c) Uncountable, not $\boldsymbol{R}$, and such that it contains a countable subset which is dense in X and a countable subset which is not dense in X.
17. Finish the proof of Corollary 3.2.11.
18. Finish the proof of Proposition 3.2.14.
19. Let X be a topological space and let S be a subset of X. Prove that $(S^\circ)^\circ = S^\circ$ and that $\bar{\bar{S}} = S$.
20. Let X be a topological space. Let S and T be subsets of X.

(a) Prove that $(S \cap T)^\circ = S^\circ \cap T^\circ$.
(b) Prove that $\overline{S \cup T} = \bar{S} \cup \bar{T}$.
(c) Give examples to show that in general $(S \cup T)^\circ \neq S^\circ \cup T^\circ$ and $\overline{S \cap T} \neq \bar{S} \cap \bar{T}$.

21. Let X be a topological space.
(a) Suppose X has the property that every subset of X which consists of exactly one point is closed. Prove that if S is any subset of X then S is closed in X.
(b) Give an example of topological space X and a subset S of X such that S is not closed in X.

4
Continuous Functions

In this chapter we define the notion of a continuous function from one topological space to another, examine the properties of such functions, and use them to compare topological spaces. We then study the important topological properties of connectedness and compactness and the preservation of these by continuous functions. As usual, all of our general topological results will be applied to the set $\boldsymbol{R}$ of real numbers, and conversely we will use our knowledge of the system $\boldsymbol{R}$ to suggest topological generalizations.

4.1 CONTINUITY

We have stated that we intend to compare topological spaces by means of continuous functions, but what is a continuous function from one topological space to another? The reader is undoubtedly familiar to some extent with continuous functions from $\boldsymbol{R}$ to $\boldsymbol{R}$. Let us examine such functions to see what it is they do in terms of the natural topology on $\boldsymbol{R}$.

A common intuitive definition of a continuous function from $\boldsymbol{R}$ to $\boldsymbol{R}$ that is sometimes used in elementary courses is this: a function f from $\boldsymbol{R}$ to $\boldsymbol{R}$ is continuous if one can draw its graph without lifting one's pencil from the paper. What does this mean in terms of the natural topology on $\boldsymbol{R}$? To be more specific, what does it mean in terms of the basic open sets

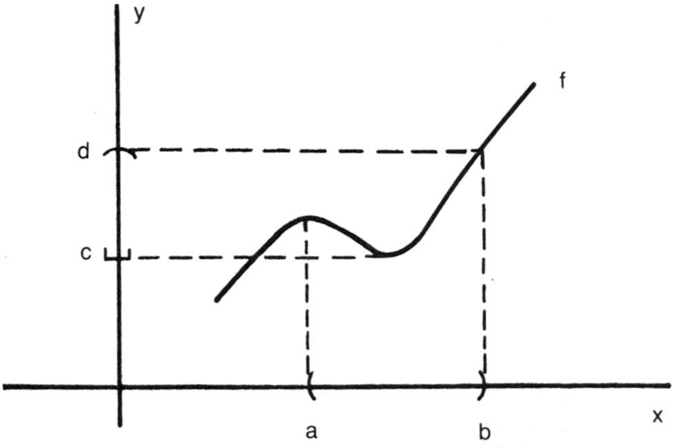

Figure 4.1 The continuous image of an open interval need not be an open set.

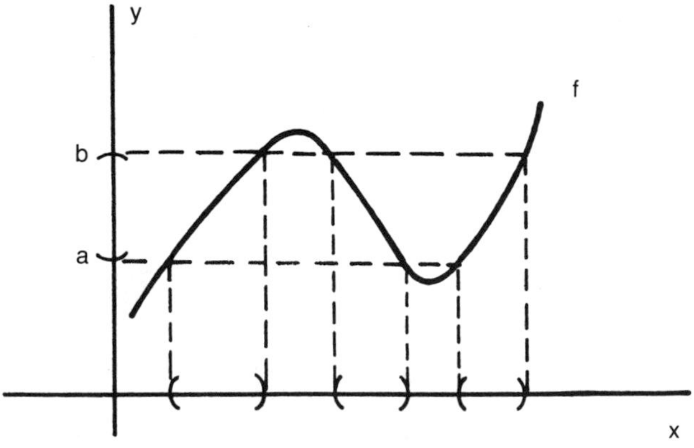

Figure 4.2 The preimage of an open interval under a continuous function is an open set.

in the natural topology, namely, the open intervals? As Figure 4.1 shows, a function from $\boldsymbol{R}$ to $\boldsymbol{R}$ which is continuous according to our intuitive definition need not carry open intervals onto open intervals, or even onto open sets. Thus in Figure 4.1 the image under the continuous function f of the open interval (a, b) is the half-open interval $[c, d)$. Hence we may conclude that the continuous image of an open set is not always an open set.

Now suppose we consider preimages under continuous functions. If we begin with an open set in the range of a continuous function and take its preimage, we will obtain a set in the domain of the function. Figure 4.2 shows a continuous function f from $\boldsymbol{R}$ to $\boldsymbol{R}$. Note that in the figure the preimage under f of the open interval (a, b) in the range is a union of open intervals, and hence an open set, in the domain. This is always so for a continuous function from $\boldsymbol{R}$ to $\boldsymbol{R}$: under such a function, the preimage of an open set is an open set. On the other hand, if a function from $\boldsymbol{R}$ to $\boldsymbol{R}$ is *not* continuous, then there will be some open set in its range whose preimage under the function is not open in its domain. For example, Figure 4.3 shows a function g from $\boldsymbol{R}$ to $\boldsymbol{R}$ which is not continuous: note that the preimage under g of the open interval (a, b) is not an open set.

The above remarks suggest that if we want continuity in topological spaces to be a generalization of our intuitive notion of continuity on $\boldsymbol{R}$,

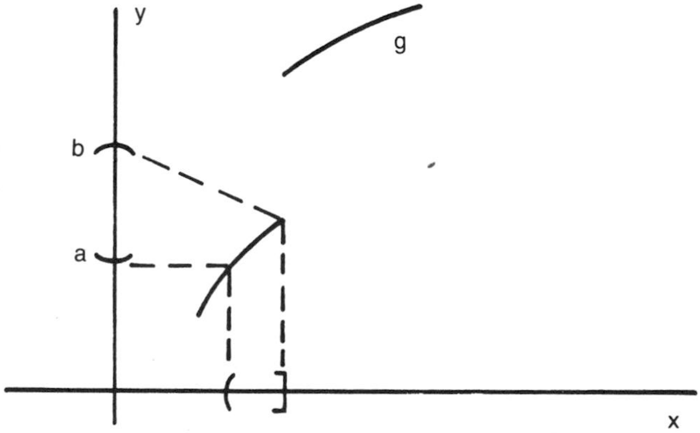

$g^{-1}((a,b))$ is a half-open interval

Figure 4.3 The preimage of some open interval under a discontinuous function will not be an open set.

then we should consider a function from one topological space to another to be continuous provided that the preimage of any open set in its range is an open set in its domain.

Definition 4.1.1 Let (X, τ) and (Y, σ) be topological spaces. A function $f\colon X \to Y$ is *continuous* if whenever $G \in \sigma$ its preimage $f^{-1}(G) \in \tau$. If f is not continuous, it is said to be *discontinuous*.

Before we proceed to some examples, let us agree that from now on the set $\boldsymbol{R}$ of real numbers will always have the natural topology unless we explicitly state otherwise. This will save us constant repetition of the phrase "let $\boldsymbol{R}$ have the natural topology."

Examples 4.1.2 1. Let $X = \{x, y, z\}$ and $\tau = \{\emptyset, \{x\}, \{x, y\}, X\}$. Let $Y = \{\alpha, \beta, \gamma\}$ and $\sigma = \{\emptyset, \{\alpha\}, Y\}$. Let $f\colon X \to Y$ and $g\colon X \to Y$ be given by

$$f(x) = \alpha \qquad f(y) = f(z) = \beta$$

and

$$g(x) = g(z) = \alpha \qquad g(y) = \beta$$

We have

$$f^{-1}(\emptyset) = \emptyset \qquad f^{-1}(\{\alpha\}) = \{x\} \qquad \text{and} \qquad f^{-1}(Y) = X$$

Therefore the preimage under f of every open set in Y is an open set in X, so f is a continuous function from X to Y. For g, on the other hand,

$$g^{-1}(\emptyset) = \emptyset \qquad g^{-1}(Y) = X \qquad \text{and} \qquad g^{-1}(\{\alpha\}) = \{x, z\}$$

and since $\{\alpha\}$ is open in Y but its preimage under g is not open in X, g is a discontinuous function from X to Y.

2. Continuity depends on the topology as well as on the function. A function which is continuous for one topology may not be continuous for another. For instance, suppose we define X, Y, σ, f, and g as in Example 1 above, but change the topology on X by setting $\tau = \{\emptyset, \{x, z\}, X\}$; then f is a discontinuous function from X to Y and g a continuous one. (Check this.)

3. Every constant function is continuous. To see this, suppose X and Y are topological spaces and $f\colon X \to Y$ is given by $f(x) = c$ for all $x \in X$. If

G is an open set in Y, then

$$f^{-1}(G) = \begin{cases} \emptyset & \text{if } c \notin G \\ X & \text{if } c \in G \end{cases}$$

Therefore $f^{-1}(G)$ is always open in X, so f is continuous.

4. Let $f\colon \boldsymbol{R} \to \boldsymbol{R}$ be given by $f(x) = x + 1$ for all $x \in \boldsymbol{R}$. Let us show that f is continuous. It will suffice to show that under f the preimage of every open interval is an open set in the natural topology. (Why?) But this is so, for if (a, b) is an open interval $f^{-1}((a, b)) = (a - 1, b - 1)$.

5. Let $\chi_{\boldsymbol{Q}}\colon \boldsymbol{R} \to \boldsymbol{R}$ be defined as follows:

$$\chi_{\boldsymbol{Q}}(x) = \begin{cases} 1 & \text{if } x \in \boldsymbol{Q} \\ 0 & \text{if } x \in \boldsymbol{R} - \boldsymbol{Q} \end{cases}$$

The function $\chi_{\boldsymbol{Q}}$ is called the *characteristic function of the rationals*. It is discontinuous because every open interval (a, b) contains both rational and irrational numbers and hence

$$\chi_{\boldsymbol{Q}}^{-1}((a, b)) = \{0, 1\}$$

and $\{0, 1\}$ is not an open set in the natural topology on $\boldsymbol{R}$.

6. A function $f\colon X \to Y$ may be discontinuous, but its restriction to a subset of X supplied with the subspace topology may be continuous. For instance, suppose $f\colon \boldsymbol{R} \to \boldsymbol{R}$ is given by

$$f(x) = \begin{cases} 1 & \text{if } x \in [0, 1] \\ 0 & \text{if } x \notin [0, 1] \end{cases}$$

Then f is discontinuous, since $f^{-1}((\frac{1}{2}, \frac{3}{2})) = [0, 1]$, which is not an open set. However, if we restrict f to the subset $[0, 1]$ supplied with the subspace topology induced by the natural topology, the restriction of f to $[0, 1]$ is a constant function, which is continuous by virtue of Example 3 above.

The definition of a continuous function from one topological space to another is given in terms of open sets, but of course there must be an equivalent formulation in terms of closed sets.

Proposition 4.1.3 Let X and Y be topological spaces. A function $f\colon X \to Y$ is continuous if and only if whenever F is a closed set in Y its preimage $f^{-1}(F)$ is a closed set in X.

Proof: Suppose f is a continuous function. If F is a closed set in Y, then $Y - F$ is an open set in Y and therefore

$$f^{-1}(Y - F) = f^{-1}(Y) - f^{-1}(F) = X - f^{-1}(F)$$

must be an open set in X. But $X - f^{-1}(F)$ open implies that $f^{-1}(F)$ is closed.

Conversely, suppose that $f^{-1}(F)$ is closed in X whenever F is closed in Y. If G is open in Y, then $Y - G$ is closed in Y, and therefore

$$f^{-1}(Y - G) = f^{-1}(Y) - f^{-1}(G) = X - f^{-1}(G)$$

must be closed in X. But this implies that $f^{-1}(G)$ is open in X. Hence f is continuous. □

The next proposition is sometimes useful when a function must be tested for continuity. It is really just a restatement of the definition of continuity in terms of the domain of the function rather than its range.

Proposition 4.1.4 Let X and Y be topological spaces. A function $f: X \to Y$ is continuous if and only if $f(\bar{S}) \subset \overline{f(S)}$ for all subsets S of X.

Proof: Suppose f is continuous. Then since $\overline{f(S)}$ is closed in Y, we have $f^{-1}(\overline{f(S)})$ closed in X. But $S \subset f^{-1}(\overline{f(S)})$, and thus $f^{-1}(\overline{f(S)})$ closed implies that $\bar{S} \subset f^{-1}(\overline{f(S)})$. Applying f to both sides of this last inclusion, we obtain $f(\bar{S}) \subset \overline{f(S)}$.

To prove the converse, let us suppose that f is a function from X to Y such that $f(\bar{S}) \subset \overline{f(S)}$ for every subset S of X. Let F be a closed set in Y and let $S = f^{-1}(F)$. We will be done if we can show that S is closed in X. We have

$$f(\bar{S}) \subset \overline{f(S)} = \overline{f(f^{-1}(F))} = \bar{F} = F$$

But $f(\bar{S}) \subset F$ implies that $\bar{S} \subset f^{-1}(F) = S$. We have thus shown that $\bar{S} \subset S$, and since $S \subset \bar{S}$ always holds, we conclude that $S = \bar{S}$. Therefore S is a closed set in X. □

Thus far all our characterizations of continuity have been global in nature. The next proposition affords us a local characterization of continuity.

Proposition 4.1.5 Let X and Y be topological spaces. A function $f: X \to Y$ is continuous if and only if it satisfies the following property for each $x \in X$:

> Whenever G is an open set in Y such that $f(x) \in G$, there is some open set H in X such that $x \in H$ and $f(H) \subset G$.

Proof: Let f be continuous and let $x \in X$. If G is open in Y and $f(x) \in G$, then $f^{-1}(G)$ is open in X and $x \in f^{-1}(G)$, so we may take $H = f^{-1}(G)$.

Now suppose that f satisfies the property stated in the proposition. Let G be any set open in Y and consider $f^{-1}(G)$. If $f^{-1}(G)$ is empty, then it is certainly an open set; if it is not empty, then for each $x \in f^{-1}(G)$ there is some open set H_x such that $x \in H_x$ and $f(H_x) \subset G$. But $f(H_x) \subset G$ implies that $H_x \subset f^{-1}(G)$, and it follows that

$$f^{-1}(G) = \bigcup_x H_x$$

where x ranges over all elements of $f^{-1}(G)$. This shows that $f^{-1}(G)$ is a union of open sets in X and hence is itself open in X, and therefore that f is continuous. □

Corollary 4.1.6 Let S be a subset of $\boldsymbol{R}$ and let S have the subspace topology induced by the natural topology on $\boldsymbol{R}$. A function $f: S \to \boldsymbol{R}$ is continuous if and only if it satisfies the following property for all $x \in S$:

> Whenever (c, d) is an open interval containing x, there is some open interval (a, b) such that $x \in (a, b)$ and $f(S \cap (a, b)) \subset (c, d)$.

The proof of the Corollary is left as an exercise. (See Exercise 6 at the end of this section.)

Proposition 4.1.5 not only gives us a local characterization of continuity, but it also suggests how we may define continuity at a point: we will say that a function f from X to Y is continuous at a point $x \in X$ if it satisfies the property of the proposition at x.

Definition 4.1.7 Let X and Y be topological spaces and let $x \in X$. A function $f: X \to Y$ is *continuous at* x if it satisfies the following property:

> Whenever G is an open set in Y such that $f(x) \in G$, there is some open set H in X that $x \in H$ and $f(H) \subset G$.

If f is not continuous at x, it is *discontinuous at* x.

Note that in the light of the preceding definition and Proposition 4.1.5, we may conclude that a function is continuous if and only if it is continuous at every point of its domain.

Examples 4.1.8 1. Let $X = \{x, y, z\}$ and $\tau = \{\emptyset, \{x\}, X\}$. Let $Y = \{\alpha, \beta, \gamma\}$ and $\sigma = \{\emptyset, \{\alpha, \beta\}, Y\}$. Let $f\colon X \to Y$ be given by

$$f(x) = \alpha \qquad f(y) = \beta \qquad \text{and} \qquad f(z) = \gamma$$

Clearly, f is continuous at x: for any open set G which contains $f(x) = \alpha$ we always have $x \in \{x\}$ and $f(\{x\}) \subset G$, and therefore we may always choose the open set H of Definition 4.1.7 to be the set $\{x\}$. Similarly, f is continuous at z because we may always choose $H = X$. However, f is discontinuous at y, because $\{\alpha, \beta\}$ is an open set which contains $f(y) = \beta$, and there is no open set H such that $y \in H$ and $f(H) \subset \{\alpha, \beta\}$. Thus the function f is not a continuous function from X to Y.

2. Let $f\colon \boldsymbol{R} \to \boldsymbol{R}$ be given by $f(x) = 2x - 1$ for all $x \in \boldsymbol{R}$. We will show that f is continuous at each $x \in \boldsymbol{R}$ and hence that it is a continuous function from $\boldsymbol{R}$ to $\boldsymbol{R}$. To show that f is continuous at $x \in \boldsymbol{R}$, it suffices to show that whenever $f(x) = 2x + 1 \in (c, d)$, there is some open interval (a, b) such that $x \in (a, b)$ and $f((a, b)) \subset (c, d)$. But if $a = (c + 1)/2$ and $b = (d + 1)/2$, this is the case, so we are done.

3. Let $f\colon \boldsymbol{R} \to \boldsymbol{R}$ be given by

$$f(x) = \begin{cases} 1 & \text{if } x > 0 \\ -1 & \text{if } x \leqslant 0 \end{cases}$$

Then f is discontinuous at $x = 0$: for $f(0) = -1 \in (-2, 0)$, but there is no open interval (a, b) such that $0 \in (a, b)$ and $f((a, b)) \subset (-2, 0)$. On the other hand, f is continuous at x for all $x \neq 0$. For instance, if $x > 0$ and $f(x) = 1 \in (c, d)$, then $(x/2, 3x/2)$ is an open interval such that $x \in (x/2, 3x/2)$ and

$$f\left(\left(\frac{x}{2}, \frac{3x}{2}\right)\right) = \{1\} \subset (c, d)$$

A similar argument holds if $x < 0$.

4. Let $\chi_{\boldsymbol{Q}}$ be the characteristic function of the rationals. Thus

$$\chi_{\boldsymbol{Q}}(x) = \begin{cases} 1 & \text{if } x \in \boldsymbol{Q} \\ 0 & \text{if } x \in \boldsymbol{R} - \boldsymbol{Q} \end{cases}$$

Since every open interval (a, b) contains both rational and irrational numbers, it follows that

$$\chi_Q((a, b)) = \{0, 1\}$$

for every open interval (a, b). Therefore if x is rational, so that $\chi_Q(x) = 1$, then there can be no open interval (a, b) such that

$$\chi_Q((a, b)) \subset (\tfrac{1}{2}, \tfrac{3}{2})$$

Hence the function is discontinuous at x if x is rational. Similarly, there can be no open interval (a, b) such that

$$\chi_Q((a, b)) \subset (-\tfrac{1}{2}, \tfrac{1}{2})$$

so the function is discontinuous at x if x is irrational. Therefore the characteristic function of the rationals χ_Q is discontinuous at every real number.

Now that we have various characterizations of continuity at our disposal, we turn to methods of combining known continuous functions to obtain new ones. Our first result says that the composition of continuous functions is a continuous function.

Proposition 4.1.9 Let X, Y, and W be topological spaces and let $f: X \to Y$ and $g: Y \to W$. If f is continuous at $x \in X$ and g is continuous at $f(x) \in Y$, then $g \circ f$ is continuous at x.

Proof: Let G be an open set in W such that $(g \circ f)(x)$ is an element of G. Since g is continuous at $f(x)$, there is some open set H in Y such that $f(x) \in H$ and $g(H) \subset G$. Since f is continuous at x, there is some open set K in X such that $x \in K$ and $f(K) \subset H$. But then K is open, $x \in K$, and since

$$(g \circ f)(K) = g(f(K)) \subset g(H) \subset G$$

we have $g \circ f$ continuous at x. □

Corollary 4.1.10 Let X, Y, and W be topological spaces. If $f: X \to Y$ and $g: Y \to W$ are continuous functions, then the composite function $g \circ f: X \to W$ is also continuous.

Since we do not in general have the arithmetic operations of addition, subtraction, multiplication, and division defined on topological spaces, we cannot always arithmetically combine functions on such spaces. We can do so for functions on $\boldsymbol{R}$, however. The next definition shows how functions on $\boldsymbol{R}$ can be combined arithmetically, and the proposition which follows demonstrates that arithmetic combinations of continuous functions are continuous.

Definition 4.1.11 Let S be a subset of $\boldsymbol{R}$, with the subspace topology induced by the natural topology on $\boldsymbol{R}$. Let f and g be functions from S to $\boldsymbol{R}$. Define

1. $rf: S \to \boldsymbol{R}$, for any $r \in \boldsymbol{R}$, by $(rf)(x) = r \cdot f(x)$ for all $x \in S$;
2. $f+g: S \to \boldsymbol{R}$ by $(f+g)(x) = f(x) + g(x)$ for all $x \in S$;
3. $f-g: S \to \boldsymbol{R}$ by $(f-g)(x) = f(x) - g(x)$ for all $x \in S$;
4. $fg: S \to \boldsymbol{R}$ by $(fg)(x) = f(x) \cdot g(x)$ for all $x \in S$;
5. $f/g: S \to \boldsymbol{R}$ by $(f/g)(x) = f(x)/g(x)$ for all $x \in S$, provided that $g(x) \neq 0$ for all $x \in S$.

Proposition 4.1.12 Let S be any subset of $\boldsymbol{R}$ and let S have the subspace topology induced by the natural topology on $\boldsymbol{R}$. If $f: S \to \boldsymbol{R}$ and $g: S \to \boldsymbol{R}$ are functions which are continuous at $x \in S$, then

1. The function rf is continuous at x for every $r \in \boldsymbol{R}$;
2. The functions $f+g$ and $f-g$ are continuous at x;
3. The function fg is continuous at x;
4. The function f/g is continuous at x if g is never zero on S.

Proof: We prove (1) and (2), leaving (3) and (4) as exercises. (See Exercises 9 and 10 at the end of this section.)

To prove (1) it will suffice to show that if $(rf)(x) = r \cdot f(x)$ is an element of an open interval (c, d), then there is some open interval (a, b) such that $x \in (a, b)$ and $(rf)(S \cap (a, b)) \subset (c, d)$. (Why?) If $r = 0$, then $(rf)(x) = 0$, and if $0 \in (c, d)$, we may take (a, b) to be any open interval containing x and obtain $(rf)(S \cap (a, b)) = \{0\} \subset (c, d)$. Thus (1) is true if $r = 0$.

Now suppose that $r > 0$. If $r \cdot f(x) \in (c, d)$, then $f(x) \in (c/r, d/r)$, and since f is continuous at x, there is some open interval (a, b) such that $x \in (a, b)$ and $f(S \cap ((a, b)) \subset (c/r, d/r)$. But this in turn implies that

$$(rf)(S \cap (a, b)) \subset (c, d)$$

Hence (1) holds if $r > 0$, and a similar argument proves the case where $r < 0$.

Now we show that $f + g$ is continuous at x. Suppose that $f(x) + g(x)$ is an element of the open interval (c, d). Let us choose real numbers

$$c_1, \quad c_2, \quad d_1, \quad d_2$$

such that

$$c_1 < f(x) < d_1 \qquad c_2 < g(x) < d_2$$

and

$$c < c_1 + c_2 < d_1 + d_2 < d$$

Thus $f(x) \in (c_1, d_1)$ and $g(x) \in (c_2, d_2)$. Since f is continuous at x, there is an open interval (a_1, b_1) such that

$$x \in (a_1, b_1) \qquad \text{and} \qquad f(S \cap (a_1, b_1)) \subset (c_1, d_1)$$

Since g is continuous at x, there is an open interval (a_2, b_2) such that

$$x \in (a_2, b_2) \qquad \text{and} \qquad g(S \cap (a_2, b_2)) \subset (c_2, d_2)$$

Let

$$(a, b) = (a_1, b_1) \cap (a_2, b_2)$$

Then (a, b) is an open interval, $x \in (a, b)$, and

$$(f + g)(S \cap (a, b)) \subset (c, d)$$

Therefore $f + g$ is continuous at x.

Now consider $f - g$: by part (1) of the proposition, $(-1)g$ is continuous at x, and therefore by what we have just proved, so is $f + (-1)g = f - g$. □

Corollary 4.1.13 Let S be any subset of $\boldsymbol{R}$ and let S have the subspace topology induced by the natural topology on $\boldsymbol{R}$. If $f: S \to \boldsymbol{R}$ and $g: S \to \boldsymbol{R}$ are continuous functions, then

1. The function $rf: S \to \boldsymbol{R}$ is continuous, for every $r \in \boldsymbol{R}$;
2. The functions $f + g: S \to \boldsymbol{R}$ and $f - g: S \to \boldsymbol{R}$ are continuous;

3. The function $fg: S \to \boldsymbol{R}$ is continuous;
4. The function $f/g: S \to \boldsymbol{R}$ is continuous if $g(x) \neq 0$ for all $x \in S$.

We conclude this section with a brief discussion of homeomorphic topological spaces. The following definition plays much the same role for topology as the definition of set equivalence (Definition 1.3.1, Section 1.3 of Chapter 1) does for set theory.

Definition 4.1.14 A function f from a topological space X to a topological space Y is a *homeomorphism* from X to Y if it is continuous and its inverse function f^{-1} exists and is a continuous function from Y to X. If there is a homeomorphism from X to Y, we say that X is *homeomorphic* to Y.

Implicit in the definition of a homeomorphism is the fact that both the function and its inverse must be one-to-one and onto. Furthermore, if a function is a homeomorphism from X to Y, then its inverse is a homeomorphism from Y to X. Note also that if X and Y are homeomorphic topological spaces, then they are equivalent sets as defined in Chapter 1, Section 1.3. Therefore sets which are not equivalent cannot be homeomorphic, no matter how they are topologized. For instance, $\boldsymbol{R}$ and $\boldsymbol{Q}$ cannot be homeomorphic because they are not equivalent. However, equivalent sets may not be homeomorphic even if they are topologized in what appear to be similar ways. For instance, the results of the next section will allow us to prove that $\boldsymbol{R}$ with the natural topology and $[0, 1]$ with the subspace topology induced by the natural topology are not homeomorphic even though they are equivalent sets. (See Exercise 15, Section 4.2 of this chapter.)

Our final proposition in this section compares the topologies on homeomorphic spaces.

Proposition 4.1.15 Let X and Y be topological spaces. If $f: X \to Y$ is a homeomorphism from X to Y, then

1. A set G is open in X if and only if its image $f(G)$ is open in Y;
2. A set F is closed in X if and only if its image $f(F)$ is closed in Y.

Proof: If G is open in X then, since $f^{-1}: Y \to X$ is continuous, we must have $(f^{-1})^{-1}(G) = f(G)$ open in Y. Conversely, if $f(G)$ is open in Y, then the continuity of f implies that $f^{-1}(f(G)) = G$ must be open in X. This proves (1). We leave (2) as an exercise. (See Exercise 17 at the end of this section.) □

The import of Proposition 4.1.15 is that if X and Y are homeomorphic topological spaces, then there is a one-to-one correspondence between their open sets (and between their closed sets). Thus homeomorphic spaces are topologically indistinguishable from one another, being topologically identical in much the same way that equivalent sets are set-theoretically identical.

EXERCISES

1. Let X and Y be topological spaces.
   (a) Prove that if X has the discrete topology, then any function from X to Y is continuous.
   (b) Prove that if Y has the indiscrete topology, then any function from X to Y is continuous.
   (c) Suppose X has the indiscrete topology, Y has the discrete topology, and f is a continuous function from X to Y. What can you say about f?
2. Let X and Y be topological spaces and let f be a continuous function from X to Y. Let S be a subset of X, and let S have the subspace topology induced by the topology on X and $f(S)$ the subspace topology induced by the topology on Y. Prove that the restriction of f to S is a continuous function from S to $f(S)$.
3. Let X be a set and S a subset of X. Define the *characteristic function χ_S of S* as follows:

$$\chi_S(x) = \begin{cases} 1 & \text{if } x \in S \\ 0 & \text{if } x \in X - S \end{cases}$$

   Thus the characteristic function of S is a function from X to the set $\{0, 1\}$. Suppose X is a topological space and let $\{0, 1\}$ have the discrete topology. Prove that a function from X to $\{0, 1\}$ is continuous if and only if it is the characteristic function of a subset of X which is both open and closed in X.
4. Let X and Y be topological spaces. Prove that a function $f: X \to Y$ is continuous if and only if it satisfies the following property: whenever S is a subset of X and x is a limit point of S, then either $f(x) \in f(S)$ or $f(x)$ is a limit point of $f(S)$.
5. A linear function on $\boldsymbol{R}$ is a function of the form $f(x) = ax + b$, for all $x \in \boldsymbol{R}$, where a and b are fixed real numbers. Using only Definition 4.1.1, prove that every linear function on $\boldsymbol{R}$ is continuous.

6. Prove Corollary 4.1.6.
7. Use Corollary 4.1.6 to prove that the absolute value function $f(x) = |x|$ is continuous on $\boldsymbol{R}$.
8. Let $f: \boldsymbol{R} \to \boldsymbol{R}$ be given by $f(x) = [x]$, where $[x]$ denotes the greatest integer less than or equal to x. Prove that f is continuous at x if and only if $x \notin \boldsymbol{Z}$.
9. Let S be a subset of $\boldsymbol{R}$ with the subspace topology induced by the natural topology on $\boldsymbol{R}$. Let $x \in S$ and let $f: S \to \boldsymbol{R}$ and $g: S \to \boldsymbol{R}$ be continuous at x.
   (a) Prove that if $g(x) = 0$, then fg is continuous at x.
   (b) Define a function h from S to $\boldsymbol{R}$ be setting
   $$h = f(g - g(x)) + g(x)(f - f(x))$$
   Prove that if h is continuous at x, then so is fg.
   (c) Use (a) and (b) to prove (3) of Proposition 4.1.12.
10. Prove (4) of Proposition 4.1.12.
11. Let $n \in \boldsymbol{N}$. Let $f: \boldsymbol{R} \to \boldsymbol{R}$ be given by $f(x) = x^n$, for all $x \in \boldsymbol{R}$. Prove that f is a continuous function on $\boldsymbol{R}$.
12. A polynomial function on $\boldsymbol{R}$ is a function whose defining equation is of the form $f(x) = a_0 x^n + a_1 x^{n-1} + \cdots + a_n$, where $n \in \boldsymbol{N}$ and $a_0, a_1, \ldots, a_n$ are fixed real numbers with $a_0 \neq 0$. Prove that every polynomial function on $\boldsymbol{R}$ is continuous.
13. Let $f: \boldsymbol{R} \to \boldsymbol{R}$.
   (a) Suppose f is continuous when $\boldsymbol{R}$ has the natural topology. Is it true that f must be continuous when $\boldsymbol{R}$ has the open ray topology? Justify your answer.
   (b) Suppose that f is continuous when $\boldsymbol{R}$ has the open ray topology. Is it true that f must be continuous when $\boldsymbol{R}$ has the natural topology? Justify your answer.
14. Does Exercise 13 suggest to you a general result concerning continuity and comparable topologies? If so, state and prove it.
15. Let $(0, 1)$ have the subspace topology induced by the natural topology on $\boldsymbol{R}$ and let $f: (0, 1) \to \boldsymbol{R}$ be given as follows: $f(x) = 0$ if x is irrational; if x is rational, write $x = p/q$, where $p \in \boldsymbol{N}$, $q \in \boldsymbol{N}$ and p, q have no common factors, and set $f(x) = 1/q$. Prove that f is continuous at x if and only if x is irrational.
16. Let f be a continuous function on $\boldsymbol{R}$ and suppose that f is not identically zero on $\boldsymbol{R}$, that is, that there is some $y \in \boldsymbol{R}$ such that $f(y) \neq 0$. Prove that there exists a closed interval $[a, b]$, with $a < b$, such that $f(x) \neq 0$ for any $x \in [a, b]$.

17. Prove (2) of Proposition 4.1.15.
18. Let X, Y, and W be topological spaces. Prove that
    (a) X is homeomorphic to X;
    (b) If X is homeomorphic to Y then Y is homeomorphic to X;
    (c) If X is homeomorphic to Y and Y is homeomorphic to W then X is homeomorphic to W.
19. Prove that a function f from a topological space X to a topological space Y is a homeomorphism if and only if f is continuous, one-to-one, onto, and has the property that $f(G)$ is open in Y whenever G is open in X.
20. Give an example of topological spaces X and Y and a one-to-one onto function f from X to Y such that f is continuous but not a homeomorphism, and show that the inverse of f is not continuous.
21. Let f be a function on $\boldsymbol{R}$ and let $x \in \boldsymbol{R}$. We say that f is *continuous from the right at* x if whenever $f(x) \in (c, d)$ there is some $b \in \boldsymbol{R}$ such that $f([x, b)) \subset (c, d)$. Similarly, f is *continuous from the left at* x if whenever $f(x) \in (c, d)$ there is some $a \in \boldsymbol{R}$ such that $f((a, x]) \subset (c, d)$.
    (a) Let $f: \boldsymbol{R} \to \boldsymbol{R}$ be given by

$$f(x) = \begin{cases} -1 & \text{if } x < 0 \\ 1 & \text{if } x \geqslant 0 \end{cases}$$

    Show that f is continuous from the right at 0 but not continuous from the left at 0.
    (b) Let f be the greatest integer function of Exercise 8 above. Show that f is continuous from the right at every $x \in \boldsymbol{R}$ but continuous from the left at x if and only if $x \notin \boldsymbol{Z}$.
22. Let a, b be real numbers with $a < b$ and let $[a, b]$ have the subspace topology induced by the natural topology on $\boldsymbol{R}$. Show that a function from $[a, b]$ to $\boldsymbol{R}$ is continuous if and only if it is continuous on (a, b), continuous from the right at a, and continuous from the left at b.
23. Let the Euclidean plane $\boldsymbol{R}^2$ have the product topology. Define

$$p_x: \boldsymbol{R}^2 \to \boldsymbol{R} \qquad \text{by } p_x((x, y)) = x \qquad \text{for all } (x, y) \in \boldsymbol{R}^2$$

and

$$p_y: \boldsymbol{R}^2 \to \boldsymbol{R} \qquad \text{by } p_y((x, y)) = y \qquad \text{for all } (x, y) \in \boldsymbol{R}^2$$

(The functions p_x and p_y are the projections of the plane onto the x- and y-axes, respectively.) Prove that

(a) The projections p_x and p_y are continuous functions;

(b) If X is a topological space and $f: X \to \boldsymbol{R}^2$, then f is continuous if and only if $p_x \circ f$ and $p_y \circ f$ are continuous functions from X to $\boldsymbol{R}$.

4.2 CONNECTEDNESS AND COMPACTNESS

In this section we consider two important topological properties, connectedness and compactness, which are preserved by continuous functions. We will make a particularly close study of these properties and their relationship to the natural topology on the set $\boldsymbol{R}$ of real numbers.

Let us begin by raising again the question we asked at the start of this chapter: what do continuous functions do to open intervals in $\boldsymbol{R}$? As we saw in Figure 4.1, if f is continuous on the open interval (a, b) then $f((a, b))$ can be a half-open interval, which is neither open nor closed in the natural topology. We might guess, then, that if the continuous image of an open interval is not necessarily open, at least it must be an interval. But this is not so: Figure 4.4 shows that the continuous image of an open interval can be a ray. It is true, however, that these are the only possibilities: the continuous image of an open interval must be either an interval or a ray. As we shall see, the reason this is so is that continuous functions preserve a topological property known as connectedness.

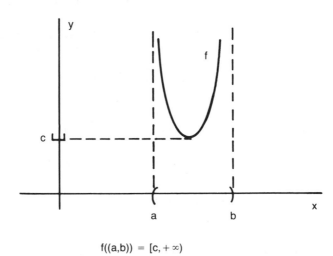

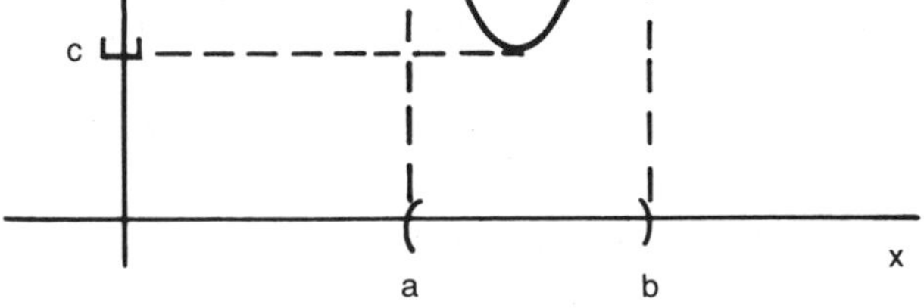

$f((a,b)) = [c, +\infty)$

Figure 4.4 The continuous image of an open interval need not be an interval.

Definition 4.2.1 A subset S of a topological space X is *disconnected* in X if there exist open sets G and H in X such that $G \cap S$ and $H \cap S$ are nonempty disjoint sets and

$$S = (G \cap S) \cup (H \cap S)$$

(see Figure 4.5). The set S is *connected* in X if it is not disconnected in X.

Note that the subset S referred to in the definition need not be a proper subset of the topological space X; if $S = X$, we simply refer to the space X as being disconnected or connected.

Examples 4.2.2 1. Let $X = \{x, y, z\}$ and let $\tau = \{\emptyset, \{x\}, \{y, z\}, X\}$. The space X is disconnected because if $G = \{x\}$ and $H = \{y, z\}$, then G and H are open in X, $G \cap X = G$ and $H \cap X = H$ are nonempty disjoint sets, and

$$X = (G \cap X) \cup (H \cap X)$$

2. Let $X = \{x, y, z\}$ and $\tau = \{\emptyset, \{x\}, \{x, y\}, \{x, z\}, X\}$. Let $S = \{y, z\}$. The subset S is disconnected in X, since the sets $G = \{x, y\}$ and $H = \{x, z\}$ are open sets in X such that $G \cap S$ and $H \cap S$ are nonempty and disjoint and

$$S = (G \cap S) \cup (H \cap S)$$

(Note that the sets G and H are not disjoint in X: the definition only requires that $G \cap S$ and $H \cap S$ be disjoint, not that G and H be disjoint.)

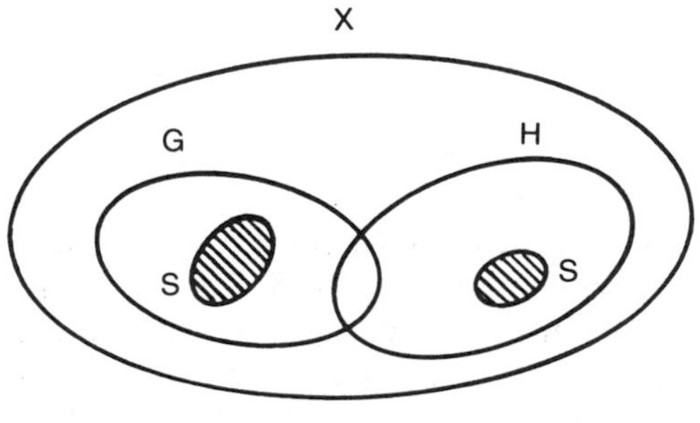

S is disconnected in X

Figure 4.5 A disconnected set.

Now let $T = \{x, y\}$; then T is connected in X, for if

$$T = (G \cap T) \cup (H \cap T)$$

with G and H open sets and $G \cap T$, $H \cap T$ nonempty and disjoint, then $x \in T$ implies that x is an element of just one of the open sets G and H, and this is impossible since every nonempty open set in X contains x. For the same reason, the space X is a connected space.

3. The set $\boldsymbol{Q}$ of rational numbers is disconnected in the space $\boldsymbol{R}$ of real numbers. To see this, let

$$G = (-\infty, \sqrt{2}) \qquad \text{and} \qquad H = (\sqrt{2}, +\infty)$$

Then G and H are open sets, $G \cap \boldsymbol{Q}$ and $H \cap \boldsymbol{Q}$ are nonempty and disjoint, and

$$\boldsymbol{Q} = (G \cap \boldsymbol{Q}) \cup (H \cap \boldsymbol{Q}).$$

4. The empty set cannot be disconnected (why not?) and therefore is a connected subset of every topological space.

We will soon prove that intervals and rays are connected subsets of $\boldsymbol{R}$, and that in fact these are the only nonempty connected subsets of $\boldsymbol{R}$.

Our first proposition concerning the phenomenon of connectedness provides us with an alternate definition of disconnectedness, one which is often easier to use than that given in Definition 4.2.1.

Proposition 4.2.3 Let X be a topological space. A subset S of X is disconnected in X if and only if S contains a nonempty proper subset T such that T is both open and closed in the subspace topology induced on S by the topology on X.

Proof: Suppose S is disconnected in X, so that

$$S = (G \cap S) \cup (H \cap S)$$

for some sets G and H such that G, H are open in X and $G \cap S$, $H \cap S$ are nonempty and disjoint. Let $T = G \cap S$. Then T is a nonempty proper subset of S, T is open in the subspace topology on S because it is the intersection of S with the open set G, and T is closed in the subspace topology on S because its complement in S is $H \cap S$, which is also open in the subspace topology on S.

Conversely, suppose that T is a nonempty proper subset of S which is both open and closed in the subspace topology on S. Since $S = T \cup (S - T)$ and since T and $S - T$ are both open in the subspace topology on S, there exist sets G and H which are open in X such that $T = G \cap S$ and $S - T = H \cap S$. Therefore

$$S = (G \cap S) \cup (H \cap S)$$

where G, H are open in X and $G \cap S$ and $H \cap S$ are nonempty disjoint sets. □

We next use Proposition 4.2.3 to show that continuous functions preserve connectedness.

Proposition 4.2.4 Let X and Y be topological spaces. If S is a connected subset of X and f is a continuous function from X to Y, then $f(S)$ is a connected subset of Y.

Proof: Suppose S is connected in X. Since f is a continuous function from X to Y, f restricted to S is a continuous function from S with the subspace topology induced by X to $f(S)$ with the subspace topology induced by Y. (See Exercise 2 of Exercises 4.1.) If $f(S)$ is disconnected in Y, then $f(S)$ must contain a nonempty proper subset T which is both open and closed in the subspace topology on $f(S)$. But since $f\colon S \to f(S)$ is continuous, $f^{-1}(T)$ is a nonempty proper subset of S which is both open and closed in the subspace topology on S; hence S must be disconnected, a contradiction. Therefore $f(S)$ is connected in Y. □

As a consequence of Proposition 4.2.4, we can prove the important Intermediate Value Theorem, which states that if f is a continuous function from a topological space X into $\boldsymbol{R}$, then on any connected subset of X the function f will assume every value intermediate to any two of its values. Figure 4.6 depicts the Intermediate Value Theorem when $X = \boldsymbol{R}$ and the connected subset of X is the interval $[0, 1]$.

Theorem 4.2.5 (Intermediate Value Theorem) Let X be a topological space and let $f\colon X \to \boldsymbol{R}$ be a continuous function. Let S be a connected subset of X. If $x \in S$ and $y \in S$ and $f(x) \neq f(y)$, then for every real number t between $f(x)$ and $f(y)$ there is some element $z \in S$ such that $f(z) = t$.

Proof: By Proposition 4.2.4, $f(S)$ is connected in $\boldsymbol{R}$. We must show that if t is any real number between $f(x)$ and $f(y)$, then $t \in f(S)$. The rays

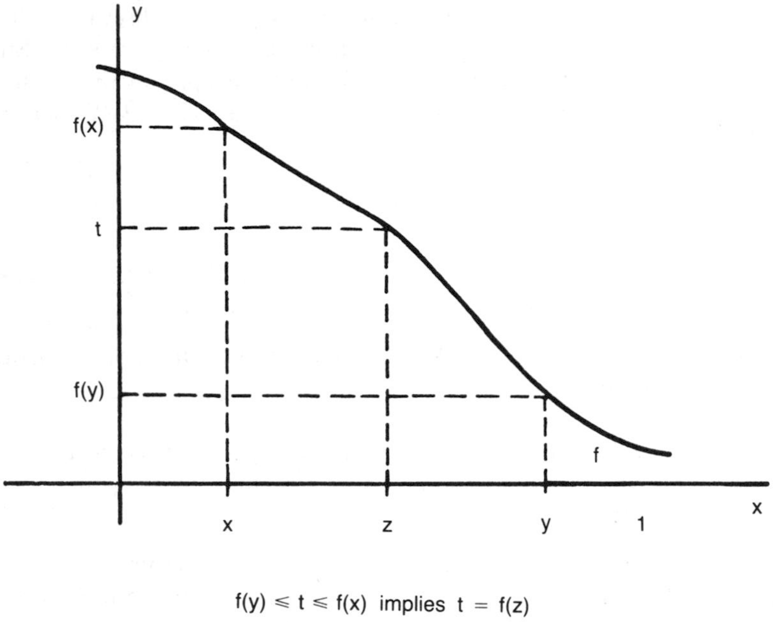

Figure 4.6 The Intermediate Value Theorem.

$(-\infty, t)$ and $(t, +\infty)$ are open in $\boldsymbol{R}$ and $(-\infty, t) \cap f(S)$, $(t, +\infty) \cap f(S)$ are disjoint sets which are nonempty because $f(x)$ is an element of one of them and $f(y)$ is an element of the other. If $t \notin f(S)$, we have

$$f(S) = ((-\infty, t) \cap f(S)) \cup ((t, +\infty) \cap f(S))$$

and hence $f(S)$ is disconnected in $\boldsymbol{R}$, a contradiction. Therefore $t \in f(S)$. □

Note that the conclusion of the Intermediate Value Theorem need not hold if the subset S is not connected. For instance, let $f: \boldsymbol{R} \to \boldsymbol{R}$ be defined by $f(x) = x$ for all $x \in \boldsymbol{R}$ and consider the disconnected subset $\boldsymbol{Q}$ of $\boldsymbol{R}$. The function f is certainly continuous, and $\sqrt{2}$ is between $f(1) = 1$ and $f(2) = 2$, but there is no $z \in \boldsymbol{Q}$ such that $f(z) = \sqrt{2}$.

Now we turn to an examination of the connected subsets of $\boldsymbol{R}$. As previously mentioned, the only nonempty connected subsets of $\boldsymbol{R}$ are the intervals and rays.

Proposition 4.2.6 A nonempty subset I of $\boldsymbol{R}$ is connected if and only if it is either an interval or a ray.

Proof: We prove first that if I is either an interval or a ray, then it is connected in $\boldsymbol{R}$. If I is a closed interval which contains only one point, it is trivial that I is connected, so we will assume that I contains more than one point. Note that if $x \in I$ and $y \in I$ with $x \leqslant y$ then $[x, y] \subset I$.

Suppose that I is disconnected in $\boldsymbol{R}$, so that $I = J \cup (I - J)$, where J is some nonempty proper subset of I which is both open and closed in the subspace topology induced on I by the natural topology on $\boldsymbol{R}$. Let $x \in J$ and $y \in I - J$. Since $x \neq y$, we may assume that $x < y$. Let

$$S = \{x \in J \mid x \leqslant z < y\}$$

Then $x \in S$, so S is nonempty, and since S is bounded above by y we conclude that $s = \sup S$ exists. Note that $x \leqslant s \leqslant y$, so $s \in I$, and hence either $s \in J$ or $s \in I - J$. We shall show that neither of these cases can occur, and therefore that I must be connected.

Suppose $s \in I - J$: $S \subset J$ implies $\bar{S} \subset \bar{J}$, and since $s = \sup S \in \bar{S}$, we must have $s \in \bar{J}$. But J is closed in the subspace topology on I, so $s \in I$ and $s \in \bar{J}$ together imply that $s \in J$. Therefore we have $s \notin J$ and $s \in J$, which is impossible.

Now suppose that $s \in J$: then $s \notin I - J$, and thus $s \neq y$, so $s < y$. Consider the interval $(s, y] \subset I$. If $J \cap (s, y]$ is not empty, then $s \neq \sup S$; hence $J \cap (s, y] = \emptyset$ and this implies that $(s, y] \subset (I - J)$. Since s is a limit point of $(s, y]$, it is therefore a limit point of $I - J$, and since $I - J$ is a closed set, it follows that $s \in I - J$. Thus we have $s \notin I - J$ and $s \in I - J$, and we have shown that I is connected.

To finish the proof we must demonstrate that if I is a nonempty connected subset of $\boldsymbol{R}$, then it must be an interval or a ray. If we can find $x \in I$, $y \in I$, and $z \in \boldsymbol{R} - I$ such that $x < z < y$, then

$$I = ((-\infty, z) \cap I) \cup ((z, +\infty) \cap I)$$

is disconnected in $\boldsymbol{R}$, a contradiction. Therefore we conclude that the connected set I has the property that whenever $x \in I$, $y \in I$ and $x < y$, then $[x, y] \subset I$. But any subset of $\boldsymbol{R}$ which has this property is either an interval or a ray. (See Exercise 7 of Section 2.1, Chapter 2.) □

Corollary 4.2.7 The set $\boldsymbol{R}$ of real numbers contains no nonempty proper subset which is both open and closed in the natural topology on $\boldsymbol{R}$.

Corollary 4.2.8 Let I be a subset of $\boldsymbol{R}$ and let f be a continuous function from I to $\boldsymbol{R}$. If I is an interval, then $f(I)$ is either an interval or a ray; if I is a ray, then $f(I)$ is either an interval or a ray.

We now know what continuous functions on $\boldsymbol{R}$ do to open intervals: they carry them onto either intervals or rays, and, in the light of Figures 4.1 and 4.4, this is as specific as it is possible to be. However, we can be more specific about what continuous functions do to closed intervals. As we shall see, continuous functions carry closed intervals onto closed intervals, and the reason this is so is that, in addition to connectedness, continuous functions also preserve a topological property known as compactness.

Definition 4.2.9 Let X be a topological space and let S be a subset of X. Let A be an index set and for each $\alpha \in A$ let G_α be an open set in X. The collection of sets $\{G_\alpha \mid \alpha \in A\}$ is an *open covering for* S if

$$S \subset \bigcup_{\alpha \in A} G_\alpha$$

An open covering $\{G_\alpha \mid \alpha \in A\}$ for S *has a finite subcovering* if there is a finite subset B of A such that

$$S \subset \bigcup_{\beta \in B} G_\beta$$

The subset S of X is *compact* in X if every open covering for S has a finite subcovering.

Note that the subset S of the definition need not be a proper subset of X. If $S = X$, then we refer to X as a compact topological space.

Examples 4.2.10 1. Let X be any topological space. If S is a finite subset of X, then S is compact in X. To prove this, let $\{G_\alpha \mid \alpha \in A\}$ be an open covering for S. If $S = \emptyset$, then we may choose any $\beta \in A$ and $\{G_\beta\}$ will certainly be a finite subcovering for S. If S is not empty, then $S = \{x_1, \ldots, x_n\}$, for some $n \in \boldsymbol{N}$. For each $i \in \boldsymbol{N}$, $1 \leqslant i \leqslant n$, there is some $\alpha_i \in A$ such that $x_i \in G_{\alpha_i}$. But then $\{G_{\alpha_i} \mid 1 \leqslant i \leqslant n\}$ is a finite subcovering for S. We have thus shown that every open covering of S has a finite subcovering, so S is indeed compact in X.

2. It follows from the previous example that every subset of a finite topological space is compact in the space. Thus, as a special case, every finite topological space is a compact space.

3. The space $\boldsymbol{R}$ of real numbers is not a compact space. The collection of intervals $\{(-n, n) \mid n \in \boldsymbol{N}\}$ is an open covering for $\boldsymbol{R}$, and if this covering had a finite subcovering, say $\{(-n_1, n_1), \ldots, (-n_k, n_k)\}$, then we would have

$$\boldsymbol{R} = \bigcup_{1 \leqslant i \leqslant k} (-n_i, n_i)$$

But if $m = \max\{n_1, \ldots, n_k\}$ this would imply that $\boldsymbol{R} = (-m, m)$, which is impossible. Since we have produced an open covering of $\boldsymbol{R}$ which has no finite subcovering, it follows that $\boldsymbol{R}$ is not compact.

4. Open intervals are not compact in $\boldsymbol{R}$: an open covering of (a, b) which has no finite subcovering is $\{(a + (1/n), b) \mid n \in \boldsymbol{n}\}$. Similarly, rays are not compact in $\boldsymbol{R}$. We leave the proofs of these facts as an exercise. (See Exercise 11 at the end of this section.)

We will soon show that every closed interval is compact in $\boldsymbol{R}$. Before doing so, however, let us prove the important fact that continuous functions preserve compactness.

Proposition 4.2.11 Let f be a continuous function from a topological space X to a topological space Y. If S is a compact subset of X, then $f(S)$ is a compact subset of Y.

Proof: Let $\{G_\alpha \mid \alpha \in A\}$ be an open covering for $f(S)$ in Y. Since f is continuous, the family $\{f^{-1}(G_\alpha) \mid \alpha \in A\}$ is an open covering for S in X, and therefore it must have a finite subcovering, say

$$\{f^{-1}(G_{\alpha_1}), \ldots, f^{-1}(G_{\alpha_n})\}$$

But then

$$S \subset \bigcup_{1 \leqslant i \leqslant n} f^{-1}(G_{\alpha_i}) \qquad \text{implies that} \qquad f(S) \subset \bigcup_{1 \leqslant i \leqslant n} G_{\alpha_i}$$

Therefore the open covering $\{G_\alpha \mid \alpha \in A\}$ for $f(S)$ has a finite subcovering, namely $\{G_{\alpha_1}, \ldots, G_{\alpha_n}\}$. □

Now we show that closed intervals are compact subsets of $\boldsymbol{R}$.

Proposition 4.2.12 Every closed interval in $\boldsymbol{R}$ is a compact subset of $\boldsymbol{R}$.

Proof: Let $\{G_\alpha \mid \alpha \in A\}$ be an open covering for the closed interval $[a, b]$. It follows that for each $x \in [a, b]$, $\{G_\alpha \mid \alpha \in A\}$ is also an open covering for the interval $[a, x]$. Let S be the set of all elements x of $[a, b]$ such that the open covering $\{G_\alpha \mid \alpha \in A\}$ for $[a, x]$ has a finite subcovering.

Since there is some $\beta \in A$ such that $a \in G_\beta$, we have $\{G_\beta\}$ a finite subcovering for the interval $[a, a]$. Hence $a \in S$, so S is nonempty, and since S is clearly bounded above by b, this implies that $s = \sup S$ exists. Note that $a \leqslant s \leqslant b$.

Since there is some $\gamma \in A$ such that $s \in G_\gamma$, and since G_γ is an open set in $\boldsymbol{R}$, there exists an open interval (c, d) such that $s \in (c, d) \subset G_\gamma$. Now $s = \sup S$ is either an element of S or a limit point of S, and in either case, there exists an element x of S such that $c < x \leqslant s$. But $x \in S$ implies that there is a finite subcovering, say $\{G_{\alpha_1}, \ldots, G_{\alpha_n}\}$, for the interval $[a, x]$. Thus $\{G_{\alpha_1}, \ldots, G_{\alpha_n}, G_\gamma\}$ must be a finite subcovering for $[a, s]$. If $s < b$, then there is some $y \in (c, d) \cap (a, b)$ such that $s < y$. But then

$$\{G_{\alpha_1}, \ldots, G_{\alpha_n}, G_\gamma\}$$

is a finite subcovering for $[a, y]$, and this implies that $y \in S$, contradicting the fact that $s = \sup S$. Therefore $s < b$ is impossible, so $s = b$, and

$$\{G_{\alpha_1}, \ldots, G_{\alpha_n}, G_\gamma\}$$

is a finite subcovering for $[a, b]$. □

We have now shown that closed intervals are compact subsets of $\boldsymbol{R}$. Of course, not all closed subsets of $\boldsymbol{R}$ are compact, and an examination of the proof of Proposition 4.2.12 shows that the boundedness of a closed interval is crucial to its compactness. However, it is true that a compact subset of $\boldsymbol{R}$ must be closed. (In fact, it is true that a subset of $\boldsymbol{R}$ is compact if and only if it is closed and bounded: this is the important Heine–Borel Theorem, which we will soon prove.) It is important to realize, however, that in a general topological space compact subsets need not be closed. The following example exhibits a topological space and a compact subset of that space which is not closed in the topology.

Example 4.2.13 Let R have the open ray topology. The interval $[0, 1]$ is neither open nor closed in the open ray topology (why?), but it is compact. To see this, suppose that $\{G_\alpha \mid \alpha \in A\}$ is an open covering for $[0, 1]$

in the open ray topology. Since $0 \in G_\beta$ for some $\beta \in A$, and since G_β is an open ray, it follows that either $G_\beta = \boldsymbol{R}$ or $G_\beta = (a, +\infty)$ for some $a < 0$. In either case, $[0, 1] \subset G_\beta$, so the open covering $\{G_\alpha \mid \alpha \in A\}$ has a finite subcovering $\{G_\beta\}$.

Let us consider for a moment the difference between $\boldsymbol{R}$ with the natural topology and $\boldsymbol{R}$ with the open ray topology. The open ray topology is much smaller than the natural topology, so small in fact that it is not possible in the open ray topology to separate distinct points by surrounding them with open sets, although this can always be done in the natural topology. Thus, in the natural topology if $x \neq y$ we can always find open sets G and H such that $x \in G$, $y \in H$, and $G \cap H = \emptyset$, whereas in the open ray topology this is not the case, for if $x < y$, then any open ray which contains x must also contain y. See Figure 4.7. This situation occurs because the open ray topology has comparatively few open sets, while the natural topology is rich in open sets. A topological space which is like the reals with the natural topology in that it has enough open sets to separate points is called a Hausdorff space.

Definition 4.2.14 A topological space is a *Hausdorff space* if whenever $x \in X$, $y \in X$, and $x \neq y$, then there exist open sets G and H in X such that $x \in G$, $y \in H$, and $G \cap H = \emptyset$.

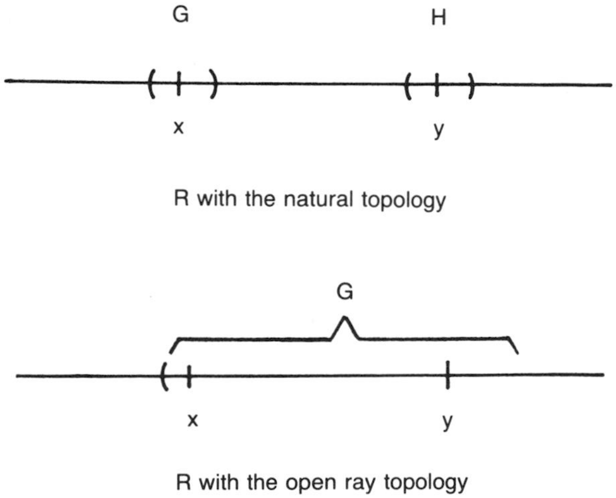

Figure 4.7 Separation of points depends on the topology.

Examples 4.2.15 1. Let $X = \{x, y, z\}$ and $\tau = \{\emptyset, \{x\}, \{x, y\}, X\}$; then X is not a Hausdorff space because there are no open sets G, H such that $x \in G$, $y \in H$, and $G \cap H = \emptyset$.

2. Every discrete topological space is a Hausdorff space. Every indiscrete topological space which has more than one point is not a Hausdorff space.

3. The set $\boldsymbol{R}$ of real numbers with the natural topology is a Hausdorff space. The set $\boldsymbol{R}$ with the open ray topology is not a Hausdorff space.

The distinction between Hausdorff and non-Hausdorff spaces is a fundamental one. Many things which are true in Hausdorff spaces are not true in non-Hausdorff spaces, and one of these is that in Hausdorff spaces, compact subsets must be closed.

Proposition 4.2.16 A compact subset of a Hausdorff space is closed.

Proof: Let S be a compact subset of the Hausdorff space X. If S is empty or if $S = X$, we are done, so we assume that S is a nonempty proper subset of X. We will prove that S is closed by showing that $X - S$ is open.

Let $y \in X - S$. Since X is a Hausdorff space, for each $x \in S$ there exist open sets G_x and H_x such that $x \in G_x$, $y \in H_x$, and $G_x \cap H_x = \emptyset$. Clearly

$$S \subset \bigcup_{x \in S} G_x$$

and because S is compact the open covering $\{G_x \mid x \in S\}$ has a finite subcovering, say $\{G_{x_1}, \ldots, G_{x_n}\}$. Therefore

$$S \subset \bigcup_{1 \leqslant i \leqslant n} G_{x_i} \qquad \text{and} \qquad y \in \bigcap_{1 \leqslant i \leqslant n} H_{x_i}$$

Note that $\bigcap_{1 \leqslant i \leqslant n} H_{x_i}$ is an open set.

For each i, $1 \leqslant i \leqslant n$, $G_{x_i} \cap H_{x_i} = \emptyset$ implies that

$$H_{x_i} \subset (X - G_{x_i}).$$

Hence

$$y \in \bigcap_{1 \leqslant i \leqslant n} H_{x_i} \subset \bigcap_{1 \leqslant i \leqslant n} (X - G_{x_i}) = \left(X - \bigcap_{1 \leqslant i \leqslant n} G_{x_i}\right) \subset (X - S)$$

We have thus shown that for each $y \in X - S$ there exists an open set which contains y and is contained in $X - S$, and it follows that $X - S$ is an open set in X. □

Now we can prove the Heine–Borel Theorem, which completely characterizes the compact subsets of $\boldsymbol{R}$ in the natural topology.

Theorem 4.2.17 (Heine–Borel) A subset S of $\boldsymbol{R}$ is compact if and only if it is closed and bounded.

Proof: Suppose S is a compact subset of $\boldsymbol{R}$. Since $\boldsymbol{R}$ is a Hausdorff space, S is closed. Furthermore, S compact implies that the open covering $\{(-n, n) \mid n \in \boldsymbol{N}\}$ has a finite subcovering, say $\{(-n_i, n_i) \mid 1 \leqslant i \leqslant k\}$. Let $m = \max\{n_1, \ldots, n_k\}$. Then $S \subset (-m, m)$, which shows that S is bounded.

Conversely, suppose that S is closed and bounded in $\boldsymbol{R}$, and let $\{G_\alpha \mid \alpha \in A\}$ be an open covering for S. Because S is bounded, there exists a closed interval $[a, b]$ such that $S \subset [a, b]$, and it follows that $\{G_\alpha \mid \alpha \in A\} \cup (\boldsymbol{R} - S)$ is an open covering for $[a, b]$. But since closed intervals are compact, this open covering must have a finite subcovering, say $\{G_{\alpha_1}, \ldots, G_{\alpha_n}, \boldsymbol{R} - S\}$. It follows that $\{G_{\alpha_1}, \ldots, G_{\alpha_n}\}$ is a finite subcovering for S. Hence S is compact. □

The following consequence of the Heine–Borel Theorem, known as the Extreme Value Theorem, is very important. It says that a continuous function into $\boldsymbol{R}$ attains its maximum and its minimum on any nonempty compact subset of its domain. Figure 4.8 illustrates the theorem for a function from $\boldsymbol{R}$ into $\boldsymbol{R}$ and the compact subset [0, 1].

Theorem 4.2.18 (Extreme Value Theorem) If $f: X \to \boldsymbol{R}$ is continuous and S is a nonempty compact subset of X, then f attains its maximum and minimum on S; that is, there exist elements s and t in S such that $f(s) \leqslant f(x) \leqslant f(t)$ for all $x \in S$.

Proof: Since f is continuous and S is compact in X, $f(S)$ is compact in $\boldsymbol{R}$. Thus, by the Heine–Borel Theorem, $f(S)$ is a closed and bounded subset of $\boldsymbol{R}$. Since S is nonempty so is $f(S)$, and hence $\sup f(S)$ and $\inf f(S)$ exist. Note that $\inf f(S) \leqslant f(x) \leqslant \sup f(S)$ for all $x \in S$. Since $f(S)$ is closed in $\boldsymbol{R}$, it contains all its limit points, and from this it follows that $\sup f(S)$ and $\inf f(S)$ are elements of $f(S)$. But this means there exist elements s and t in S such that $f(s) = \inf f(S)$, $f(t) = \sup f(S)$, and we are done. □

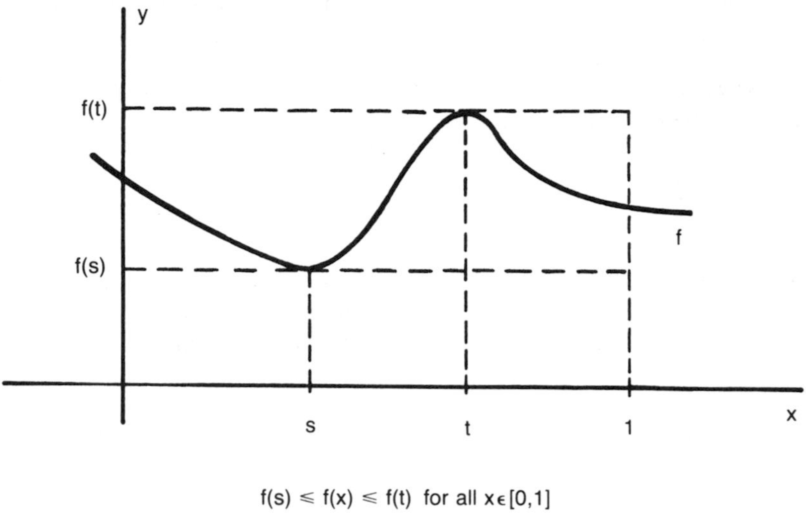

Figure 4.8 The Extreme Value Theorem.

Corollary 4.2.19 If S is a nonempty, closed, and bounded subset of $\boldsymbol{R}$ and $f: S \to \boldsymbol{R}$ is continuous, then f attains its maximum and minimum on S.

Note that in the preceding corollary both closedness and boundedness are necessary. For example, let $f(x) = x$ for all $x \in \boldsymbol{R}$. This function is certainly continuous on $\boldsymbol{R}$, but it attains neither its maximum nor its minimum on the closed (but unbounded) set $\boldsymbol{R}$; similarly, it attains neither its maximum nor its minimum on the bounded (but not closed) set $(0, 1)$.

We now bring this section to a close by proving that the continuous image of a closed interval in $\boldsymbol{R}$ is a closed interval.

Proposition 4.2.20 If f is a continuous real-valued function defined on the closed interval $[a, b]$, then $f([a, b])$ is a closed interval.

Proof: Since $[a, b]$ is connected and compact and f is continuous, it follows that $f([a, b])$ is connected and compact. But $f([a, b])$ connected in $\boldsymbol{R}$ implies that it is either an interval or a ray, and $f([a, b)]$ compact in $\boldsymbol{R}$ implies that it is closed and bounded. Since closed rays are unbounded, $f([a, b])$ must therefore be a closed interval. □

EXERCISES

1. Let X be a topological space and let S be a nonempty subset of X. Prove:
   (a) If X has the indiscrete topology, then S is connected;
   (b) If X has the discrete topology, then S is connected if and only if S has exactly one point.
2. Let X be a topological space and let $\{0, 1\}$ have the discrete topology. Prove that X is a connected space if and only if there does not exist a continuous function from X onto $\{0, 1\}$.
3. Prove that the set of all irrational real numbers is disconnected in $\boldsymbol{R}$.
4. Let $[0, 1]$ have the subspace topology induced by the natural topology on $\boldsymbol{R}$. Prove that the Cantor ternary set is disconnected in $[0, 1]$.
5. Let X be a topological space and let S be connected in X. Prove that
   (a) The closure $\bar{S}$ is connected in X;
   (b) If S is dense in X, then X is a connected space.
6. (a) Let f be a nonconstant function from $\boldsymbol{R}$ to $\boldsymbol{R}$. Show that f cannot be continuous if either $f(\boldsymbol{R}) \subset \boldsymbol{Q}$ or $f(\boldsymbol{R}) \subset (\boldsymbol{R} - \boldsymbol{Q})$.
   (b) Show that there is no continuous function $f: \boldsymbol{R} \to \boldsymbol{R}$ such that $f(x)$ is rational whenever x is irrational and $f(x)$ is irrational whenever x is rational.
7. A polynomial $p(x) = a_0x^n + \cdots + a_n$, where $a_0 \neq 0$, is said to be of *degree n*. Prove that a polynomial of odd degree has at least one real root; that is, prove that if n is odd then there is at least one real number r such that $p(r) = 0$.
8. Prove that if $x \in [0, +\infty)$ and $n \in \boldsymbol{N}$, then $\sqrt[n]{x}$ exists in $[0, +\infty)$.
9. Let $[0, 1]$ have the subspace topology induced by the natural topology on $\boldsymbol{R}$ and let $f: [0, 1] \to [0, 1]$ be continuous. Prove that f has a fixed point; that is, prove that there is some $x \in [0, 1]$ such that $f(x) = x$.
10. Let X be a topological space and let S be a subset of X. Prove the following:
    (a) If X has the indiscrete topology, then S is compact in X;
    (b) If X has the discrete topology, then S is compact in X if and only if S is a finite set.
11. Prove from the definition of compactness that open intervals and rays are not compact in $\boldsymbol{R}$.
12. Let $f: \boldsymbol{R} \to \boldsymbol{R}$ be continuous. What can be said about $f((a, +\infty))$? About $f([a, +\infty))$?

13. Prove that if X is a compact topological space and S is a closed set in X, then S is compact in X.
14. Let $[0, 1]$ have the subspace topology induced by the natural topology on $\boldsymbol{R}$. Prove that the Cantor ternary set is compact in $[0, 1]$.
15. Let $[0, 1]$ have the subspace topology induced by the natural topology on $\boldsymbol{R}$. Prove that $\boldsymbol{R}$ is not homeomorphic to $[0, 1]$.
16. Let X be a compact topological space and let Y be a Hausdorff space. Let $f: X \to Y$ be one-to-one, onto, and continuous. Prove that f is a homeomorphism.
17. Let $n \in \boldsymbol{N}$. Let $f: [0, +\infty) \to [0, +\infty)$ be given by $f(x) = \sqrt[n]{x}$ for all $x \in [0, +\infty)$. Prove that f is a continuous function.
18. Prove the Cantor Intersection Theorem: let $\{F_n \mid n \in \boldsymbol{N}\}$ be a collection of nonempty closed sets in $\boldsymbol{R}$ such that
    (a) F_1 is bounded; and
    (b) $F_{n+1} \subset F_n$ for all $n \in \boldsymbol{N}$.
    Then $\bigcap_{n \in N} F_n \neq \emptyset$.
    (Hint: if the intersection is empty, then $\{\boldsymbol{R} - F_n \mid n \in \boldsymbol{N}\}$ is an open covering for F_1.)
19. Let X be a topological space. A collection $\{F_\alpha \mid \alpha \in A\}$ of closed sets in X has the *finite intersection property* if the intersection of any finite number of the sets in the collection is nonempty. Prove that X is a compact topological space if and only if whenever $\{F_\alpha \mid \alpha \in A\}$ is a collection of closed sets in X having the finite intersection property, then $\bigcap_{\alpha \in A} F_\alpha \neq \emptyset$.
20. Topological spaces can be classified according to the manner in which their open sets can be used to separate points. The following axioms are known as *separation axioms*.
    Let X be a topological space; then

    0. X is a *T_0-space* if whenever $x \in X$ and $y \in X$ with $x \neq y$ there is an open set in X which contains one of the points x or y but not the other.
    1. X is a *T_1-space* if whenever $x \in X$ and $y \in X$ with $x \neq y$ there are open sets G and H in X such that $x \in G$, $y \in H$, $x \notin H$, and $y \notin G$.
    2. X is a *T_2-space* if whenever $x \in X$ and $y \in X$ with $x \neq y$ there are open sets G and H in X such that $x \in G$, $y \in H$, and $G \cap H = \emptyset$. (Thus T_2-space is merely another name for Hausdorff space.)
    3. X is a *regular space* if whenever F is a closed set in X and $x \in X - F$ there are open sets G and H in X such that $x \in G$, $F \subset H$, and $G \cap H = \emptyset$. If X is both a regular space and a T_1-space, it is a *T_3-space*.

4. X is a *normal space* if whenever E and F are disjoint closed sets in X there are open sets G and H in X such that $E \subset G$, $F \subset H$, and $G \cap H = \emptyset$. If X is both a normal space and a T_1-space it is a *T_4-space*.

(a) Give an example of a T_4-space.
(b) Give an example of a space which is not a T_0-space.
(c) The five separation axioms are progressively stronger: prove that if X is a T_i-space, $1 \leqslant i \leqslant 4$, then X is a T_{i-1}-space.

For each i, $0 \leqslant i \leqslant 3$, there exist spaces which are T_i-spaces but not T_{i+1}-spaces.

(d) Give an example of an infinite topological space which is a T_0-space but not a T_1-space.
(e) Prove that if X is a T_1-space, then every subset of X which consists of a single point is closed in X.
(f) Give an example of a topological space which is a T_1-space but not a T_2-space.

It is not so easy to find examples of spaces which are T_2 but not T_3 and T_3 but not T_4. The interested reader should consult Kelley's *General Topology* or another textbook in topology. It is not too difficult, however, to show that $\boldsymbol{R}$ is a T_4 space.

(g) Let X be a topological space. Suppose that whenever E and F are disjoint closed sets in X there is a continuous function $f: X \to \boldsymbol{R}$ such that $f(E) \subset (-\infty, 0)$ and $f(F) \subset (0, +\infty)$. Prove that X is a normal space.
(h) Use (g) to show that $\boldsymbol{R}$ is a normal space, and hence a T_4-space. (Hint: define functions f, g from $\boldsymbol{R}$ to $\boldsymbol{R}$ by

$$f(x) = \inf\{|x - y| \mid y \in E\} \qquad \text{for all } x \in \boldsymbol{R}$$

and

$$g(x) = \inf\{|x - y| \mid y \in F\} \qquad \text{for all } x \in \boldsymbol{R}$$

and consider $f - g$.)

5
Sequences and Series

In the previous chapter we utilized continuous functions to study topological spaces, paying particular attention to the space $\boldsymbol{R}$ of real numbers. In this chapter we will continue this process, but concentrating our attention now on the continuous functions known as sequences. The first section of the chapter examines the properties of sequences in general topological spaces, while the second considers sequences in the space of real numbers. The third section is devoted to the special type of real sequence known as an infinite series, and the fourth examines the relationship between real sequences and functional limits.

5.1 SEQUENCES

A sequence in a topological space X is a function from the set $\boldsymbol{N}$ of natural numbers with the discrete topology into X.

Definition 5.1.1 Let X be a topological space and let $\boldsymbol{N}$ be the set of natural numbers provided with the discrete topology. A function $f: \boldsymbol{N} \to X$ is called a *sequence in X*. If $n \in \boldsymbol{N}$, the element $f(n) \in X$ is called the nth *term* of the sequence and is denoted by x_n.

Note that every sequence is a continuous function, since any function whose domain has the discrete topology must be continuous.

Sequences are important for what they tell us about the range space X, and hence it is common practice to suppress all mention of the function f and focus attention only on the image (i.e., the terms) of the sequence. We shall adopt this convention and in addition shall use the notation $\{x_n\}$ for a sequence. Therefore rather than speaking of

"the sequence $f: \boldsymbol{N} \to X$ given by $f(n) = x_n$ for all $n \in \boldsymbol{N}$"

we will refer to

"the sequence $\{x_n\}$ in X"

Our next definition states what convergence means for a sequence in a general topological space.

Definition 5.1.2 Let $\{x_n\}$ be a sequence in the topological space X and let $x \in X$. If for every open set G containing x there is some $m \in \boldsymbol{N}$ such that $x_n \in G$ for all $n \geqslant m$, then we say that the sequence *converges to* x, and that x is a *limit* of the sequence. If the sequence has a limit in X, it is a *convergent* sequence; if it has no such limit, it is a *divergent* sequence.

Thus a sequence converges to x if for every open set G containing x there is some term of the sequence such that this term and all succeeding terms are in G. Another way to say this is that the sequence converges to x if every open set which contains x also contains all but finitely many terms of the sequence (see Figure 5.1).

Examples 5.1.3 1. Let $X = \{x, y, z\}$ and $\tau = \{\emptyset, \{x\}, \{x, y\}, X\}$. Let $\{x_n\}$ be the sequence in X defined by $x_n = x$ for all $n \in \boldsymbol{N}$. This sequence converges to x, since if G is an open set containing x then we have either $G = \{x\}$, $G = \{x, y\}$, or $G = X$; hence $x_n = x \in G$ for all $n \geqslant 1$. This sequence also converges to y, for if G is an open set containing y, then either $G = \{x, y\}$ or $G = X$; in either case $x_n = x$ is an element of G. Similarly, the sequence converges to z. This example shows that a convergent sequence need not have a unique limit.

2. Consider the sequence $\{x_n\} = \{(-1)^n\}$ in $\boldsymbol{R}$. This sequence is divergent because for any $x \in \boldsymbol{R}$ there exists an open interval (a, b) containing x such that either 1 or -1 does not belong to (a, b); therefore for any $m \in \boldsymbol{N}$ there is some $n \geqslant m$ such that $x_n \notin (a, b)$. This proves that the sequence cannot converge to x for any $x \in \boldsymbol{R}$, and hence that it is divergent.

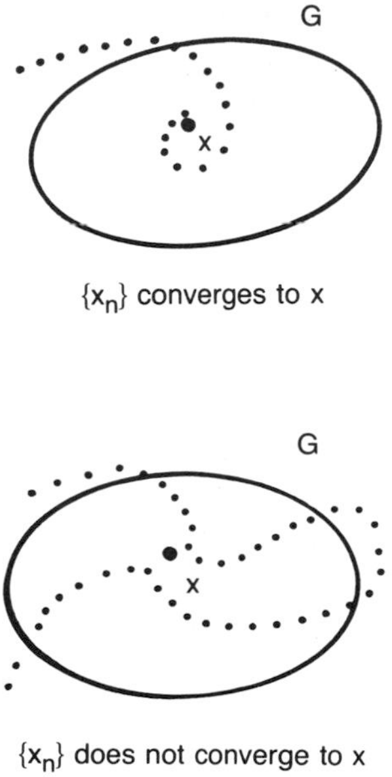

Figure 5.1 Convergence of sequences.

3. Let $\{x_n\} = \{1/n\}$ in $\boldsymbol{R}$. We shall demonstrate that zero is the unique limit of this sequence. Suppose that G is an open set in $\boldsymbol{R}$ such that $0 \in G$. Let (a, b) be an open interval such that $0 \in (a, b) \subset G$. From Corollary 2.1.13, there is some $m \in \boldsymbol{N}$ such that $0 < 1/m < b$. But then $1/n \leqslant 1/m$ for all $n \geqslant m$, and thus $x_n \in G$ for all $n \geqslant m$. Therefore the sequence converges to 0.

Now we will show that if $x \neq 0$, then the sequence cannot converge to x. First, if $x < 0$, then $G = (3x/2, x/2)$ is an open set containing x and $x_n = 1/n$ is not an element of G for any $n \in \boldsymbol{N}$ because $x/2 < 0$. Thus the sequence cannot converge to x if $x < 0$. If $x > 0$, then there is some $k \in \boldsymbol{N}$ such that $0 < 1/k < x$, and then $G = (1/k, x + 1)$ is an open set which contains x and $1/n$ is not an element of G for all $n \geqslant k$, so the sequence cannot converge to x.

4. Let $x \in \boldsymbol{R}$ and consider the sequence $\{x + 1/n\}$ in $\boldsymbol{R}$. This sequence converges to the unique limit x in $\boldsymbol{R}$. The proof is similar to that of the previous example. (See Exercise 4 at the end of this section.)

5. In this example we show that every real number is the limit of a sequence of rational numbers. Let $x \in \boldsymbol{R}$. By Theorem 2.1.12, for each $n \in \boldsymbol{N}$ there is some rational number q_n such that $x < q_n < x + 1/n$. We

claim that the sequence $\{q_n\}$ converges to x. To show this, it suffices to show that if (a, b) is an open interval containing x, then there is some $m \in \boldsymbol{N}$ such that $q_n \in (a, b)$ for all $n \geqslant m$. But by the previous example, the sequence $(x + 1/n)$ converges to x, so there is some $m \in \boldsymbol{N}$ such that $x + 1/n$ is an element of (a, b) for all $n \geqslant m$, and thus q_n must be an element of (a, b) for all $n \geqslant m$. A similar argument shows that every real number is the limit of a sequence of irrational numbers. (See Exercise 5 at the end of this section.)

6. Let $\boldsymbol{R}$ have the open ray topology; then the sequence $\{1/n\}$ converges to x for all $x \leqslant 0$. (See Exercise 6 at the end of this section.)

Examples 1 and 6 of Examples 5.1.3 show that in general a convergent sequence need not have a unique limit. Note that the topological spaces involved in these examples are not Hausdorff spaces. On the other hand, all the sequences we have seen so far which do have unique limits have been sequences in Hausdorff spaces. This is another reason why Hausdorff spaces are important: a convergent sequence in a Hausdorff space must have a unique limit.

Proposition 5.1.4 Every convergent sequence in a Hausdorff space has a unique limit.

Proof: Let $\{x_n\}$ be a convergent sequence in a Hausdorff space X, and assume that this sequence converges to the elements x and y of X, where $x \neq y$. Since X is Hausdorff, there exists open sets G and H such that $x \in G$, $y \in H$, and $G \cap H = \emptyset$. Since x is a limit of the sequence, there is some $m \in \boldsymbol{N}$ such that $x_n \in G$ for all $n \geqslant m$, and since y is also a limit of the sequence, there is some $k \in \boldsymbol{N}$ such that $x_n \in H$ for all $n \geqslant k$. But if $p = \max\{m, k\}$, then $x_p \in G \cap H$, a contradiction. □

Corollary 5.1.5 Every convergent sequence in $\boldsymbol{R}$ (with the natural topology) has a unique limit.

Proposition 5.1.4 allows us to use the limit notation in Hausdorff spaces.

Definition 5.1.6 Let X be a Hausdorff space. If the sequence $\{x_n\}$ converges to $x \in X$, we write $\operatorname{Lim}_{n \to \infty} x_n = x$.

Next we show that continuous functions preserve the limits of sequences. This does not depend on whether or not the spaces involved are Hausdorff.

Proposition 5.1.7 Let f be a continuous function from the topological space X to the topological space Y. If $\{x_n\}$ is a sequence which converges to x in X, then the sequence $\{f(x_n)\}$ converges to $f(x)$ in Y.

Proof: Let G be an open set in Y such that $f(x) \in G$; then $f^{-1}(G)$ is an open set in X. Since $x \in f^{-1}(G)$ and $\{x_n\}$ converges to x, there is some $m \in N$ such that $x_n \in f^{-1}(G)$ for all $n \geqslant m$. But then $f(x_n) \in G$ for all $n \geqslant m$, so $\{f(x_n)\}$ converges to $f(x)$. □

The converse of the preceding proposition is not true. That is, it is in general not the case that if $f: X \to Y$ is such that the sequence $\{f(x_n)\}$ converges to $f(x)$ whenever $\{x_n\}$ converges to x, then f must be a continuous function. In fact, this is not even true if X and Y are both Hausdorff spaces. It is true for functions from $\boldsymbol{R}$ to $\boldsymbol{R}$, however, and this is an important fact about continuity on the reals.

Proposition 5.1.8 A function $f: \boldsymbol{R} \to \boldsymbol{R}$ is continuous if and only if for every convergent sequence $\{x_n\}$ in $\boldsymbol{R}$,

$$\lim_{n \to \infty} x_n = x \text{ implies that } \qquad \lim_{n \to \infty} f(x_n) = f(x)$$

Proof: If f is continuous, the conclusion follows from Proposition 5.1.7. We must prove the converse: if f is a function on $\boldsymbol{R}$ which has the property that

$$\lim_{n \to \infty} f(x_n) = f(x) \text{ whenever } \qquad \lim_{n \to \infty} x_n = x$$

then f is continuous. We establish this by showing that $f(\bar{S}) \subset \overline{f(S)}$ for all subsets S of $\boldsymbol{R}$ (see Proposition 4.1.4).

For any subset S of $\boldsymbol{R}$, we have $\bar{S} = S \cup \dot{S}$, and since $f(S) \subset \overline{f(S)}$, we need only show that $f(\dot{S}) \subset \overline{f(S)}$. This is trivial if $S = \emptyset$, so we assume that S is nonempty. Let $x \in \dot{S}$; then x is a limit point of S and we must prove that $f(x) \in \overline{f(S)}$. In other words, we must demonstrate that if x is a limit point of S, then $f(x)$ is either a point of $f(S)$ or a limit point of $f(S)$.

Suppose x is a limit point of S. If $f(x)$ is a point of $f(S)$, we are done, so assume that $f(x)$ does not belong to $f(S)$ and consider the open interval $(x - 1, x + 1)$. Since x is a limit point of S, there exists $x_1 \in (x - 1, x + 1)$ such that $x_1 \in S$ and $x_1 \neq x$. Now suppose that we have found points

$$x_1, \ldots, x_n$$

such that

$$x_k \in \left(x - \frac{1}{k}, x + \frac{1}{k}\right) \qquad x_k \in S \qquad \text{and} \qquad x_k \neq x \qquad \text{for } 1 \leqslant k \leqslant n$$

Since x is a limit point of S, there exists

$$x_{n+1} \in \left(x - \frac{1}{n+1}, x + \frac{1}{n+1}\right)$$

such that $x_{n+1} \in S$ and $x_{n+1} \neq x$. Thus we define inductively a sequence $\{x_n\}$ such that for each $n \in \boldsymbol{N}$,

$$x_n \in \left(x - \frac{1}{n}, x + \frac{1}{n}\right) \qquad x_n \in S \qquad \text{and} \qquad x_n \neq x$$

It is easy to check that this sequence converges to x, and therefore by hypothesis the sequence $\{f(x_n)\}$ converges to $f(x)$.

Now we are ready to show that $f(x)$ is a limit point of $f(S)$. Let G be an open set containing $f(x)$. Since the sequence $\{f(x_n)\}$ converges to $f(x)$, there is some $m \in \boldsymbol{N}$ such that $f(x_n) \in G$ for all $n \geqslant m$. Hence $f(x_m)$ is an element of G. But $x_m \in S$ implies that $f(x_m) \in f(S)$, and $f(x_m) \neq f(x)$ because $f(x) \notin f(S)$ by assumption. We have thus proved that every open set which contains $f(x)$ must contain a point of $f(S)$ distinct from $f(x)$ and this establishes that $f(x)$ is a limit point of $f(S)$. □

Corollary 5.1.9 Let S be a subset of $\boldsymbol{R}$ and $x \in S$. A function $f: S \to \boldsymbol{R}$ is continuous at x if and only if for every sequence $\{x_n\}$ in S which converges to x, the sequence $\{f(x_n)\}$ converges to $f(x)$ in $\boldsymbol{R}$.

Notice that the proof of Proposition 5.1.8 depended heavily on the fact that for an arbitrary $x \in \boldsymbol{R}$ it is possible to define a family of open sets $\{G_n \mid n \in \boldsymbol{N}\}$ such that $G_{n+1} \subset G_n$ and $x \in G_n$ for all $n \in \boldsymbol{N}$. (In the proof, we took $G_n = (x - 1/n, x + 1/n)$.) The reason that Proposition 5.1.8 does not hold in general is that it is not always possible to define such a family of open sets for every element in a topological space.

Examples 5.1.10 Let $f: \boldsymbol{R} \to \boldsymbol{R}$ be given by

$$f(x) = \begin{cases} 1 & \text{if } x > 0 \\ -1 & \text{if } x \leqslant 0 \end{cases}$$

We see that f is discontinuous at 0, since the sequence $\{1/n\}$ converges to 0 while the sequence $\{f(1/n)\} = \{1\}$ converges to 1. However, f is continuous at x for all $x \neq 0$. For instance, suppose $x > 0$; then if $\{x_n\}$ converges to x, there must be some $m \in \boldsymbol{N}$ such that $x_n > 0$ for all $n \geqslant m$. But then $f(x_n) = 1$ for all $n \geqslant m$, so the sequence $\{f(x_n)\}$ converges to $f(x) = 1$, and hence f is continuous at x. A similar argument shows that f is continuous at x if $x < 0$.

The existence of the arithmetic operations in the system $\boldsymbol{R}$ allows us to form new sequences from old ones. Thus, if $\{x_n\}$ and $\{y_n\}$ are sequences in $\boldsymbol{R}$, we may form the sequences $\{x_n + y_n\}$, $\{x_n - y_n\}$, $\{x_n y_n\}$, and $\{rx_n\}$, for any $r \in \boldsymbol{R}$. Furthermore, if $y_n \neq 0$ for all $n \in \boldsymbol{N}$, we may also form the sequence $\{x_n/y_n\}$. Our next proposition describes the manner in which the limits of these newly constructed sequences are related to the limits of the original ones.

Proposition 5.1.11 If $\{x_n\}$ and $\{y_n\}$ are convergent sequences in $\boldsymbol{R}$, then

1. For all $r \in \boldsymbol{R}$, the sequence $\{rx_n\}$ converges and

$$\operatorname*{Lim}_{n\to\infty} rx_n = r \cdot \operatorname*{Lim}_{n\to\infty} x_n$$

2. The sequences $\{x_n + y_n\}$ and $\{x_n - y_n\}$ converge and

$$\operatorname*{Lim}_{n\to\infty} [x_n + y_n] = \operatorname*{Lim}_{n\to\infty} x_n + \operatorname*{Lim}_{n\to\infty} y_n$$

$$\operatorname*{Lim}_{n\to\infty} [x_n - y_n] = \operatorname*{Lim}_{n\to\infty} x_n - \operatorname*{Lim}_{n\to\infty} y_n$$

3. The sequence $\{x_n y_n\}$ converges and

$$\operatorname*{Lim}_{n\to\infty} [x_n y_n] = [\operatorname*{Lim}_{n\to\infty} x_n][\operatorname*{Lim}_{n\to\infty} y_n]$$

4. If $y_n \neq 0$ for all $n \in \boldsymbol{N}$ and $\operatorname{Lim}_{n\to\infty} y_n \neq 0$, the sequence $\{x_n/y_n\}$ converges and

$$\operatorname*{Lim}_{n\to\infty} \left[\frac{x_n}{y_n}\right] = \frac{\operatorname{Lim}_{n\to\infty} x_n}{\operatorname{Lim}_{n\to\infty} y_n}$$

Proof: We shall prove (1) and part of (2), leaving the rest as an exercise.

Let $r \in \boldsymbol{R}$. We know that the function f defined by $f(t) = rt$, for all $t \in R$, is continuous on $\boldsymbol{R}$, and hence by Proposition 5.1.7,

$$\lim_{n \to \infty} x_n = x \qquad \text{implies that} \qquad \lim_{n \to \infty} f(x_n) = f(x)$$

But $f(x_n) = rx_n$ and $f(x) = rx$, so

$$\lim_{n \to \infty} x_n = x \qquad \text{implies that} \qquad \lim_{n \to \infty} rx_n = rx$$

and we have established that (1) holds.

Now suppose that $\{x_n\}$ converges to x and $\{y_n\}$ converges to y. We wish to show that $\{x_n + y_n\}$ converges to $x + y$. It will suffice to show that if (a, b) is an open interval containing $x + y$, then there is some $m \in \boldsymbol{N}$ such that (a, b) contains $x_n + y_n$ for all $n \geqslant m$.

Now, it is always possible to choose real numbers c_1, d_1, c_2, d_2, such that $x \in (c_1, d_1)$, $y \in (c_2, d_2)$, and

$$a < c_1 + c_2 < d_1 + d_2 < b$$

Since $\{x_n\}$ converges to x, there is some $k_1 \in \boldsymbol{N}$ such that

$$x_n \in (c_1, d_1) \qquad \text{for all } n \geqslant k_1$$

Similarly, there is some $k_2 \in \boldsymbol{N}$ such that

$$y_n \in (c_2, d_2) \qquad \text{for all } n \geqslant k_2$$

If $m = \max\{k_1, k_2\}$, then $n \in \boldsymbol{N}$, $n \geqslant m$ implies that

$$x_n + y_n \in (c_1 + c_2, d_1 + d_2)$$

But since $(c_1 + c_2, d_1 + d_2) \subset (a, b)$, this shows that $x_n + y_n$ is an element of (a, b) for all $n \geqslant m$, and we are done. □

In a practical sense, the definition of a convergent sequence is not very useful for deciding whether or not a particular sequence converges, for in order to apply the definition, we must know (or guess) beforehand what its limit is. Thus, in using the definition to show that a sequence $\{x_n\}$ converges, we must somehow decide that x is a limit of the sequence and then verify that this is so by showing that for every open set G which contains x there is some $m \in \boldsymbol{N}$ such that $x_n \in G$ for all $n \geqslant m$. It would be convenient to have a characterization of convergent sequences which would allow us to examine a particular sequence and determine whether or not it is convergent without having to predetermine its limit, if any. Such a characterization exists for sequences of real numbers, and we will develop it in the next section.

EXERCISES

1. Let X be an indiscrete topological space. Prove that any sequence in X converges to every point of X.
2. Let X be a discrete topological space. Prove that a sequence $\{x_n\}$ in X converges if and only if there is some $m \in \boldsymbol{N}$ such that $x_n = x_m$ for all $n \geqslant m$.
3. Is the converse of Proposition 5.1.4 true? That is, if X is a topological space which is not Hausdorff, does it follow that there must be a sequence in X which converges to more than one point? Justify your answer.
4. Let $x \in \boldsymbol{R}$ and let $x_n = x + 1/n$, for all $n \in \boldsymbol{N}$. Prove that $\{x_n\}$ converges to x.
5. Prove that every real number is the limit of a sequence of irrational numbers.
6. Let $\boldsymbol{R}$ have the open ray topology. Show that the sequence $\{1/n\}$ converges to x for all $x \leqslant 0$.
7. Let X be a topological space. Show that if $\{x_n\}$ is a sequence in X which converges to x and $x_n \neq x$ for any $n \in \boldsymbol{N}$, then x is a limit point of the set of points $\{x_n \mid n \in \boldsymbol{N}\}$.
8. Is the converse of Exercise 7 true? That is, if a sequence in a topological space has a limit point which is not a point of the sequence, then does the sequence converge to the limit point? Justify your answer.
9. Let S be a nonempty subset of $\boldsymbol{R}$. Prove that $x \in \bar{S}$ if and only if there is a sequence of points of S which converges to x.
10. Let f: $(0, 1) \rightarrow \boldsymbol{R}$ be given by $f(x) = 0$ if x is irrational, $f(x) = 1/q$ if $x = p/q$ where p, q are integers having no common factor. Show that f is discontinuous at every rational point and continuous at every irrational point of $(0, 1)$.
11. Prove that if $\{x_n\}$ and $\{y_n\}$ are convergent sequences in $\boldsymbol{R}$ with $x_n \leqslant y_n$ for all $n \in \boldsymbol{N}$, then $\operatorname{Lim}_{n\rightarrow\infty} x_n \leqslant \operatorname{Lim}_{n\rightarrow\infty} y_n$.

12. Let $\{x_n\}$ be a convergent sequence in $\boldsymbol{R}$. Prove that

$$\lim_{n \to \infty} x_n = x \qquad \text{if and only if } \lim_{n \to \infty} [x_n - x] = 0$$

13. Finish the proof of Proposition 5.1.11.
14. Let the Euclidean plane $\boldsymbol{R}^2$ have the product topology. Prove that a sequence $\{(x_n, y_n)\}$ in $\boldsymbol{R}^2$ converges to (x, y) in $\boldsymbol{R}^2$ if and only if $\operatorname{Lim}_{n \to \infty} x_n = x$ and $\operatorname{Lim}_{n \to \infty} y_n = y$ in $\boldsymbol{R}$.

5.2 REAL SEQUENCES

In this section we characterize the convergent sequences in the system of real numbers. Our procedure will be as follows: we will show how to construct from every real sequence two associated sequences of a special kind, and will then prove that the original sequence converges if and only if both of these associated sequences converge to the same limit.

A sequence $\{x_n\}$ of real numbers is a function from $\boldsymbol{N}$ to $\boldsymbol{R}$ whose image is the set of real numbers $\{x_n \mid n \in \boldsymbol{N}\}$. This subset of $\boldsymbol{R}$ may or may not be bounded.

Definition 5.2.1 Let $\{x_n\}$ be a sequence of real numbers. If its image $\{x_n \mid n \in \boldsymbol{N}\}$ is bounded above in $\boldsymbol{R}$, we say that the *sequence is bounded above*, and refer to the least upper bound of the image as the *least upper bound of the sequence*. Similarly, the *sequence is bounded below* if its image is bounded below, and the *greatest lower bound of the sequence* is the greatest lower bound of the image. A sequence of real numbers whose image is a bounded subset of $\boldsymbol{R}$ is said to be a *bounded sequence*.

If a sequence is convergent, we know that its terms must get closer and closer to its limit. But for a sequence of real numbers, this means that the absolute value of the terms cannot grow without bound; in other words, a convergent real sequence must be a bounded sequence.

Proposition 5.2.2 Every convergent sequence of real numbers is a bounded sequence.

Proof: Let $\{x_n\}$ be a sequence of real numbers which converges to the real number x; then there exists $m \in \boldsymbol{N}$ such that $x_n \in (x - 1, x + 1)$ for all $n \geqslant m$. Let

$$a = \min\{x_1, \ldots, x_{m-1}, x - 1\}$$

and

$$b = \max \{x_1, \ldots, x_{m-1}, x + 1\}$$

Since $a \leqslant x_n \leqslant b$ for all $n \in N$, the image $\{x_n \mid n \in N\}$ of the sequence is a bounded subset of R. □

Proposition 5.2.2 tells us that a convergent sequence in R must be bounded, but of course the converse is not true: a bounded sequence need not be convergent. (For example, the sequence $\{(-1)^n\}$ is bounded but not convergent in R.) However, the converse is true for the class of sequences known as monotonic sequences.

Definition 5.2.3 A sequence $\{x_n\}$ in R is *monotonically increasing* if $x_n \leqslant x_{n+1}$ for all $n \in N$, and *monotonically decreasing* if $x_n \geqslant x_{n+1}$ for all $n \in N$. A sequence is referred to as *monotonic* if it is either monotonically increasing or monotonically decreasing.

Example 5.2.4 The sequence $\{2^n\}$ is monotonically increasing, the sequence $\{1/n\}$ is monotonically decreasing, and the constant sequence $\{1\}$ is both monotonically increasing and monotonically decreasing; therefore these are all monotonic sequences. On the other hand, the sequence $\{(-1)^n\}$ is neither monotonically increasing nor monotonically decreasing, and hence is not monotonic.

Proposition 5.2.5 Every monotonically increasing sequence in R which is bounded above converges to its least upper bound and every monotonically decreasing sequence in R which is bounded below converges to its greatest lower bound.

Proof: Let $\{x_n\}$ be a monotonically increasing sequence which is bounded above, and let $x = \sup \{x_n \mid n \in N\}$; then x is either an element or a limit point of the set $\{x_n \mid n \in N\}$. We will show that in either case the sequence converges to x.

Suppose x is an element of $\{x_n \mid n \in N\}$, so that $x = x_m$ for some $m \in N$. Since x is an upper bound for the set and the sequence is monotonically increasing, it follows that $x_n = x$ for all $n \geqslant m$. But this surely implies that the sequence converges to x.

Now suppose that x is a limit point of $\{x_n \mid n \in N\}$, and let (a, b) be an open interval which contains x. Because x is a limit point, (a, b) contains some element x_m of the set, and since x is an upper bound for the monotonically increasing sequence, it follows that $x_m \leqslant x_n \leqslant x$ for all $n \geqslant m$, and hence that $x_n \in (a, b)$ for all $n \geqslant m$. Therefore the sequence converges to x in this case also. We have thus shown that every monotonically increasing sequence which is bounded above converges to its least upper bound, and the proof for monotonically decreasing sequences is similar. □

Corollary 5.2.6 A monotonic sequence in $\boldsymbol{R}$ converges if and only if it is bounded.

The preceding corollary completely characterizes convergent monotonic sequences in $\boldsymbol{R}$. Our goal is a similar characterization for all convergent real sequences. We shall accomplish this by making use of our characterization of convergent monotonic sequences. We will first show that from any bounded sequence we can construct two bounded monotonic sequences, then prove that the original sequence converges if and only if the two monotonic sequences both converge to the same limit.

Before beginning the program outlined above, we introduce some convenient notation for divergent monotonic sequences. If a monotonically increasing sequence is not bounded above, it diverges because its terms grow without bound, in which case we say that the sequence diverges to *plus infinity*. Similarly, a monotonically decreasing sequence which is not bounded below is said to diverge to *minus infinity*.

Definition 5.2.7 If $\{x_n\}$ is a monotonically increasing sequence which is unbounded above, we write

$$\operatorname*{Lim}_{n \to \infty} x_n = +\infty$$

and say that the sequence *diverges to plus infinity*. If $\{x_n\}$ is a monotonically decreasing sequence which is unbounded below, we write

$$\operatorname*{Lim}_{n \to \infty} x_n = -\infty$$

and say that the sequence *diverges to minus infinity*.

Next we show that it is possible to construct from every bounded real sequence a monotonically increasing sequence and a monotonically decreasing sequence.

Proposition 5.2.8 Let $\{x_n\}$ be a sequence in $\boldsymbol{R}$.

1. If $\{x_n\}$ is bounded above, for each $k \in \boldsymbol{N}$ let

$$s_k = \sup \{x_n \mid n \geqslant k\}$$

   The sequence $\{s_k\}$ is monotonically decreasing and it converges if $\{x_n\}$ is bounded below.
2. If $\{x_n\}$ is bounded below, for each $k \in \boldsymbol{N}$ let

$$t_k = \inf \{x_n \mid n \geqslant k\}$$

   The sequence $\{t_k\}$ is monotonically increasing and it converges if $\{x_n\}$ is bounded above.

Proof: We prove (1), leaving (2) as an exercise. (See Exercise 5 at the end of this section.)

If the sequence $\{x_n\}$ is bounded above, then certainly for each $k \in \boldsymbol{N}$ the set $\{x_n \mid n \geqslant k\}$ is bounded above, and hence s_k exists for each $k \in \boldsymbol{N}$. Since

$$\{x_n \mid n \geqslant k+1\} \subset \{x_n \mid n \geqslant k\}$$

we have

$$s_{k+1} \leqslant s_k$$

for all $k \in \boldsymbol{N}$, and thus $\{s_k\}$ is a monotonically decreasing sequence. If $\{x_n\}$ is bounded below, then so is $\{s_k\}$, and hence being a bounded monotonically decreasing sequence, it must converge to its greatest lower bound. □

Examples 5.2.9 1. Let $\{x_n\} = \{1, \frac{1}{2}, \frac{1}{3}, \ldots\}$. This sequence is bounded above and below, so we may form both of the sequences $\{s_k\}$ and $\{t_k\}$:

$$s_1 = \sup \{1, \tfrac{1}{2}, \tfrac{1}{3}, , \ldots\} = 1 \qquad t_1 = \inf \{1, \tfrac{1}{2}, \tfrac{1}{3}, \ldots\} = 0$$

$$s_2 = \sup \{\tfrac{1}{2}, \tfrac{1}{3}, \ldots\} = \tfrac{1}{2} \qquad t_2 = \inf \{\tfrac{1}{2}, \tfrac{1}{3}, \ldots\} = 0$$

$$s_3 = \sup \{\tfrac{1}{3}, \tfrac{1}{4}, \ldots\} = \tfrac{1}{3} \qquad t_3 = \inf \{\tfrac{1}{3}, \tfrac{1}{4}, \ldots\} = 0$$

and so on. Therefore $\{s_k\} = \{1/k\}$, which is a monotonically decreasing sequence which converges to 0, and $\{t_k\} = \{0\}$, which is a monotonically increasing sequence which converges to 0. Note that both $\{s_k\}$ and $\{t_k\}$

converge to the same limit and that the original sequence $\{x_n\}$ also converges to this limit.

2. Let $\{x_n\} = \{1, -1, \frac{1}{2}, -2, \frac{1}{3}, -3, \ldots\}$. This sequence is bounded above, so the sequence $\{s_k\}$ exists, and we have

$$\{s_k\} = \{1, \tfrac{1}{2}, \tfrac{1}{2}, \tfrac{1}{3}, \tfrac{1}{3}, \ldots\}$$

Note that $\text{Lim}_{k \to \infty} s_k = 0$. Since $\{x_n\}$ is not bounded below, the sequence $\{t_k\}$ does not exist.

Now we are ready to define the concepts of the limit superior and limit inferior of a real sequence. These concepts will be useful in our characterization of convergent sequences in $\boldsymbol{R}$.

Definition 5.2.10 Let $\{x_n\}$ be a sequence in $\boldsymbol{R}$. Define the *limit superior* of $\{x_n\}$, denoted by $\text{Lim sup } x_n$, and the *limit inferior* of $\{x_n\}$, denoted by $\text{Lim inf } x_n$, as follows:

1. If $\{x_n\}$ is bounded above, let $\{s_k\}$ be the monotonically decreasing sequence defined in Proposition 5.2.8 (1) and set

$$\text{Lim sup } x_n = \text{Lim}_{k \to \infty} s_k$$

If $\{x_n\}$ is not bounded above, set $\text{Lim sup } x_n = +\infty$.

2. If $\{x_n\}$ is bounded below, let $\{t_k\}$ be the monotonically increasing sequence of Proposition 5.2.8 (2) and set

$$\text{Lim inf } x_n = \text{Lim}_{k \to \infty} t_k$$

If $\{x_n\}$ is not bounded below, set $\text{Lim inf } x_n = -\infty$.

Examples 5.2.11 1. Let $\{x_n\} = \{1/n\}$. This sequence is bounded above and below, and by virtue of Example 1 of Examples 5.2.9,

$$\text{Lim sup } x_n = \text{Lim}_{k \to \infty} s_k = \text{Lim}_{k \to \infty} \frac{1}{k} = 0$$

$$\text{Lim inf } x_n = \text{Lim}_{k \to \infty} t_k = \text{Lim}_{k \to \infty} 0 = 0$$

Note that $\text{Lim sup } x_n = \text{Lim inf } x_n = 0$ and that $\{x_n\}$ also converges to 0.

2. Let $\{x_n\} = \{(-1)^n\}$. This sequence is bounded and

$$s_k = \sup \{(-1)^n \mid n \geqslant k\} = 1$$

$$t_k = \inf \{(-1)^n \mid n \geqslant k\} = -1$$

Thus

$$\text{Lim sup } x_n = \lim_{k\to\infty} s_k = \lim_{k\to\infty} 1 = 1$$

and

$$\text{Lim inf } x_n = \lim_{k\to\infty} t_k = \lim_{k\to\infty} -1 = -1$$

Note that Lim sup $x_n \neq$ Lim inf x_n and that the sequence $\{x_n\}$ diverges.

3. Let $\{x_n\} = \{1, -1, \frac{1}{2}, -2, \frac{1}{3}, -3, \ldots\}$. By virtue of Example 2 of Examples 5.2.9,

$$\text{Lim sup } x_n = \lim_{k\to\infty} s_k = 0$$

while, since $\{x_n\}$ is not bounded below,

$$\text{Lim inf } x_n = -\infty$$

4. Consider the sequence $\{x_n\} = \{(-1)^n n\}$. This sequence is unbounded above and unbounded below, so

$$\text{Lim sup } x_n = +\infty \qquad \text{and} \qquad \text{Lim inf } x_n = -\infty$$

5. Consider the sequence $\{x_n\} = \{n\}$. This sequence is unbounded above, hence

$$\text{Lim sup } x_n = +\infty$$

It is bounded below, however, and since $t_k = k$, for all $k \in \boldsymbol{N}$, the monotonically increasing sequence $\{t_k\}$ diverges to $+\infty$. Therefore we have

$$\text{Lim inf } x_n = \lim_{k\to\infty} t_k = +\infty$$

Similarly, the sequence $\{y_n\} = \{-n\}$ has

$$\text{Lim sup } y_n = -\infty \qquad \text{and} \qquad \text{Lim inf } y_n = -\infty$$

We will soon prove that a sequence of real numbers converges if and only if its limit superior and limit inferior are finite and equal. In general, the limit superior and limit inferior of a sequence may be thought of as measuring the "degree of divergence" of a sequence. For instance, if the limit superior and limit inferior are both infinite, as in Examples 4 and 5 of Examples 5.2.11, then no part of the sequence converges, and it is very badly divergent. If one of the limit superior or limit inferior is finite while the other is infinite, as in Example 3 above, then the sequence is less badly divergent, for at least "part" of it converges. If the limit superior and limit inferior are finite but unequal, as in Example 2, then the sequence is "almost" convergent, for its "parts" converge even though it does not converge as a whole. Finally, if the limit superior and limit inferior are finite and equal, as in Example 1, then the sequence converges.

It is easy to show that $\operatorname{Lim\,inf} x_n \leqslant \operatorname{Lim\,sup} x_n$ for any sequence $\{x_n\}$ in $\boldsymbol{R}$. We outline a proof of this fact, leaving the details as an exercise. (See Exercise 6 at the end of this section.)

Proposition 5.2.12 If $\{x_n\}$ is a sequence in $\boldsymbol{R}$, then

$$\operatorname{Lim\,inf} x_n \leqslant \operatorname{Lim\,sup} x_n$$

Proof: Suppose first that $\{x_n\}$ is bounded above and below. It is easy to check that

$$t_i = \inf\{x_n \mid n \geqslant i\} \leqslant \sup\{x_n \mid n \geqslant j\} = s_j$$

for all $i \in \boldsymbol{N}$, $j \in \boldsymbol{N}$, and from this it follows that

$$\operatorname{Lim\,inf} x_n = \operatorname*{Lim}_{k \to \infty} t_i = \sup\{t_i\} \leqslant \inf\{s_j\} = \operatorname*{Lim}_{j \to \infty} = \operatorname{Lim\,sup} x_n$$

On the other hand, if $\{x_n\}$ is not bounded above, then

$$\operatorname{Lim\,sup} x_n = +\infty$$

so surely

$$\operatorname{Lim\,inf} x_n \leqslant \operatorname{Lim\,sup} x_n$$

in this case, and similarly if $\{x_n\}$ is not bounded below. □

Now we are ready to prove the theorem which characterizes convergent sequences in $\boldsymbol{R}$. The theorem not only tells us exactly which real sequences are convergent but also what the limit of a convergent sequence must be.

Theorem 5.2.13 If $\{x_n\}$ is a sequence in $\boldsymbol{R}$, then $\{x_n\}$ converges to $x \in \boldsymbol{R}$ if and only if

$$\text{Lim sup } x_n = \text{Lim inf } x_n = x$$

Proof: If $\{x_n\}$ converges to x, then $\{x_n\}$ must be a bounded sequence, and therefore the monotonic sequences $\{s_k\}$ and $\{t_k\}$ of Proposition 5.2.8 exist and are convergent. Hence

$$\text{Lim sup } x_n = \underset{k \to \infty}{\text{Lim}}\ s_k = \inf \{s_k\} = s$$

and

$$\text{Lim inf } x_n = \underset{k \to \infty}{\text{Lim}}\ t_k = \sup \{t_k\} = t$$

both exist as finite real numbers. Note that for all $k \in \boldsymbol{N}$,

$$s \leqslant s_k \qquad \text{and} \qquad t \geqslant t_k$$

If $\text{Lim}_{n \to \infty} x_n = x < s$, then there exists an open interval (a, b) such that $x \in (a, b)$ and $b < s$. But then there is some $m \in \boldsymbol{N}$ such that $x_n \in (a, b)$ for all $n \leqslant m$, and therefore

$$s_m = \sup \{x_n \mid n \geqslant m\} \leqslant b < s$$

which contradicts the fact that s is a lower bound for the sequence $\{s_k\}$. Thus we must have $x \geqslant s$. A similar argument shows that $x \leqslant t$, and hence

$$s \leqslant x \leqslant t$$

But $t \leqslant s$ by Proposition 5.2.12. Therefore $x = s = t$.

Now we prove the converse. Suppose that $\{x_n\}$ is a real sequence such that $\text{Lim sup } x_n = \text{Lim inf } x_n = x$, where x is a finite real number. We must show that $\{x_n\}$ converges to x. Let (a, b) be an open interval which contains x. Since

$$x = \text{Lim sup } x_n = \lim_{k \to \infty} s_k$$

there exists $m_1 \in \boldsymbol{N}$ such that $s_k \in (a, b)$ for all $k \geqslant m_1$. Similarly, because $x = \text{Lim inf } x_n$, there exists $m_2 \in \boldsymbol{N}$ such that $t_k \in (a, b)$ for all $k \geqslant m_2$. If $m = \max\{m_1, m_2\}$ then for all $k \geqslant m$, we have $s_k \in (a, b)$ and $t_k \in (a, b)$. But $s_k = \sup\{x_n \mid n \geqslant k\}$ implies that $s_k \geqslant x_k$ for all $k \in \boldsymbol{N}$, and similarly $t_k \leqslant x_k$ for all $k \in \boldsymbol{N}$. Hence $t_k \leqslant x_k \leqslant s_k$ for all $k \geqslant m$, and it follows that $x_k \in (a, b)$ for all $k \geqslant m$. Therefore $\{x_n\}$ converges to x, and we are done. □

Theorem 5.2.13 shows that the question of convergence for an arbitrary real sequence can be reduced to a question of convergence for two associated monotonic sequences: the original sequence converges if and only if the associated monotonic sequences both converge to the same limit. Since convergence for monotonic sequences is merely a question of the boundedness of their images as subsets of the reals, this is indeed a result of considerable power.

EXERCISES

1. For each of the following real sequences, prove convergence or divergence. If the sequence converges, find its limit.

   (a) $\left\{\dfrac{n+1}{n}\right\}$

   (b) $\left\{\dfrac{n}{n+1}\right\}$

   (c) $\{n^2 - n\}$

   (d) $\left\{\dfrac{2n+1}{n^2+1}\right\}$

   (e) $\left\{\dfrac{2-3n^2}{n+2}\right\}$

   (f) $\{r/n\}$, any $r \in \boldsymbol{R}$

2. Let $r \in \boldsymbol{R}$, $r \geqslant 0$. Under what circumstances does the sequence $\{r^n\}$ converge, and what is its limit? Prove your answers.
3. Let $\{x_n\}$ be a real sequence. Show that

$$\lim_{k \to \infty} x_n = 0 \qquad \text{if and only if } \lim_{k \to \infty} |x_n| = 0$$

4. Let $\{x_n\}$ be a real sequence which converges to the real number x. Prove that for any $m \in \boldsymbol{N}$,
   (a) The sequence $\{x_n^m\}$ converges to x^m;
   (b) If it is defined, the sequence $\{\sqrt[m]{x_n}\}$ converges to $\sqrt[m]{x}$.

5. Prove (2) of Proposition 5.2.8.
6. Prove Proposition 5.2.12.
7. For each of the following sequences, find $\operatorname{Lim\,sup} x_n$ and $\operatorname{Lim\,inf} x_n$, state whether the sequence converges or diverges, and find its limit if it converges.

   (a) $\{n(n-1)\}$
   (b) $\left\{\dfrac{n-1}{2n}\right\}$
   (c) $\left\{\dfrac{n^2+1}{n^3}\right\}$
   (d) $\{1, 2, 3, 4, 1, 2, 3, 4, \ldots\}$
   (e) $\{1, 1, 2, 1, 2, 3, 1, 2, 3, 4, \ldots\}$
   (f) $\{0, -1, 0, -2, 0, -3, \ldots\}$
   (g) $\{2, 0, -2, 0, 2, 0, -2, 0, \ldots\}$

8. Let $\{x_n\}$ be a real sequence. A sequence $\{y_n\}$ is said to be a *subsequence* of $\{x_n\}$ if both of the following conditions hold:
   (i) For every $m \in \boldsymbol{N}$, there is some $n \in \boldsymbol{N}$ such that $y_m = x_n$;
   (ii) If $y_m = x_n$ and $y_{m+1} = x_k$, then $n < k$.
   Prove that if $\{x_n\}$ has a subsequence which is bounded below by b then $\operatorname{Lim\,sup} x_n \geqslant b$, while if $\{x_n\}$ has no subsequence which is bounded below, then $\operatorname{Lim\,sup} x_n = -\infty$. State and prove a similar result for $\operatorname{Lim\,inf} x_n$.
9. Let $\{x_n\}$ be a bounded real sequence. Let $\operatorname{Lim\,sup} x_n = s$ and $\operatorname{Lim\,inf} x_n = t$, and let ε be any positive real number. Prove:
   (a) There are infinitely many values of n for which $x_n > s - \varepsilon$;
   (b) There are infinitely many values of n for which $x_n < t + \varepsilon$;
   (c) $x_n < s + \varepsilon$ for all but finitely many values of n;
   (d) $x_n > t - \varepsilon$ for all but finitely many values of n.
10. Prove that every bounded real sequence has a convergent subsequence by showing that if $\{x_n\}$ is such a sequence, then it has a subsequence which converges to $\operatorname{Lim\,sup} x_n$. State and prove a similar result for $\operatorname{Lim\,inf} x_n$.
11. Let $\{x_n\}$ and $\{y_n\}$ be bounded real sequences, with $x_n \leqslant y_n$ for all $n \in \boldsymbol{N}$. Show that

$$\operatorname{Lim\,sup} x_n \leqslant \operatorname{Lim\,sup} y_n$$

   and

$$\operatorname{Lim\,inf} x_n \leqslant \operatorname{Lim\,inf} y_n$$

12. Let $\{x_n\}$ and $\{y_n\}$ be bounded real sequences, and consider the sequence $\{x_n + y_n\}$.

(a) Prove that

$$\text{Lim sup}\,[x_n + y_n] \leqslant \text{Lim sup}\, x_n + \text{Lim sup}\, y_n$$

and

$$\text{Lim inf}\,[x_n + y_n] \geqslant \text{Lim inf}\, x_n + \text{Lim inf}\, y_n$$

(b) Give examples to show that equality need not hold in (a).

5.3 INFINITE SERIES

In this section we consider the special kind of real sequence known as an infinite series. We begin with the definition of an infinite series.

Definition 5.3.1 Let $\{x_n\}$ be a sequence in $\boldsymbol{R}$. For each $m \in \boldsymbol{N}$ let

$$S_m = x_1 + x_2 + \cdots + x_m = \sum_{k=1}^{m} x_k$$

The real number S_m is the mth *partial sum* of the sequence $\{x_n\}$. The sequence

$$\{S_m\} = \left\{ \sum_{k=1}^{m} x_k \,\middle|\, m \in \boldsymbol{N} \right\}$$

of mth partial sums of $\{x_n\}$ is called the *infinite series with* nth *term* x_n and is denoted by

$$\sum_{n=1}^{\infty} x_n$$

Note that an infinite series $\Sigma_{n=1}^{\infty} x_n$ is *not* an "infinite sum." The notation $\Sigma_{n=1}^{\infty} x_n$ merely stands for a special sequence, namely the sequence of mth partial sums of the sequence $\{x_n\}$.

Examples 5.3.2 1. The infinite series $\Sigma_{n=1}^{\infty} (-1)^n$ is the sequence $\{S_m\}$, where

$$S_m = \sum_{k=1}^{m} (-1)^k$$

Clearly, $S_m = -1$ if m is an odd natural number and $S_m = 0$ if m is an even natural number. Therefore $\Sigma_{n=1}^{\infty}(-1)^n$ is the sequence $\{-1, 0, -1, 0, \ldots\}$.

2. The infinite series

$$\sum_{n=1}^{\infty} \frac{1}{2^n}$$

is the sequence of partial sums $\{\Sigma_{k=1}^{m}(1/2^k)\}$. But

$$\sum_{k=1}^{m} \frac{1}{2^k} = 1 - \frac{1}{2^m}$$

so $\Sigma_{n=1}^{\infty}(1/2^n)$ is the sequence $\{1-(1/2^m)\} = \{\frac{1}{2}, \frac{3}{4}, \frac{7}{8}, \ldots\}$.

Since an infinite series is just a special kind of sequence, we can speak of its convergence or divergence.

Definition 5.3.3 The infinite series $\Sigma_{n=1}^{\infty} x_n$ is said to *converge* if its sequence of partial sums $\{\Sigma_{k=1}^{m} x_k\}$ converges, and to *diverge* if its sequence of partial sums diverges. If the sequence of partial sums converges to $x \in \boldsymbol{R}$, we say that the infinite series *converges to* x. Although the series $\Sigma_{n=1}^{\infty} x_n$ is a sequence, not a number, it is traditional to write

$$\sum_{n=1}^{\infty} x_n = x$$

to indicate that the series converges to x.

Examples 5.3.4 1. The infinite series $\Sigma_{n=1}^{\infty}(-1)^n$ of Example 1, Examples 5.3.2, diverges because its sequence of partial sums is $\{-1, 0, -1, 0, \ldots\}$, which is clearly divergent.

2. The series $\Sigma_{n=1}^{\infty}(1/2^n)$ of Example 2, Examples 5.3.2, converges to 1 because its sequence of partial sums is $\{1-(1/2^m)\}$, which converges to 1.

3. The infinite series $\Sigma_{n=1}^{\infty}(1/n)$ is known as the *harmonic series.* We will show that the harmonic series diverges. If $\{S_m\}$ is the sequence of partial sums of the harmonic series, we have

$$S_2 = 1 + \tfrac{1}{2} > 1$$

$$S_4 = S_2 + \tfrac{1}{3} + \tfrac{1}{4} > S_2 + 2(\tfrac{1}{4}) = S_2 + \tfrac{1}{2}$$

$$S_8 = S_4 + \tfrac{1}{5} + \tfrac{1}{6} + \tfrac{1}{7} + \tfrac{1}{8} > S_4 + 4(\tfrac{1}{8}) = S_4 + \tfrac{1}{2}$$

$$S_{16} = S_8 + \tfrac{1}{9} + \cdots + \tfrac{1}{16} > S_8 + 8(\tfrac{1}{16}) = S_8 + \tfrac{1}{2}$$

and so on. Thus the sequence of partial sums is clearly unbounded, and hence divergent, and therefore the harmonic series diverges.

Now we prove some results about infinite series. Our first result will completely characterize the type of series known as a geometric series.

Definition 5.3.5 A series of the form

$$\sum_{n=0}^{\infty} ar^n = a + ar + ar^2 + ar^3 + \cdots$$

is called a *geometric series with ratio r.*

Proposition 5.3.6 If $|r| < 1$, the geometric series $\Sigma_{n=0}^{\infty} ar^n$ converges to $a/(1-r)$. If $a \neq 0$ and $|r| \geqslant 1$, the series diverges.

Proof: It is easy to show that if $r \neq 1$ the mth partial sum of the geometric series is

$$S_m = \frac{a(1-r^m)}{1-r}$$

and the proposition follows from this fact and the examination of a few special cases. The details are left to the reader. (See Exercise 1 at the end of this section.) □

The geometric series are virtually the only important series for which it is possible to determine at a glance whether or not a particular series is convergent. Other types of series generally require more sophisticated tests to determine whether or not they converge, and there are many such tests, some of them quite specialized. In the remainder of this section, we present a few of the more general convergence tests; others are given in the exercises at the end of the section.

Our first test is one which is frequently useful in establishing the divergence of a series.

Proposition 5.3.7 If the series $\Sigma_{n=1}^{\infty} x_n$ converges, then $\text{Lim}_{n\to\infty} x_n = 0$.

Proof: If the series $\Sigma_{n=1}^{\infty} x_n$ converges, then its sequence of partial sums $\{S_m\} = \{S_1, S_2, S_3, \ldots\}$ converges to some real number x. But then the sequence $\{T_m\} = \{0, S_1, S_2, \ldots\}$ also converges to x. (Why?) Since

$x_m = S_m - T_m$ for all $m \in \boldsymbol{N}$, we have

$$\lim_{m \to \infty} x_m = \lim_{m \to \infty} [S_m - T_m] = \lim_{m \to \infty} S_m - \lim_{m \to \infty} T_m = x - x = 0. \square$$

Proposition 5.3.7 provides a divergence test for series because it implies that if $\lim_{n \to \infty} x_n \neq 0$, then the series $\sum_{n=1}^{\infty} x_n$ must diverge. Note that the converse of the proposition is *not* true: a series $\sum_{n=1} x_n$ may diverge even though $\lim_{n \to \infty} x_n = 0$, as the example of the harmonic series shows.

Example 5.3.8 The series $\sum_{n=1}^{\infty} (n+1)/n$ diverges, since

$$\lim_{n \to \infty} \frac{n+1}{n} = 1$$

The next result may also be thought of as providing a test for divergence, though not a particularly easy one to apply. It is included here because it is needed to prove the more useful comparison test.

Proposition 5.3.9 (Boundedness Test) If $\sum_{n=1}^{\infty} x_n$ is an infinite series with $x_n \geqslant 0$ for all $n \in \boldsymbol{N}$, then the series converges if and only if its sequence of partial sums is bounded above.

Proof: The infinite series converges if and only if its sequence of partial sums converges. But since $x_n \geqslant 0$ for all $n \in \boldsymbol{N}$, the sequence of partial sums is monotonically increasing and therefore converges if and only if it is bounded above. □

The next result, known as the *comparison test*, is perhaps the most useful of all convergence tests.

Proposition 5.3.10 (Comparison Test) If $\sum_{n=1}^{\infty} x_n$ and $\sum_{n=1}^{\infty} y_n$ are infinite series with $0 \leqslant x_n \leqslant y_n$ for all $n \in \boldsymbol{N}$, then

1. If $\sum_{n=1}^{\infty} y_n$ converges, so does $\sum_{n=1}^{\infty} x_n$;
2. If $\sum_{n=1}^{\infty} x_n$ diverges, so does $\sum_{n=1}^{\infty} y_n$.

Proof: If $\sum_{n=1}^{\infty} y_n$ converges, then by Proposition 5.3.9, the sequence of partial sums $\{\sum_{k=1}^{m} y_k\}$ is bounded above. But $0 \leqslant x_n \leqslant y_n$ for all $n \in \boldsymbol{N}$ then implies that the sequence of partial sums $\{\sum_{k=1}^{m} x_k\}$ is bounded above, and hence the series $\sum_{n=1}^{\infty} x_n$ must converge. This proves (1), and the proof of (2) is similar. (See Exercise 2 at the end of this section.) □

Examples 5.3.11 1. Let p be a real number and consider the *p-series* $\Sigma_{n=1}^{\infty}(1/n^p)$. Since $0 < 1/n \leqslant 1/n^p$ when $p < 1$ and since $\Sigma_{n=1}^{\infty}(1/n)$ diverges, we see that when $p < 1$, the p-series diverges by comparison with the harmonic series.

2. Since $0 < 1/n^n \leqslant 1/2^n$ for all $n \geqslant 2$ and since $\Sigma_{n=2}^{\infty}(1/2^n)$ converges (why?), the series $\Sigma_{n=2}^{\infty}(1/n^n)$ converges by comparison. Therefore the series $\Sigma_{n=1}^{\infty}(1/n^n) = 1 + \Sigma_{n=2}^{\infty}(1/n^n)$ also converges. (Why?)

The final convergence test we shall present is the well-known *ratio test*.

Proposition 5.3.12 (Ratio Test) Let $\Sigma_{n=1}^{\infty} x_n$ be a series of positive real numbers.

1. If $\displaystyle\operatorname{Lim}_{n\to\infty} \frac{x_{n+1}}{x_n} < 1$, then $\displaystyle\sum_{n=1}^{\infty} x_n$ converges;
2. If $\displaystyle\operatorname{Lim}_{n\to\infty} \frac{x_{n+1}}{x_n} > 1$, then $\displaystyle\sum_{n=1}^{\infty} x_n$ diverges.

Proof: 1. If $\operatorname{Lim}_{n\to\infty}(x_{n+1}/x_n) = L < 1$, let r be a positive real number such that $L < r < 1$. Since there is some $m \in \boldsymbol{N}$ such that $(x_{n+1}/x_n) < r$ for all $n \geqslant m$, we have

$$x_{m+1} < rx_m$$

$$x_{m+2} < rx_{m+1} < r^2 x_m$$

$$x_{m+3} < rx_{m+2} < r^3 x_m$$

and so on. Therefore, by induction

$$x_{m+k} < r^k x_m \qquad \text{for all } k \in \boldsymbol{N}$$

But the infinite series

$$\sum_{k=1}^{\infty} x_m r^k$$

is a geometric series with ratio r, and since $|r| < 1$, it converges. Thus the series

$$\sum_{k=1}^{\infty} x_{m+k}$$

converges (by comparison wth the geometric series). But if $\Sigma_{k=1}^{\infty} x_{m+k}$ converges, so does $\Sigma_{n=1}^{\infty} x_n$. (See Exercise 3 at the end of this section.)

2. If $\operatorname{Lim}_{n\to\infty} (x_{n+1}/x_n) > 1$, then there is some $m \in N$ such that $(x_{n+1}/x_n) > 1$ for all $n \geqslant m$, and thus

$$x_m < x_{m+1} < x_{m+2} < \cdots$$

Since $\Sigma_{n=1}^{\infty} x_n$ is a series of positive real numbers, this shows that $\operatorname{Lim}_{n\to\infty} x_n \neq 0$, and hence the series diverges by Proposition 5.3.7. □

Note that if $\operatorname{Lim}_{n\to\infty} (x_{n+1}/x_n) = 1$, then the ratio test says nothing about convergence or divergence: in such a case the series may either converge or diverge.

Examples 5.3.13 1. The series $\Sigma_{n=1}^{\infty} (1/n!)$ converges by the ratio test because

$$\operatorname*{Lim}_{n\to\infty} \frac{x_{n+1}}{x_n} = \operatorname*{Lim}_{n\to\infty} \frac{1/(n+1)!}{1/n!} = \operatorname*{Lim}_{n\to\infty} \frac{1}{n+1} = 0$$

2. The series $\Sigma_{n=1}^{\infty} (2^n/n)$ diverges by the ratio test because

$$\operatorname*{Lim}_{n\to\infty} \frac{x_{n+1}}{x_n} = \operatorname*{Lim}_{n\to\infty} \frac{2^{n+1}/(n+1)}{2^n/n} = \operatorname*{Lim}_{n\to\infty} 2 \frac{n}{n+1} = 2$$

EXERCISES

1. Prove Proposition 5.3.6.
2. Prove (2) of Proposition 5.3.10.
3. Prove that a series $\Sigma_{n=1}^{\infty} x_n$ converges if there is some $m \in N$ such that $\Sigma_{k=1}^{\infty} x_{m+k}$ converges.
4. For each of the following series, prove convergence or divergence.

(a) $\displaystyle\sum_{n=1}^{\infty} (-1)^n \frac{n+1}{n}$

(b) $\displaystyle\sum_{n=1}^{\infty} \frac{3}{\sqrt{n}}$

(c) $\displaystyle\sum_{n=1}^{\infty} \frac{n^3}{n!}$

(d) $\displaystyle\sum_{n=1}^{\infty} 3(5)^{1-n}$

(e) $\displaystyle\sum_{n=1}^{\infty} \frac{n^3}{2^n}$

(f) $\displaystyle\sum_{n=1}^{\infty} \frac{1}{1+2^{n-1}}$

5. (a) Show that every series of the form $\Sigma_{n=1}^{\infty}(a_n/10^n)$, where

$$a_n \in \{0, 1, 2, \ldots, 9\} \qquad \text{for all } n \in \boldsymbol{N}$$

converges to some $x \in [0, 1]$.

(b) Show that if $x \in [0, 1]$ has the decimal representation

$$x = 0.a_1 a_2 a_3 \ldots$$

then

$$\sum_{n=1}^{\infty} \frac{a_n}{10^n} = x$$

6. (a) Prove that if $\Sigma_{n=1}^{\infty} x_n = x$ and $\Sigma_{n=1}^{\infty} y_n = y$, then

$$\sum_{n=1}^{\infty} (x_n + y_n) = x + y$$

(b) Prove that if $\Sigma_{n=1}^{\infty} x_n = x$ and r is any real number, then

$$\sum_{n=1}^{\infty} r x_n = r x$$

7. (a) Generalize Exercise 3 above by showing that if one infinite series is created by eliminating finitely many terms from another one, then either both series converge or both diverge.

(b) Show that the hypotheses of the comparison test may be weakened by replacing the condition

$$0 \leqslant x_n \leqslant y_n \qquad \text{for all } n \in \boldsymbol{N}$$

with

$$0 \leqslant x_n \leqslant y_n \qquad \text{for all but finitely many indices } n$$

8. Show that if $\Sigma_{n=1}^{\infty} |x_n|$ converges, then $\Sigma_{n=1}^{\infty} x_n$ converges.

9. The ratio test may be generalized as follows:

Let $\Sigma_{n=1}^{\infty} x_n$ be a series of nonzero real numbers.

(i) If $\operatorname{Lim\,sup} \left| \dfrac{x_{n+1}}{x_n} \right| < 1$, then $\displaystyle\sum_{n=1}^{\infty} x_n$ converges;

(ii) If $\operatorname{Lim\,inf} \left| \dfrac{x_{n+1}}{x_n} \right| > 1$, then $\displaystyle\sum_{n=1}^{\infty} x_n$ diverges.

Prove this.

10. The *root test* for convergence of a nonnegative series may be stated as follows:

Let $\Sigma_{n=1}^{\infty} x_n$ be a series of nonnegative real numbers.

(i) If $\lim_{n\to\infty} \sqrt[n]{x_n} < 1$, then $\sum_{n=1}^{\infty} x_n$ converges;

(ii) If $\lim_{n\to\infty} \sqrt[n]{x_n} > 1$, then $\sum_{n=1}^{\infty} x_n$ diverges.

(a) Apply the root test to the following series:

$$\sum_{n=1}^{\infty} \frac{1}{n^n} \qquad \sum_{n=1}^{\infty} \left(\frac{n+1}{2n+1}\right)^n \qquad \sum_{n=1}^{\infty} \frac{3^n}{n^3}$$

(b) Prove the root test.

11. Consider a series of the form $\Sigma_{n=1}^{\infty} (-1)^{n+1} x_n$, where x_n is positive for all $n \in \boldsymbol{N}$. Such a series is called an *alternating series*. The following is a convergence test for alternating series:

The alternating series $\Sigma_{n=1}^{\infty} (-1)^{n+1} x_n$ converges if

(i) $x_n \geqslant x_{n+1}$ for all $n \in \boldsymbol{N}$, and

(ii) $\lim_{n\to\infty} x_n = 0$.

(a) Apply the test to the following series:

$$\sum_{n=1}^{\infty} (-1)^{n+1} \frac{1}{n} \qquad \sum_{n=1}^{\infty} (-1)^{n+1} \frac{n+1}{n} \qquad \sum_{n=1}^{\infty} (-1)^{n+1} \frac{\ln n}{n+1}$$

($\ln n$ is the natural logarithm, or logarithm to the base e, of n.)

(b) Prove the test.

12. The *integral test* for the convergence of positive series may be stated as follows: Let $\Sigma_{n=1}^{\infty} x_n$ be a series of positive real numbers, and let $f: [1, +\infty) \to [0, +\infty)$ be a continuous function such that $f(n) = x_n$ for all $n \in \boldsymbol{N}$ and $f(x) \leqslant f(y)$ if $x \geqslant y$. The series $\Sigma_{n=1}^{\infty} x_n$ converges if and only if the improper integral $\int_1^{+\infty} f(x)\, dx$ exists.

(a) Apply the integral test to the following series:

$$\sum_{n=1}^{\infty} \frac{1}{n} \qquad \sum_{n=1}^{\infty} \frac{1}{n^p} \ (p \text{ any real number}) \qquad \sum_{n=2}^{\infty} \frac{1}{n(\ln n)}$$

(b) Prove the integral test.

5.4 FUNCTIONAL LIMITS

Let f be a function from $\boldsymbol{R}$ to $\boldsymbol{R}$. The reader has undoubtedly encountered expressions of the form $\operatorname{Lim}_{x \to c} f(x) = x_0$. An expression such as this, which is known as a *functional limit*, is meant to convey the information that the functional values $f(x)$ approach x_0 as x approaches c. In this section we present a brief discussion of functional limits. We begin with the definition of such a limit, show how questions about the limit of a function f can be reduced to questions about the limits of sequences of the form $\{f(x_n)\}$, and conclude with a characterization of continuous functions on $\boldsymbol{R}$ which utilizes functional limits.

Our definition of functional limit is given in terms of the natural topology on $\boldsymbol{R}$.

Definition 5.4.1 Let S be a subset of $\boldsymbol{R}$, $f: S \to \boldsymbol{R}$, and let c be a limit point of S. The real number x_0 is *the limit of f as x approaches c*, written

$$\operatorname*{Lim}_{x \to c} f(x) = x_0$$

if for every open set G containing x_0 there exists an open set H containing c such that $f(x) \in G$ for all $x \in S \cap H - \{c\}$.

Suppose that $\operatorname{Lim}_{x \to c} f(x) = x_0$. The definition says that if we surround the point x_0 with an open set G, then there must be an open set H about c such that f carries the elements of $S \cap H$ (with the possible exception of c, if $c \in S$) into the set G. Therefore by choosing G to be smaller and smaller, we can force the functional values of f to approach x_0 more and more closely. Note that the value of f at c is immaterial, for we never consider $f(c)$; indeed, f need not even be defined at c, for we do not require that c be an element of the domain S of f, but only that it be a limit point of S. (It is necessary to require that c be a limit point of S in order to ensure that the open set H contains some point of the domain S.) We also remark that use of the phrase "*the* limit of f as x approaches c" in the definition is justified, for the limit of f as x approaches c is unique if it exists. As usual, this depends on the fact that $\boldsymbol{R}$ is a Hausdorff space, and we leave the proof to the reader. (See Exercise 1 at the end of this section.)

Examples 5.4.2 1. Let $f: \boldsymbol{R} \to \boldsymbol{R}$ be given by

$$f(x) = \begin{cases} 2 & \text{if } x = 1 \\ x & \text{if } x \neq 1 \end{cases}$$

We claim that $\text{Lim}_{x\to 1} f(x) = 1$. To prove this, we must show that if G is an open set which contains 1, then there exists an open set H which contains 1 and such that $f(x) \in G$ for all $x \in H - \{1\}$. But since $f(x) = x$ for all $x \neq 1$, we may take $H = G$. Note that $\text{Lim}_{x\to 1} f(x) = 1$ even though $f(1) \neq 1$.

2. Let $f\colon (0, 1) \to \boldsymbol{R}$ be given by $f(x) = 2x + 1$ for all $x \subset (0, 1)$. We claim that $\text{Lim}_{x\to c} f(x) = 2c + 1$ for all $c \in [0, 1]$. To see this, let G be an open set containing $2c + 1$. There is some open interval (a, b) contained in G such that $2c + 1 \in (a, b)$. If $H = ((a - 1)/2, (b - 1)/2)$, then H is open, $c \in H$, and $f(x) \in (a, b) \subset G$ for all $x \in (0, 1) \cap H - \{c\}$. Note that if $c = 0$ or $c = 1$, then $f(c)$ is not defined, but $\text{Lim}_{x\to c} f(x)$ exists nonetheless.

3. Let $f\colon \boldsymbol{R} \to \boldsymbol{R}$ be given by

$$f(x) = \begin{cases} 1 & \text{if } x > 0 \\ -1 & \text{if } x \leqslant 0 \end{cases}$$

If $c > 0$, then $\text{Lim}_{x\to c} f(x) = 1$, for if G is an open set containing 1, we may choose $H = (a, b)$, where $0 < a < c < b$. A similar argument shows that if $c < 0$, then $\text{Lim}_{x\to c} f(x) = -1$. We claim, however, that $\text{Lim}_{x\to 0} f(x)$ does not exist. To see this, suppose $\text{Lim}_{x\to 0} f(x) = x_0$ does exist. If G is an open set containing x_0, there must exist an open set H containing 0 such that $f(x) \in G$ for all $x \in H - \{0\}$. Choose G to be any open set which does not contain both 1 and -1. Since an open set H which contains 0 must contain both a positive number x_1 and a negative number x_2, and since $f(x_1) = 1$ and $f(x_2) = -1$, it is not possible to have $f(x) \in G$ for all $x \in H - \{0\}$, and this establishes the claim.

Now, as promised, we demonstrate the connection between the limit of a function f as x approaches c and the limits of certain sequences of the form $\{f(x_n)\}$.

Proposition 5.4.3 Let S be a subset of $\boldsymbol{R}$, $f\colon S \to \boldsymbol{R}$, and let c be a limit point of S; then $\text{Lim}_{x\to c} f(x) = x_0$ if and only if $\text{Lim}_{n\to\infty} f(x_n) = x_0$ for every sequence $\{x_n\}$ in S which converges to c and for which $x_n \neq c$ for all $n \in \boldsymbol{N}$.

Proof: Suppose $\text{Lim}_{x\to c} f(x) = x_0$ and let $\{x_n\}$ be a sequence of points of S which converges to c and such that $x_n \neq c$ for all $n \in \boldsymbol{N}$. If G is an open set containing x_0, there is an open set H containing c such that $f(x) \in G$ for all $x \in S \cap H - \{c\}$. Since $\text{Lim}_{n\to\infty} x_n = c$, there is some $m \in \boldsymbol{N}$ such that $x_n \in H$ for all $n \geqslant m$. Since $x_n \neq c$ for all $n \in \boldsymbol{N}$ and since $\{x_n\}$ is a sequence of points of S, it follows that $x_n \in S \cap H - \{c\}$ for all $n \geqslant m$. But then $f(x_n) \in G$ for all $n \geqslant m$, and this shows that the sequence $\{f(x_n)\}$ converges to x_0.

To prove the converse, we show that if $\text{Lim}_{x \to c} f(x) \neq x_0$, then there is some sequence $\{x_n\}$ in S such that $x_n \neq c$ for all $n \in \boldsymbol{N}$ and $\text{Lim}_{n \to \infty} x_n = c$, but $\text{Lim}_{n \to \infty} f(x_n) \neq x_0$. We argue as follows: if $\text{Lim}_{x \to c} f(x) \neq x_0$, there exists and open set G containing x_0 such that for every open set H containing c there is some point $x \in S \cap H - \{c\}$ for which $f(x) \notin G$. In particular, this is so if we choose $H = H_n = (c - 1/n, c + 1/n)$, $n \in \boldsymbol{N}$. Thus for every $n \in \boldsymbol{N}$ there is some point $x_n \in S \cap H_n - \{c\}$ such that $f(x_n) \notin G$. Clearly, $\{x_n\}$ is a sequence of points of S which converges to c, $x_n \neq c$ for all $n \in \boldsymbol{N}$, and $\{f(x_n)\}$ cannot converge to x_0 because $f(x_n) \notin G$ for any $n \in \boldsymbol{N}$. □

Proposition 5.4.3 says that the study of functional limits in $\boldsymbol{R}$ reduces to the study of certain convergent sequences in the domain of the function. Thus the concept of the limit of a function on $\boldsymbol{R}$ does not really yield anything new, for it is merely sequences "in disguise": whatever can be accomplished by using limits of functions can also be accomplished using limits of sequences. Many of our earlier results concerning functions and sequences can be used to derive corresponding results concerning functional limits. As an example, we present a result which is sometimes used as the definition of continuity on the real line.

Proposition 5.4.4 Let S be a subset of $\boldsymbol{R}$ and let $c \in S$ be a limit point of S. A function $f: S \to \boldsymbol{R}$ is continuous at c if and only if $\text{Lim}_{x \to c} f(x) = f(c)$.

Proof: If f is continuous at c, then for every sequence $\{x_n\}$ in S which converges to c, we must have $\{f(x_n)\}$ converging to $f(c)$ by Proposition 5.1.7. But then Proposition 5.4.3 shows that $\text{Lim}_{x \to c} f(x) = f(c)$. Conversely, if $\text{Lim}_{x \to c} f(x) = f(c)$, then for every open set G containing $f(c)$ there is an open set H containing c such that $f(x) \in G$ for all $x \in S \cap H - \{c\}$. But since $f(c) \in G$, we conclude that for every open set G containing $f(c)$ there is an open set H containing c such that $f(x) \in G$ for all $x \in S \cap H$, and this is the definition of continuity of f at the point c. □

We conclude this section with a statement of the familiar properties of limits. The proof is left to the reader. (See Exercise 3 at the end of this section.)

Proposition 5.4.5 Let S be a subset of $\boldsymbol{R}$, c be a limit point of S, and $f\colon S \to \boldsymbol{R}$, $g\colon S \to \boldsymbol{R}$. If $\operatorname{Lim}_{x \to c} f(x)$ and $\operatorname{Lim}_{x \to c} g(x)$ exist, then

1. For all $r \in \boldsymbol{R}$, $\operatorname*{Lim}_{x \to c} rf(x)$ exists and $\operatorname*{Lim}_{x \to c} rf(x) = r \cdot \operatorname*{Lim}_{x \to c} f(x)$;
2. $\operatorname*{Lim}_{x \to c} (f + g)(x)$ exists and $\operatorname*{Lim}_{x \to c} (f + g)(x) = \operatorname*{Lim}_{x \to c} f(x) + \operatorname*{Lim}_{x \to c} g(x)$;
3. $\operatorname*{Lim}_{x \to c} (fg)(x)$ exists and $\operatorname*{Lim}_{x \to c} (fg)(x) = [\operatorname*{Lim}_{x \to c} f(x)][\operatorname*{Lim}_{x \to c} g(x)]$;
4. If $\operatorname*{Lim}_{x \to c} g(x) \neq 0$, then $\operatorname*{Lim}_{x \to c} (f/g)(x)$ exists and

$$\operatorname*{Lim}_{x \to c} \left(\frac{f}{g}\right)(x) = \frac{\operatorname{Lim}_{x \to c} f(x)}{\operatorname{Lim}_{x \to c} g(x)}$$

EXERCISES

1. Let S be a subset of $\boldsymbol{R}$, c be a limit point of S, and $f\colon S \to \boldsymbol{R}$. Prove that if $\operatorname{Lim}_{x \to c} f(x)$ exists it is unique.
2. For each of the following, find the indicated limit or show that it does not exist. Justify your answers.

   (a) $\operatorname*{Lim}_{x \to c} |x|$, where c is any real number;

   (b) $\operatorname*{Lim}_{x \to c} x^n$, for $n \in \boldsymbol{N}$ and c any real number;

   (c) $\operatorname*{Lim}_{x \to m} [x]$, where $m \in \boldsymbol{Z}$ and $[x]$ is the greatest integer less than or equal to x;

   (d) $\displaystyle\operatorname*{Lim}_{x \to 2} \frac{x^2 + x + 1}{x - 1}$;

   (e) $\displaystyle\operatorname*{Lim}_{x \to 1} \frac{x^2 + x + 1}{x - 1}$.
3. Prove Proposition 5.4.5

6
Metric Spaces

Throughout this text we have been using the real number system $\boldsymbol{R}$ as both an example of a topological space and as a guide toward further topological generalization. In this latter connection, there are several important properties of $\boldsymbol{R}$ which we have not yet considered from a topological point of view, and foremost among these is the property that between any two points of $\boldsymbol{R}$ there is a well-defined distance. (Recall that the distance between the real numbers x and y is defined to be $|x - y|$, the absolute value of their difference.) In this chapter we will extend the concept of distance between points to sets other than $\boldsymbol{R}$ by means of functions known as metrics. We will see that if it is possible to define a metric function on a set, then it is possible to topologize the set in a very interesting and fruitful manner, thus creating a topological space called a *metric space*. Metric spaces share many of the properties of the real number system $\boldsymbol{R}$ and are of considerable importance in analysis.

6.1 THE METRIC TOPOLOGY

Suppose x and y are real numbers. Then (see Corollary 2.1.4, Section 2.1 of Chapter 2)

1. The distance between x and y is a nonnegative real number:

$$|x - y| \geqslant 0$$

2. The distance between x and y is 0 if and only if $x = y$:

$$|x - y| = 0 \qquad \text{if and only if } x = y$$

3. The distance between x and y is the same as the distance between y and x:

$$|x - y| = |y - x|$$

   Furthermore, if z is also a real number, then

4. The triangle inequality holds for x, y, and z:

$$|x - y| \leqslant |x - z| + |y - z|$$

These properties of distance in $\boldsymbol{R}$ motivate the following definition of a distance function, or metric, on an arbitrary set M.

Definition 6.1.1 Let M be a set. A function $\rho\colon M \times M \to \boldsymbol{R}$ is a *metric* on M if

1. $\rho((x, y)) \geqslant 0$ for all $x \in M$, $y \in M$;
2. $\rho((x, y)) = 0$ if and only if $x = y$ in M;
3. $\rho((x, y)) = \rho((y, x))$ for all $x \in M$, $y \in M$;
4. (triangle inequality) $\rho((x, y)) \leqslant \rho((x, z)) + \rho((y, z))$ for all $x \in M$, $y \in M$, and $z \in M$.

In order to simplify the notation, from now on we will write $\rho(x, y)$ rather than $\rho((x, y))$.

It should be clear that if ρ is a metric on M, then it *defines* a distance between any two points of M: the distance between $x \in M$ and $y \in M$, as measured by ρ, is just the nonnegative real number $\rho(x, y)$. Thus, the smaller the real number $\rho(x, y)$, the closer together the points x and y are in M, as measured by ρ.

Let us consider some examples of metrics on sets.

Examples 6.1.2 1. Let $\rho\colon \boldsymbol{R} \times \boldsymbol{R} \to \boldsymbol{R}$ be given by $\rho(x, y) = |x - y|$, for all $x \in \boldsymbol{R}$, $y \in \boldsymbol{R}$. Then ρ is a metric on $\boldsymbol{R}$, called the *absolute value metric* or *natural metric* on $\boldsymbol{R}$.

2. Let M be any set, and define $\rho\colon M \times M \to \boldsymbol{R}$ as follows:

$$\text{for all } x \in M,\ y \in M \qquad \rho(x, y) = \begin{cases} 1 & \text{if } x \neq y \\ 0 & \text{if } x = y \end{cases}$$

It is easy to check that the function ρ satisfies the conditions of Definition 6.1.1 and is therefore a metric on M. This metric is called the *discrete metric* on M.

3. Let M be any set, S a subset of M, and $\rho\colon M \times M \to \boldsymbol{R}$ a metric on M; then the restriction of the function ρ to $S \times S$ is a metric on S. For instance, the restriction of the absolute value metric on $\boldsymbol{R}$ to the closed unit interval [0, 1] is a metric on [0, 1].

4. Define $\rho\colon \boldsymbol{R}^2 \times \boldsymbol{R}^2 \to \boldsymbol{R}$ as follows: for all points (x_1, y_1) and (x_2, y_2) in $\boldsymbol{R}^2$, let

$$\rho((x_1, y_1), (x_2, y_2)) = [(x_1 - x_2)^2 + (y_1 - y_2)^2]^{1/2}$$

The function ρ is a metric on the Euclidean plane $\boldsymbol{R}^2$. We leave the proof of this fact as an exercise. (See Exercise 2 at the end of this section.) This metric is known as the *product metric* on $\boldsymbol{R}^2$.

5. Let $M = \{\text{bounded sequences in } \boldsymbol{R}\}$, and define $\rho\colon M \times M \to \boldsymbol{R}$ as follows: if $\{x_n\} \in M$ and $\{y_n\} \in M$, then

$$\rho(\{x_n\}, \{y_n\}) = \sup \{|x_n - y_n| \mid n \in \boldsymbol{N}\}$$

Let us show that ρ is a metric on M.

First we must show that $\rho(\{x_n\}, \{y_n\})$ exists for all $\{x_n\} \in M$ and $\{y_n\} \in M$. However, since $\{x_n\}$ and $\{y_n\}$ are bounded sequences in $\boldsymbol{R}$, it follows that the set $\{|x_n - y_n| \mid n \in \boldsymbol{N}\}$ is a bounded subset of $\boldsymbol{R}$ (why?) and hence $\rho(\{x_n\}, \{y_n\}) = \sup \{|x_n - y_n| \mid n \in \boldsymbol{N}\}$ exists.

If $\{x\} \in M$ and $\{y_n\} \in M$, then $|x_n - y_n| \geqslant 0$ for all $n \in \boldsymbol{N}$, so surely

$$\rho(\{x_n\}, \{y_n\}) = \sup \{|x_n - y_n| \mid n \in \boldsymbol{N}\} \geqslant 0$$

Furthermore, $\sup \{|x_n - y_n| \mid n \in \boldsymbol{N}\} = 0$ if and only if $x_n = y_n$ for all $n \in \boldsymbol{N}$, that is, if and only if $\{x_n\} = \{y_n\}$ in M. Thus the function ρ satisfies conditions (1) and (2) of Definition 6.1.1. It satisfies condition (3) because

$$\sup \{|x_n - y_n| \mid n \in \boldsymbol{N}\} = \sup \{|y_n - x_n| \mid n \in \boldsymbol{N}\}$$

and thus

$$\rho(\{x_n\}, \{y_n\}) = \rho(\{y_n\}, \{x_n\})$$

Now suppose that $\{z_n\}$ is also a bounded sequence of real numbers. By the triangle inequality for absolute value, we have

$$|x_n - y_n| \leqslant |x_n - z_n| + |y_n - z_n|$$

for all $n \in \boldsymbol{N}$, and it is easy to check that this in turn implies that

$$\sup \{|x_n - y_n| \mid n \in \boldsymbol{N}\} \leqslant \sup \{|x_n - z_n| \mid n \in \boldsymbol{N}\} + \sup \{|y_n - z_n| \mid n \in \boldsymbol{N}\}$$

Thus

$$\rho(\{x_n\}, \{y_n\}) \leqslant \rho(\{x_n\}, \{z_n\}) + \rho(\{y_n\}, \{z_n\})$$

Therefore ρ is indeed a metric on M. The set M of all bounded sequences in $\boldsymbol{R}$ is commonly denoted by l^∞, and the metric of this example is called the *supremum metric* on l^∞.

If M is a set, ρ is a metric on M, and x is a point of M, it is natural to consider the subset of M consisting of all points which lie within a specified distance of x.

Definition 6.1.3 Let M be a set and let ρ be a metric on M. If $x \in M$ and ε is a positive real number, then

1. The *open ball of radius* ε *about* x is the set

$$\beta(x, \varepsilon) = \{y \in M \mid \rho(x, y) < \varepsilon\}$$

2. The *closed ball of radius* ε *about* x is the set

$$\bar{\beta}(x, \varepsilon) = \{y \in M \mid \rho(x, y) \leqslant \varepsilon\}$$

Thus the open ball $\beta(x, \varepsilon)$ consists of all points of M which are at a distance less than ε from x, while the closed ball $\bar{\beta}(x, \varepsilon)$ consists of all points of M which are at a distance of ε or less from x. In Figure 6.1, the

Figure 6.1 Open and closed balls.

open ball consists of all points of M which lie within the interior of the circle, while the closed ball consists of the interior points together with those which lie on the circumference of the circle.

Examples 6.1.4 1. Let ρ be the absolute value metric on $\boldsymbol{R}$, let $x \in \boldsymbol{R}$, and let ε be a positive real number. We have

$$\beta(x, \varepsilon) = \{y \in \boldsymbol{R} \mid |x - y| < \varepsilon\} = (x - \varepsilon, x + \varepsilon)$$

Hence for any $x \in \boldsymbol{R}$ and any positive real number ε, the open ball $\beta(x, \varepsilon)$ is just the open interval $(x - \varepsilon, x + \varepsilon)$. Similarly, the closed ball $\bar{\beta}(x, \varepsilon)$ is just the closed interval $[x - \varepsilon, x + \varepsilon]$.

2. Let M be a set and let ρ be the discrete metric on M, so that $\rho(x, y) = 1$ if $x \neq y$ in M, and $\rho(x, y) = 0$ if $x = y$ in M. If $\varepsilon \in \boldsymbol{R}$, $\varepsilon > 1$, we have

$$\beta(x, \varepsilon) = \{y \in M \mid \rho(x, y) < \varepsilon\} = M$$

while if $0 < \varepsilon \leqslant 1$, we have

$$\beta(x, \varepsilon) = \{y \in M \mid \rho(x, y) < \varepsilon\} = \{x\}$$

Similarly, if $\varepsilon \geqslant 1$, then $\bar{\beta}(x, \varepsilon) = M$, while if $0 < \varepsilon < 1$, then $\bar{\beta}(x, \varepsilon) = \{x\}$.

3. Let ρ be the product metric on R^2 of Example 4 of Examples 6.1.2. Let (h, k) be a point in $\boldsymbol{R}^2$, and let ε be a positive real number; then

$$\beta((h, k), \varepsilon) = \{(x, y) \in \boldsymbol{R}^2 \mid (x - h)^2 + (y - k)^2 < \varepsilon^2\}$$

which is the set of all points in the plane which lie in the interior of the circle with center at (h, k) and radius ε; that is, $\beta((h, k), \varepsilon)$ is the open disc with center (h, k) and radius ε. Similarly, the closed ball $\bar{\beta}((h, k), \varepsilon)$ is the closed disc with center (h, k) and radius ε.

The preceding examples suggest a connection between a metric on a set and a certain topology on the set. For instance, consider the absolute value metric on $\boldsymbol{R}$. It is clear that a subset G of $\boldsymbol{R}$ is open in the natural topology on $\boldsymbol{R}$ if and only if for each $x \in G$ there is some positive real number ε such that $\beta(x, \varepsilon) = (x - \varepsilon, x + \varepsilon) \subset G$. (Why?) Thus we could use the natural metric on $\boldsymbol{R}$ to *define* the natural topology on $\boldsymbol{R}$: we could define a subset G of $\boldsymbol{R}$ to be open in the natural topology if for every $x \in G$ there is some positive real number ε such that $\beta(x, \varepsilon) \subset G$. In this manner, given any set M with a metric ρ, we may use ρ to define a topology on M.

Proposition 6.1.5 Let M be a set and let ρ be a metric on M. If τ_ρ is the collection of all subsets G of M having the property that for each $x \in G$, there is some positive real number ε such that $\beta(x, \varepsilon) \subset G$, then τ_ρ is a topology on M.

We leave the proof of Proposition 6.1.5 as an exercise. (See Exercise 6 at the end of this section.)

Definition 6.1.6 Let M be a set and let ρ be a metric on M. Let τ_ρ be the topology on M defined in the statement of Proposition 6.1.5. The topological space (M, τ_ρ) is called a *metric space*, and the topology τ_ρ is the *metric topology induced on M by* ρ.

We remark that it is common practice to suppress mention of the metric topology τ_ρ and thus refer to "the metric space (M, ρ)" rather than "the metric space (M, τ_ρ)."

We have seen that if ρ is a metric on a set M, then ρ induces a metric topology on M. Of course, this topology may be identical to a topology defined on M without reference to ρ.

Examples 6.1.7 1. Let ρ be the absolute value metric on $\boldsymbol{R}$. We will show that the metric topology induced on $\boldsymbol{R}$ by ρ is the natural topology, thus justifying the remarks of the paragraph preceding Proposition 6.1.5.

If G is a subset of $\boldsymbol{R}$ which is open in the metric topology, then for each $x \in G$, there is some positive real number ε such that $\beta(x, \varepsilon) \subset G$. But since $\beta(x, \varepsilon) = (x - \varepsilon, x + \varepsilon)$ is an open interval, this shows that G is open in the natural topology on $\boldsymbol{R}$. Conversely, if G is open in the natural topology on $\boldsymbol{R}$, then for each $x \in G$ there is some open interval (a, b) such that $x \in (a, b)$ and $(a, b) \subset G$. Thus if $\varepsilon = \min\{x - a, b - x\}$, then ε is a positive real number such that $\beta(x, \varepsilon) = (x - \varepsilon, x + \varepsilon) \subset (a, b) \subset G$, so G is open in the metric topology on $\boldsymbol{R}$.

We have thus shown that the metric topology induced on $\boldsymbol{R}$ by the absolute value metric is identical to the natural topology on $\boldsymbol{R}$. This result has an important consequence, for it implies that all the results we have previously obtained concerning $\boldsymbol{R}$ with the natural topology remain valid when we consider $\boldsymbol{R}$ to be a metric space with the absolute value metric.

2. Let M be a set and let ρ be the discrete metric on M. For all $x \in M$, the one-point subset $\{x\}$ is open in the metric topology because

$$\beta(x, \tfrac{1}{2}) \subset \{x\}$$

But if every one-point subset of a topological space is open, then the topology is the discrete topology. Hence the metric topology induced on a set by the discrete metric is the discrete topology.

Metric spaces are a special type of topological space, and in the next two sections of this chapter we shall study them as such. Before we conclude this section, however, we must clear up two more points concerning the metric topology by showing that open balls are open sets and closed balls are closed sets in the metric topology. The proofs of these facts are typical examples of metric space arguments which use the triangle inequality.

Proposition 6.1.8 Let (M, ρ) be a metric space. In the metric topology on M, every open ball is an open set and every closed ball is a closed set.

Proof: Let $\beta(x, \varepsilon)$ be an open ball in M. If $\beta(x, \varepsilon) = \emptyset$, then surely it is open, so assume that $\beta(x, \varepsilon) \neq \emptyset$. In order to show that $\beta(x, \varepsilon)$ is an open set in the metric space (M, ρ), we must show that if z is any element of $\beta(x, \varepsilon)$, there is some positive real number η such that $\beta(z, \eta) \subset \beta(x, \varepsilon)$. If $z = x$, we choose $\eta = \varepsilon$. If $z \neq x$, then $0 < \rho(x, z) < \varepsilon$, and we choose $\eta = \varepsilon - \rho(x, z)$. Clearly, η is a positive real number, and we claim that $\beta(z, \eta) \subset \beta(x, \varepsilon)$. To see this, note that if $y \in \beta(z, \eta)$, then $\rho(z, y) < \eta$ and by the triangle inequality we have

$$\rho(x, y) \leqslant \rho(x, z) + \rho(z, y) < \rho(x, z) + \eta = \varepsilon$$

But $\rho(x, y) < \varepsilon$ implies that $y \in \beta(x, \varepsilon)$. Therefore $\beta(z, \eta) \subset \beta(x, \varepsilon)$, and we have proved that every open ball is an open set in the metric topology.

Now we prove that every closed ball $\bar{\beta}(x, \varepsilon)$ is closed in (M, ρ). If $\bar{\beta}(x, \varepsilon) = M$, then surely it is closed, so assume that $\bar{\beta}(x, \varepsilon) \neq M$ and consider $M - \bar{\beta}(x, \varepsilon)$. It will suffice to show that $M - \bar{\beta}(x, \varepsilon)$ is open. If $z \in M - \bar{\beta}(x, \varepsilon)$, then $\rho(x, z) > \varepsilon$. Choose $\eta = \rho(x, z) - \varepsilon$. Again, η is a positive real number, and we claim that $\beta(z, \eta) \subset M - \bar{\beta}(x, \varepsilon)$. This follows because if $y \in \beta(z, \eta)$, then $\rho(y, z) < \eta$, and by the triangle inequality, $\rho(x, z) \leqslant \rho(x, y) + \rho(y, z)$. Therefore

$$\rho(x, y) \geqslant \rho(x, z) - \rho(y, z) > \rho(x, z) - \eta = \varepsilon$$

Hence $y \notin \bar{\beta}(x, \varepsilon)$ and therefore $\beta(z, \eta)$ is contained in $M - \bar{\beta}(x, \varepsilon)$. We have thus shown that for every $z \in M - \bar{\beta}(x, \varepsilon)$ there is an open ball about z which is contained in $M - \bar{\beta}(x, \varepsilon)$, and hence $M - \bar{\beta}(x, \varepsilon)$ is open in the metric topology, and we are done. □

We should note that although $\bar{\beta}(x, \varepsilon)$ is a closed set containing the open set $\beta(x, \varepsilon)$, it is not true in general that $\bar{\beta}(x, \varepsilon)$ is the closure of $\beta(x, \varepsilon)$. For example, suppose $M = \{x, y\}$ with the discrete metric. Then

$$\beta(x, 1) = \{x\} \qquad \text{and} \qquad \bar{\beta}(x, 1) = M$$

while the closure of $\beta(x, 1)$ is the set

$$\overline{\beta(x, 1)} = \{x\}$$

EXERCISES

1. Verify that the discrete metric of Example 2 of Examples 6.1.2 satisfies the conditions of Definition 6.1.1 and thus is indeed a metric.
2. Verify that the product metric of Example 4 of Examples 6.1.2 is a metric on $\boldsymbol{R}^2$.
3. Let M be the set of all sequences in $\boldsymbol{R}$ which converge to 0. If $\{x_n\}$ and $\{y_n\}$ are elements of M, set

$$\rho(\{x_n\}, \{y_n\}) = \sup \{|x_n - y_n| \mid n \in \boldsymbol{N}\}$$

Prove that ρ is a metric on M.

4. Let $\sigma\colon \boldsymbol{R}^2 \times \boldsymbol{R}^2 \to \boldsymbol{R}$ be given by

$$\sigma((x_1, y_1), (x_2, y_2)) = \max \{|x_1 - x_2|, |y_1 - y_2|\}$$

for all (x_1, y_1), (x_2, y_2) in $\boldsymbol{R}^2$. Prove that σ is a metric on $\boldsymbol{R}^2$.

5. For each $q \in \boldsymbol{Q}$, $q \neq 0$, we may write $q = (s/t)2^m$, where s, t are integers not divisible by 2 and $m \in \boldsymbol{Z}$, and define

$$|q|_2 = 2^{-m}$$

Also define $|0|_2 = 0$. (This is known as the 2-*adic valuation* on $\boldsymbol{Q}$; if we had used another prime number p in place of 2, we would have the p-*adic valuation* on $\boldsymbol{Q}$.) Prove that

(a) $|q|_2 \geqslant 0$ for all $q \in \boldsymbol{Q}$;
(b) $|q|_2 = 0$ if and only if $q = 0$;
(c) $|qr|_2 = |q|_2 |r|_2$ for all $q \in \boldsymbol{Q}$, $r \in \boldsymbol{Q}$;
(d) $|q + r|_2 \leqslant \max \{|q|_2, |r|_2\}$ for all $q \in \boldsymbol{Q}$, $r \in \boldsymbol{Q}$.

Let $\rho: \boldsymbol{Q} \times \boldsymbol{Q} \to \boldsymbol{R}$ be given by $\rho(q, r) = |q - r|_2$ for all $q \in \boldsymbol{Q}$, $r \in \boldsymbol{Q}$, and show that ρ is a metric on $\boldsymbol{Q}$.

6. Prove Proposition 6.1.5.
7. Let $M = [-1, 1] \cup \{2\}$ and define $\varphi: M \times M \to \boldsymbol{R}$ by

$$\varphi(x, y) = \begin{cases} 0 & \text{if } x = y \\ |x - y| & \text{if } x \neq y \text{ and neither } x \text{ nor } y \text{ is } 2 \\ 1 & \text{if one (but not both) of } x \text{ or } y \text{ is } 2 \end{cases}$$

Prove that φ is a metric on M.

8. Let σ be the metric of Exercise 4 above. Let $(x_1, y_1) \in \boldsymbol{R}^2$ and let ε be a positive real number. Give a geometric description of the open ball $\beta((x_1, y_1), \varepsilon)$ and the closed ball $\bar{\beta}((x_1, y_1), \varepsilon)$ in the metric space $(\boldsymbol{R}^2, \sigma)$.
9. Let τ be the product topology on $\boldsymbol{R}^2$, let ρ be the product metric of Example 4 of Examples 6.1.2, and let σ be the metric of Exercise 4 above.
   (a) Are the metric topologies induced on $\boldsymbol{R}^2$ by ρ and σ identical? Justify your answer.
   (b) Are either of the metric topologies induced on $\boldsymbol{R}^2$ by ρ or σ identical with the product topology τ on $\boldsymbol{R}^2$? Justify your answer.
10. (a) Let ρ be the 2-adic metric of Exercise 5 above. Let $q \in \boldsymbol{Q}$ and let ε be a positive real number. Describe the open ball $\beta(q, \varepsilon)$ and the closed ball $\bar{\beta}(q, \varepsilon)$ in the metric space $(\boldsymbol{Q}, \rho)$.
    (b) The absolute value metric on $\boldsymbol{R}$ induces an absolute value metric on $\boldsymbol{Q}$ by restriction. Do the absolute value metric on $\boldsymbol{Q}$ and the 2-adic metric induce the same metric topology on $\boldsymbol{Q}$?
11. Define $\psi: \boldsymbol{R} \times \boldsymbol{R} \to \boldsymbol{R}$ by

$$\psi(x, y) = \frac{|x - y|}{1 + |x - y|}$$

for all $x \in \boldsymbol{R}$, $y \in \boldsymbol{R}$. Is ψ a metric on $\boldsymbol{R}$? If so, does it induce the same metric topology on $\boldsymbol{R}$ as does the absolute value metric?

12. Let l^1 denote the set of all sequences $\{x_n\}$ in $\boldsymbol{R}$ such that the series $\sum_{n=1}^{\infty} |x_n|$ converges. Define $\rho: l^1 \times l^1 \to \boldsymbol{R}$ by

$$\rho(\{x_n\}, \{y_n\}) = \sum_{n=1}^{\infty} |x_n - y_n|$$

for all $\{x_n\} \in l^1$, $\{y_n\} \in l^1$. Prove that (l^1, ρ) is a metric space.

13. We shall use this exercise to construct the metric space $(\boldsymbol{R}^n, \rho_n)$ known as *Euclidean n-space*. The reader is asked to fill in the details. Let $n \in \boldsymbol{N}$, and set $\boldsymbol{R}^n = \{(x_1, \ldots, x_n) \mid x_k \in \boldsymbol{R}, 1 \leqslant k \leqslant n\}$. Thus, $\boldsymbol{R}^n$ is the set of all n-tuples of real numbers.

(a) Prove the Cauchy–Schwartz inequality for $\boldsymbol{R}^n$: if $(x_1, \ldots, x_n)$ and $(y_1, \ldots, y_n)$ are elements of $\boldsymbol{R}^n$, then

$$\left| \sum_{k=1}^{n} x_k y_k \right| \leqslant \left[\sum_{k=1}^{n} x_k{}^2 \right]^{1/2} \cdot \left[\sum_{k=1}^{n} y_k{}^2 \right]^{1/2}$$

Hint: Let $t \in \boldsymbol{R}$ and consider the nonnegative real number

$$\sum_{k=1}^{n} (tx_k + y_k)^2 = t^2 \sum_{k=1}^{n} x_k{}^2 + 2t \sum_{k=1}^{n} x_k y_k + \sum_{k=1}^{n} y_k{}^2$$

Set

$$t = -\left[\sum_{k=1}^{n} x_k y_k \middle/ \sum_{k=1}^{n} x_k{}^2 \right]$$

rearrange, and take square roots.

(b) Prove the Minkowski inequality for $\boldsymbol{R}^n$: if $(x_1, \ldots, x_n)$ and $(y_1, \ldots, y_n)$ are elements of $\boldsymbol{R}^n$, then

$$\left[\sum_{k=1}^{n} (x_k + y_k)^2 \right]^{1/2} \leqslant \left[\sum_{k=1}^{n} x_k{}^2 \right]^{1/2} + \left[\sum_{k=1}^{n} y_k{}^2 \right]^{1/2}$$

Hint: Expand

$$\sum_{k=1}^{n} (x_k + y_k)^2$$

and use (a).

(c) If $x = (x_1, \ldots, x_n)$ and $y = (y_1, \ldots, y_n)$ in $\boldsymbol{R}^n$, define

$$\rho_n(x, y) = \left[\sum_{k=1}^{n} (x_k - y_k)^2 \right]^{1/2}$$

Prove that ρ_n is a metric on $\boldsymbol{R}^n$.

(d) Show that ρ_1 is the absolute value metric on $\boldsymbol{R}$ and that ρ_2 is the product metric on $\boldsymbol{R}^2$. Describe open balls and closed balls in the metric space $(\boldsymbol{R}^3, \rho_3)$.

14. We shall use this exercise to construct an important metric space known as *Hilbert space*, or l^2. The reader is asked to fill in the details. Let l^2 denote the set of all sequences $\{x_n\}$ in $\boldsymbol{R}$ such that the associated series $\Sigma_{n=1}^{\infty} x_n^2$ converges.
   (a) Prove that if $\{x_n\} \in l^2$ and $\{y_n\} \in l^2$, then $\{x_n y_n\} \in l^2$. Hint: Use Exercise 13(a).
   (b) Prove the Cauchy–Schwartz inequality for l^2: if $\{x_n\} \in l^2$ and $\{y_n\} \in l^2$, then

$$\left|\sum_{n=1}^{\infty} x_n y_n\right| \leqslant \left[\sum_{n=1}^{\infty} x_n^2\right]^{1/2} \cdot \left[\sum_{n=1}^{\infty} y_n^2\right]^{1/2}$$

   (c) Prove that if $\{x_n\}$ and $\{y_n\}$ are elements of l^2, then $\{x_n + y_n\}$ is an element of l^2. Next show that the Minkowski inequality

$$\left[\sum_{n=1}^{\infty} (x_n + y_n)^2\right]^{1/2} \leqslant \left[\sum_{n=1}^{\infty} x_n^2\right]^{1/2} + \left[\sum_{n=1}^{\infty} y_n^2\right]^{1/2}$$

   holds in l^2.
   (d) If $\{x_n\} \in l^2$, define the *norm* of $\{x_n\}$, denoted by $\|x_n\|$, as follows:

$$\|x_n\| = \left[\sum_{n=1}^{\infty} x_n^2\right]^{1/2}$$

   Prove that
   (i) $\|x_n\| \geqslant 0$ for all $\{x_n\} \in l^2$;
   (ii) $\|x_n\| = 0$ if and only if $x_n = 0$ for all $n \in \boldsymbol{N}$;
   (iii) $\|cx_n\| = |c| \cdot \|x_n\|$ for all $\{x_n\} \in l^2$ and all $c \in \boldsymbol{R}$;
   (iv) $\|x_n + y_n\| \leqslant \|x_n\| + \|y_n\|$ for all $\{x_n\} \in l^2$ and $\{y_n\} \in l^2$.
   Show that the function ρ defined on $l^2 \times l^2$ by

$$\rho(\{x_n\}, \{y_n\}) = \|x_n - y_n\|$$

   is a metric on l^2.
15. Let (M, ρ) be a compact metric space. Prove that M contains a countable dense subset.

6.2. CONTINUITY IN METRIC SPACES

In this section we study continuous functions from one metric space to another. We begin with a characterization of continuity in terms of the metrics on the spaces, then define the concept of a uniformly continuous function from one metric space to another, and finally show that a function continuous on a compact metric space is uniformly continuous there.

Let (M_1, ρ_1) and (M_2, ρ_2) be metric spaces and $f: M_1 \to M_2$. According to Definition 4.1.7, f is continuous at $x \in M_1$ if and only if for every open set G in M_2 which contains $f(x)$, there is some open set H in M_1 such that $x \in H$ and $f(H) \subset G$. Our first result in this section reformulates this characterization of continuity at a point in terms of the metrics ρ_1 and ρ_2.

Proposition 6.2.1 Let (M_1, ρ_1) and (M_2, ρ_2) be metric spaces and let $f: M_1 \to M_2$. If $x \in M_1$, then f is continuous at x if and only if for every positive real number ε there is some positive number δ such that for all $y \in M_1$,

$$\rho_1(x, y) < \delta \qquad \text{implies} \qquad \rho_2(f(x), f(y)) < \varepsilon.$$

Proof: Suppose f is continuous at $x \in M$. If G is open in M_2 and $f(x)$ is contained in G, then there must be some open set H in M_1 such that $x \in H$ and $f(H) \subset G$. In particular this is so if $G = \beta(f(x), \varepsilon)$, the open ball of radius ε about $f(x)$. Thus there is some open set H in M_1 such that $x \in H$ and $f(H) \subset \beta(f(x), \varepsilon)$. But since H is open, there must be some positive real number δ such that $\beta(x, \delta) \subset H$. Now if $y \in M_1$ with $\rho_1(x, y) < \delta$, then $y \in \beta(x, \delta)$ and hence $f(y) \in \beta(f(x), \varepsilon)$, so that $\rho_2(f(x), f(y)) < \varepsilon$.

Conversely, let $x \in M_1$ and suppose that for every positive real number ε there is some positive real number δ such that whenever $y \in M_1$ and $\rho_1(x, y) < \delta$ then $\rho_2(f(x), f(y)) < \varepsilon$. If G is an open set in M_2 which contains $f(x)$, then there is some positive real number ε such that the open ball $\beta(f(x), \varepsilon) \subset G$. For this ε, there is some positive real number δ such that $y \in M_1$, $\rho_1(x, y) < \delta$ implies that $\rho_2(f(x), f(y)) < \varepsilon$. But then $f(\beta(x, \delta)) \subset \beta(f(x), \varepsilon)$, and since $\beta(x, \delta)$ is an open set in M_1 which contains x, this shows that f is continuous at x. □

Corollary 6.2.2 Let S be a subset of $\boldsymbol{R}$ and $f: S \to \boldsymbol{R}$. If $x \in S$, then f is continuous at x if and only if for every real number $\varepsilon > 0$ there is some real number $\delta > 0$ such that if $y \in S$ and $|x - y| < \delta$, then $|f(x) - f(y)| < \varepsilon$.

Proposition 6.2.1 is often used as the definition of continuity at a point for functions from one metric space to another; similarly, Corollary 6.2.2 is often used as the definition of continuity at a point for functions on $\boldsymbol{R}$. When this is done, these definitions are known as "$\delta - \varepsilon$ definitions," and when we use Proposition 6.2.1 or Corollary 6.2.2 to show that a function is continuous at a point, we will be doing "$\delta - \varepsilon$ proofs." The following examples illustrate the technique of $\delta - \varepsilon$ proofs.

Examples 6.2.3 1. Let $f\colon \boldsymbol{R} \to \boldsymbol{R}$ be given by $f(x) = 3x + 2$ for all $x \in \boldsymbol{R}$. We use Corollary 6.2.2 to show that f is continuous at x for all $x \in \boldsymbol{R}$.

Given any real number $\varepsilon > 0$, we must find a real number $\delta > 0$ such that $|x - y| < \delta$ implies $|f(x) - f(y)| < \varepsilon$. But if $|x - y| < \delta$, then

$$|f(x) - f(y)| = |(3x + 2) - (3y + 2)| = 3|x - y| < 3\delta$$

Thus if we take $\delta = \varepsilon/3$, then $|x - y| < \delta = \varepsilon/3$ implies

$$|f(x) - f(y)| = 3|x - y| < 3\delta = 3\left(\frac{\varepsilon}{3}\right) = \varepsilon$$

Therefore f is continuous at x for all $x \in \boldsymbol{R}$. Note that our choice of δ depended only on ε.

2. Let $f\colon \boldsymbol{R} \to \boldsymbol{R}$ be defined by

$$f(x) = \begin{cases} 1 & \text{if } x > 0 \\ -1 & \text{if } x \leqslant 0 \end{cases}$$

The function f is not continuous at $x = 0$. To see this, choose $\varepsilon = 1$; then there is no $\delta > 0$ such that $|0 - y| < \delta$ implies $|f(0) - f(y)| < \varepsilon$, because no matter how δ is chosen, there is some y such that $0 < y < \delta$, and hence such that

$$|0 - y| = y < \delta \text{ while } |f(0) - f(y)| = |-1 - 1| = 2 \not< \varepsilon = 1$$

3. Let $f\colon (0, 1] \to \boldsymbol{R}$ be given by $f(x) = 1/x$ for all $x \in (0, 1]$. To show that f is continuous at each $x \in (0, 1]$, we must show that for every $\varepsilon > 0$ there is some $\delta > 0$ such that if $y \in (0, 1]$ and $|x - y| < \delta$, then

$$|f(x) - f(y)| = \frac{|x - y|}{xy} < \varepsilon$$

But $|x - y| < \delta$ implies that

$$\frac{|x - y|}{xy} < \frac{\delta}{xy}$$

From this last inequality we see that we must choose δ in such a way that the quotient δ/xy is less than ε. It will not help us to choose δ to be a small positive real number if at the same time we force the product xy to be small, for then the quotient δ/xy may be very large. To keep this from happening, we will select a preliminary δ which forces xy to be bounded below and then adjust this preliminary δ to obtain $\delta/xy < \varepsilon$.

Suppose we choose $\delta \leqslant x/2$. If x and y are in the interval $(0, 1]$ and $|x-y| < \delta \leqslant x/2$, then $x/2 < y < 3x/2$ and therefore $xy > x(x/2) = x^2/2$. Choosing $\delta \leqslant x/2$ thus forces xy to be bounded below by $x^2/2$. Now let us see what effect this has on $|f(x) - f(y)|$: since $xy > x^2/2$, we have

$$|f(x) - f(y)| = \frac{|x-y|}{xy} < \frac{\delta}{xy} < \frac{\delta}{x^2/2} = \frac{2\delta}{x^2}$$

Clearly, if we now adjust δ so that it is less than or equal to $\varepsilon x^2/2$ as well as less than or equal to $x/2$, then we will have $|f(x) - f(y)| < \varepsilon$. Therefore we let $\delta = \min\{x/2, \varepsilon x^2/2\}$, and conclude that if x and y are in $(0, 1]$ with $|x-y| < \delta$, then $xy > x^2/2$ and hence

$$|f(x) - f(y)| = \frac{|x-y|}{xy} < \frac{\delta}{xy} < \frac{2\delta}{x^2} < \frac{2}{x^2}\frac{\varepsilon x^2}{2} = \varepsilon$$

Note that in this example our choice of δ depended on both ε and on x.

In Example 1 of Examples 6.2.3 we were able to find a $\delta > 0$ such that $|x-y| < \delta$ implied $|f(x) - f(y)| < \varepsilon$, and δ depended only on ε, and not on the point x under consideration. In Example 3 of Examples 6.2.3, on the other hand, the δ we obtained depended on both ε and on the point x. Continuous functions for which it is possible to find a $\delta > 0$ which depends only on ε and not on x are said to be uniformly continuous.

Definition 6.2.4 Let (M_1, ρ_1) and (M_2, ρ_2) be metric spaces. A function $f: M_1 \to M_2$ is *uniformly continuous* if for every positive real number ε there is some positive real number δ such that whenever $x \in M_1$ and $y \in M_1$ with $\rho_1(x, y) < \delta$, then $\rho_2(f(x), f(y)) < \varepsilon$.

Let us contrast Proposition 6.2.1 with Definition 6.2.4: the proposition says that f is continuous *at* x if for every positive real number δ there is some positive real number ε such that if $y \in M_1$ and $\rho_1(x, y) < \delta$, then $\rho_2(f(x), f(y)) < \varepsilon$; this is a local condition at the point x, and thus in general the number δ will depend on both the choice of ε and on the point

x being considered. The definition says that f is uniformly continuous *on the space* M_1 if for every positive real number ε there is some positive real number δ such that whenever $x \in M_1$ and $y \in M_1$ with $\rho_1(x, y) < \delta$, then $\rho_2(f(x), f(y)) < \varepsilon$; this is a global condition on M_1, and thus δ can depend only on the choice of ε, and not on the points x and y.

Uniform continuity is clearly a stronger condition than continuity: if f is a function which is uniformly continuous on a metric space M_1, then Proposition 6.2.1 shows that f is continuous at every point of M_1, hence continuous on M_1. Therefore uniform continuity implies continuity. The converse is not true, however, for (as Example 2 of Examples 6.2.5 below demonstrates), a function can be continuous on a metric space without being uniformly continuous there.

Examples 6.2.5 1. Let f: $\boldsymbol{R} \to \boldsymbol{R}$ be given by $f(x) = 3x + 2$ for all $x \in \boldsymbol{R}$. As Example 1 of Examples 6.2.3 shows, for any $\varepsilon > 0$, we may choose $\delta = \varepsilon/3$ and have $|f(x) - f(x)| < \varepsilon$ whenever $|x - y| < \delta$. Therefore the function f is uniformly continuous on $\boldsymbol{R}$.

2. Let f: $(0, 1] \to \boldsymbol{R}$ be given by $f(x) = 1/x$ for all $x \in (0, 1]$. Example 3 of Examples 6.2.3 shows that this function is continuous on $(0, 1]$, but we claim that it is not uniformly continuous there. To see this, suppose the contrary; then for any $\varepsilon > 0$ there is some $\delta > 0$ such that

$$|x - y| < \delta \qquad \text{implies} \qquad \frac{|x - y|}{xy} < \varepsilon$$

If x is any point of $(0, 1)$ such that $0 < x < \delta$, and if $y = x/2$, then

$$|x - y| = \frac{x}{2} < \delta$$

Therefore

$$\frac{|x - y|}{xy} = \frac{1}{x} < \varepsilon$$

But this is impossible, for no matter how δ is chosen, $1/x$ will be arbitrarily large for x sufficiently near zero. Thus f cannot be uniformly continuous on $(0, 1]$.

3. Let $a \in \boldsymbol{R}$, $0 < a < 1$, and let f: $[a, 1] \to \boldsymbol{R}$ be given by $f(x) = 1/x$ for all $x \in [a, 1]$. We claim that f is uniformly continuous on $[a, 1]$. For any $\varepsilon > 0$, choose $\delta = \varepsilon a^2$. If x and y are points in $[a, 1]$, then $xy > a^2$; if also

$|x - y| < \delta = \varepsilon a^2$, then

$$|f(x) - f(y)| = \frac{|x - y|}{xy} < \frac{(\varepsilon a^2)}{xy} < \frac{(\varepsilon a^2)}{a^2} = \varepsilon$$

and therefore f is uniformly continuous on $[a, 1]$.

We have just shown that the function f defined by $f(x) = 1/x$ is uniformly continuous on the metric space $[a, 1]$ if $0 < a < 1$, but is not uniformly continuous on $(0, 1]$. The reason this is so is that $[a, 1]$ is compact while $(0, 1]$ is not. The following important result tells us that a function which is continuous on a compact metric space is uniformly continuous there.

Theorem 6.2.6 Let (M_1, ρ_1) and (M_2, ρ_2) be metric spaces and let $f: M_1 \to M_2$ be continuous. If M_1 is compact, then f is uniformly continuous on M_1.

Proof: Let ε be a positive real number. Since f is continuous at each point t of M_1, there is some $\delta = \delta(t) > 0$ such that if $z \in M_1$ and $\rho_1(t, z) < \delta(t)$ then $\rho_2(f(t), f(z)) < \varepsilon/2$. (We write $\delta = \delta(t)$ to emphasize that δ depends on the point t as well as on the number ε.) The collection of open balls $\{\beta(t, \delta(t)/2) \mid t \in M_1\}$ is an open covering for M_1 and since M_1 is compact, it must have a finite subcovering, say

$$\beta\left(t_1, \frac{\delta(t_1)}{2}\right), \ldots, \beta\left(t_n, \frac{\delta(t_n)}{2}\right)$$

Let $\delta = \min\{\delta(t_1)/2, \ldots, \delta(t_n)/2\}$. Clearly δ is a positive real number, and we claim that if x and y are points of M_1 such that $\rho_1(x, y) < \delta$, then $\rho_2(f(x), f(y)) < \varepsilon$.

If $x \in M_1$, then $x \in \beta(t_k, \delta(t_k)/2)$, for some k, $1 \leqslant k \leqslant n$, and hence

$$\rho_1(x, t_k) < \frac{\delta(t_k)}{2} < \delta(t_k)$$

But this implies that

$$\rho_2(f(x), f(t_k)) < \frac{\varepsilon}{2}$$

If $y \in M_1$ and $\rho_1(x, y) < \delta$, then $\rho_1(x, y) < \delta(t_k)/2$. By the triangle inequality, we have

$$\rho_1(t_k, y) \leqslant \rho_1(t_k, x) + \rho_1(x, y) < \frac{\delta(t_k)}{2} + \frac{\delta(t_k)}{2}$$

and therefore $\rho_2(f(t_k), f(y)) < \varepsilon/2$ also. Thus

$$\rho_2(f(x), f(y)) < \rho_2(f(x), f(t_k)) + \rho_2(f(t_k), f(y)) < \frac{\varepsilon}{2} + \frac{\varepsilon}{2} = \varepsilon$$

and we are done. □

Corollary 6.2.7 Let f: $[a, b] \rightarrow \boldsymbol{R}$. If f is continuous on $[a, b]$, then it is uniformly continuous on $[a, b]$.

EXERCISES

1. Let a, b be fixed real numbers. If f: $\boldsymbol{R} \rightarrow \boldsymbol{R}$ is given by $f(x) = ax + b$ for all $x \in \boldsymbol{R}$, use Corollary 6.2.2 to show that f is continuous on $\boldsymbol{R}$.
2. For each of the following functions, use Corollary 6.2.2 to show that the function is continuous at the given values of x.
   (a) $f(x) = x^2$, for all $x \in \boldsymbol{R}$;
   (b) $f(x) = \sqrt{x}$, for $x \in \boldsymbol{R}$, $x \geqslant 0$;
   (c) $f(x) = 1/(x^2 - 1)$, for $x \in \boldsymbol{R}$, $x > 1$.
3. Let f: $\boldsymbol{R} \rightarrow \boldsymbol{R}$ be given by

   $$f(x) = \begin{cases} x & \text{if } x \neq 0 \\ 1 & \text{if } x = 0 \end{cases}$$

   Use Corollary 6.2.2 to show that f is not continuous at $x = 0$.
4. Let a be any real number and let f: $\boldsymbol{R} \rightarrow \boldsymbol{R}$ be given by

   $$f(x) = \begin{cases} \dfrac{1}{x} & \text{if } x \neq 0 \\ a & \text{if } x = 0 \end{cases}$$

   Use Corollary 6.2.2 to show that f is continuous at x if and only if $x \neq 0$.

5. Consider the identity function $f(x) = x$.
   (a) Using only Definition 6.2.4, show that this function is uniformly continuous on [0, 1].
   (b) Is this function uniformly continuous on $\boldsymbol{R}$? Justify your answer using Definition 6.2.4.
6. For each of the following, either prove that the given function is uniformly continuous on the given set or prove that it is not.
   (a) $f(x) = x^2$, on $[-n, n]$, $n \in \boldsymbol{N}$;
   (b) $f(x) = \sqrt{x}$, on $[0, +\infty)$;
   (c) $f(x) = 1/(x^2 - 1)$, on $(1, +\infty)$;
   (d) $f(x) = 1/(x^2 - 1)$, on $[a, +\infty)$, where $a > 1$.
7. Let S be a subset of $\boldsymbol{R}$ and let $f\colon S \to \boldsymbol{R}$. We say that f satisfies a *Lipschitz condition* on S if there exists $c \in \boldsymbol{R}$, $c > 0$, such that

$$|f(x) - f(y)| < c|x - y|$$

   for all elements x and y of S. Prove that if f satisfies a Lipschitz condition on S then f is uniformly continuous on S.
8. Give an example to show that the converse of the result stated in Exercise 7 above is not true.

6.3 SEQUENCES IN METRIC SPACES

In this section we study sequences in metric spaces. We will characterize convergent sequences in a metric space in terms of the metric, then define the concepts of Cauchy sequence and complete metric space, show that $\boldsymbol{R}$ is a complete metric space, prove that the intersection of nested closed balls in a complete metric space is nonempty, and finally prove the important Bolzano–Weierstrass Theorem. Before we can do any of this, however, we must show that metric spaces are Hausdorff spaces, and thus that convergent sequences in them have unique limits.

Proposition 6.3.1 Every metric space is a Hausdorff space.

Proof: If x and y are distinct points in a metric space (M, ρ) and we let $\varepsilon = \rho(x, y)$, then $\beta(x, \varepsilon/2)$ and $\beta(y, \varepsilon/2)$ are nonempty disjoint open sets which separate the points x and y. □

Corollary 6.3.2 Every convergent sequence in a metric space has a unique limit.

Now we are ready to characterize convergent sequences in a metric space in terms of the metric.

Proposition 6.3.3 Let (M, ρ) be a metric space. If $\{x_n\}$ is a sequence in M, then $\operatorname{Lim}_{n\to\infty} x_n = x$ if and only if for every positive real number ε there is some $m \in \boldsymbol{N}$ such that $\rho(x, x_n) < \varepsilon$ for all $n \geqslant m$.

Proof: Suppose $\operatorname{Lim}_{n\to\infty} x_n = x$ in M; then for every open set G in M which contains x there is some $m \in \boldsymbol{N}$ such that $x_n \in G$ for all $n \geqslant m$. If ε is a positive real number, then $\beta(x, \varepsilon)$ is an open set which contains x, and hence there is some $m \in \boldsymbol{N}$ such that $x_n \in \beta(x, \varepsilon)$ for all $n \geqslant m$. But this implies that $\rho(x, x_n) < \varepsilon$ for all $n \geqslant m$.

Conversely, let $x \in M$, let $\{x_n\}$ be a sequence in M, and suppose that for every positive real number ε there is some $m \in \boldsymbol{N}$ such that $\rho(x, x_n) < \varepsilon$ for all $n \geqslant m$. Let G be any open set which contains x. Since G is open, there is some $\varepsilon > 0$ such that $\beta(x, \varepsilon) \subset G$, and therefore for this ε, there is some $m \in \boldsymbol{N}$ such that $\rho(x, x_n) < \varepsilon$ for all $n \geqslant m$. But this implies that $x_n \in \beta(x, \varepsilon) \subset G$ for all $n \geqslant m$, and thus that $\operatorname{Lim}_{n\to\infty} x_n = x$. □

Corollary 6.3.4 If $\{x_n\}$ is a sequence in $\boldsymbol{R}$, then $\operatorname{Lim}_{n\to\infty} x_n = x$ if and only if for every positive real number ε there is some $m \in \boldsymbol{N}$ such that $|x - x_n| < \varepsilon$ for all $n \geqslant m$.

Examples 6.3.5 1. Let us use Corollary 6.3.4 to show that $\{(n+1)/n\}$ converges to 1 in $\boldsymbol{R}$. Given $\varepsilon > 0$ we must find $m \in \boldsymbol{N}$ such that $|1 - (n+1)/n| = 1/n < \varepsilon$ for all $n \geqslant m$. However, since $\varepsilon > 0$, there is some $m \in \boldsymbol{N}$ such that $0 < 1/m < \varepsilon$; hence for all $n \geqslant m$, we have $|1 - (n+1)/n| = 1/n < 1/m < \varepsilon$.

2. Consider the sequence $\{(-1)^n\}$ in $\boldsymbol{R}$. If this sequence converged to x, then we could take $\varepsilon = 1$ in Corollary 6.3.4, and therefore there would be some $m \in \boldsymbol{N}$ such that $|x - (-1)^n| < 1$ for all $n \geqslant m$. But this is impossible, for if $x \geqslant 0$ and n is any odd natural number, then $|x - (-1)^n| = x + 1 \geqslant 1$, while if $x < 0$ and n is any even natural number, then $|x - (-1)^n| = 1 - x > 1$.

Proposition 6.3.3 is often used as the definition of a convergent sequence in a metric space. Similarly, Corollary 6.3.4 is often used as the definition of a convergent sequence in $\boldsymbol{R}$.

It is possible to do more with sequences in metric spaces that it is in more general topological spaces because in metric spaces there is an explicit measure of the distance between points. Thus, in a manner of speaking, Proposition 6.3.3 says that if a sequence $\{x_n\}$ converges to x in a metric space, then the terms of the sequence are getting closer and closer to x, and it follows as a consequence of the triangle inequality that the terms of the sequence must be getting closer and closer to each other. Sequences whose terms get closer and closer to each other are called *Cauchy sequences*.

Definition 6.3.6 Let (M, ρ) be a metric space. A sequence $\{x_n\}$ in M is a *Cauchy sequence* if for every positive real number ε there is some $m \in \boldsymbol{N}$ such that $\rho(x_n, x_k) < \varepsilon$ for all $n \geqslant m$ and $k \geqslant m$.

Proposition 6.3.7 Every convergent sequence in a metric space is a Cauchy sequence.

Proof: Suppose $\{x_n\}$ converges to x in the metric space (M, ρ). By Proposition 6.3.3, for every positive real number ε there is some $m \in \boldsymbol{N}$ such that $\rho(x, x_n) < \varepsilon/2$ for all $n \geqslant m$. But then for all $n \geqslant m$ and $k \geqslant m$ we have $\rho(x_n, x_k) \leqslant \rho(x_n, x) + \rho(x_k, x) < \varepsilon/2 + \varepsilon/2 = \varepsilon$, which shows that $\{x_n\}$ is a Cauchy sequence. □

Corollary 6.3.8 If $\{x_n\}$ is a convergent sequence in $\boldsymbol{R}$, then for every $\varepsilon > 0$ there is some $m \in \boldsymbol{N}$ such that $|x_n - x_k| < \varepsilon$ for all $n \geqslant m$ and $k \geqslant m$.

Unfortunately, the converse of Proposition 6.3.7 is not true in general: in an arbitrary metric space, Cauchy sequences need not converge.

Example 6.3.9 Let $\boldsymbol{R}$ have the absolute value metric, as usual, and consider the set $\boldsymbol{Q}$ of rational numbers as a subspace of $\boldsymbol{R}$. In the metric space $\boldsymbol{Q}$, Cauchy sequences need not converge. To see this, let $\{q_n\}$ be a sequence of rational numbers which converges to $\sqrt{2}$ in $\boldsymbol{R}$. (We know that such a sequence exists. Why?) Since this sequence converges in $\boldsymbol{R}$, it must be a Cauchy sequence in $\boldsymbol{R}$, and hence it must also be a Cauchy sequence when considered as a sequence in the subspace $\boldsymbol{Q}$. But the sequence $\{q_n\}$ cannot converge in $\boldsymbol{Q}$, for if it did it would necessarily converge to a rational number, and therefore could not converge to the irrational number $\sqrt{2}$ in $\boldsymbol{R}$.

Definition 6.3.10 A metric space is *complete* if every Cauchy sequence in the space converges to a point of the space.

We have just seen that $\boldsymbol{Q}$ with the absolute value metric is not a complete metric space, so not all metric spaces are complete. It is a very important fact that $\boldsymbol{R}$ with the absolute value metric is a complete metric space, and we will prove this. Before we can do so, however, we need the following result.

Proposition 6.3.11 Every Cauchy sequence in $\boldsymbol{R}$ is bounded.

Proof If $\{x_n\}$ is a Cauchy sequence in $\boldsymbol{R}$, then there is some $m \in \boldsymbol{N}$ such that $|x_n - x_k| < 1$ for all $n \geqslant m$ and $k \geqslant m$. Therefore, for all $n \geqslant m$,

$$|x_n| = |x_n - x_m + x_m| \leqslant |x_n - x_m| + |x_m| < 1 + |x_m|$$

Let $a = \max\{|x_1|, \ldots, |x_m|\}$: for all $n \in \boldsymbol{N}$, $|x_n| < 1 + a$, and hence $\{x_n\}$ is bounded. □

Theorem 6.3.12 $\boldsymbol{R}$ is a complete metric space.

Proof: Let $\{x_n\}$ be a Cauchy sequence in $\boldsymbol{R}$. We must show that $\{x_n\}$ converges to some real number. By Proposition 6.3.11, $\{x_n\}$ is bounded; therefore $s = \text{Lim sup } x_n$ exists as a finite real number, and if $\{x_n\}$ does converge, it must converge to s. We will prove that this is so.

We claim that for every positive real number ε and every $m \in \boldsymbol{N}$ there is some $p \in \boldsymbol{N}$, $p > m$, such that $|s - x_p| < \varepsilon$. To see this, recall that $s = \text{Lim}_{k \to \infty} s_k$, where $s_k = \sup\{x_n \mid n \geqslant k\}$ for each $k \in \boldsymbol{N}$. Since the sequence $\{s_k\}$ converges to s, there is some $t \in \boldsymbol{N}$ such that

$$|s - s_k| < \frac{\varepsilon}{2} \qquad \text{for all } k \geqslant t$$

Let $m \in \boldsymbol{N}$ and choose $r \in \boldsymbol{N}$ such that $r > \max\{t, m\}$. Then $r > m$ and

$$|s - s_r| < \frac{\varepsilon}{2}$$

If $|s_r - x_p| \geqslant \varepsilon/2$ for all $p \geqslant r$, then s_r cannot be the supremum for the set $\{x_n \mid n \geqslant r\}$; but this is the definition of s_r. Therefore there is some $p \geqslant r > m$ such that $|s_r - x_p| < \varepsilon/2$. Thus we have $p > m$ and

$$|s - x_p| \leqslant |s - s_r| + |s_r - x_p| < \frac{\varepsilon}{2} + \frac{\varepsilon}{2} = \varepsilon$$

This proves the claim made at the beginning of this paragraph.

Now we are ready to show that the Cauchy sequence $\{x_n\}$ converges to s. Since $\{x_n\}$ is a Cauchy sequence, there is some $m \in \boldsymbol{N}$ such that

$$|x_n - x_k| < \frac{\varepsilon}{2} \qquad \text{for all } n \geqslant m \text{ and } k \geqslant m$$

By the claim of the preceding paragraph, however, there is some $p > m$ such that $|s - x_p| < \varepsilon/2$. Thus for all $n \geqslant p$,

$$|s - x_n| \leqslant |s - x_p| + |x_n - x_p| < \frac{\varepsilon}{2} + \frac{\varepsilon}{2} = \varepsilon$$

Hence $\{x_n\}$ converges to s and we are done. □

Corollary 6.3.13 A sequence in $\boldsymbol{R}$ converges if and only if it is a Cauchy sequence.

Complete metric spaces have many properties which incomplete ones do not. For instance, suppose that $\{\bar{\beta}(x_n, \varepsilon_n) \mid n \in \boldsymbol{N}\}$ is a collection of closed balls in a metric space such that the real sequence $\{\varepsilon_n\}$ converges to 0 and $\bar{\beta}(x_{n+1}, \varepsilon_{n+1}) \subset \bar{\beta}(x_n, \varepsilon_n)$ for all $n \in \boldsymbol{N}$. If the metric space is complete, then the intersection of these nested closed balls must consist of exactly one point, while if it is not complete the intersection may be empty.

Theorem 6.3.14 Let (M, ρ) be a complete metric space. If

$$\{\bar{\beta}(x_n, \varepsilon_n) \mid n \in \boldsymbol{N}\}$$

is a collection of closed balls in M such that $\operatorname{Lim}_{n \to \infty} \varepsilon_n = 0$ in $\boldsymbol{R}$ and

$$\bar{\beta}(x_{n+1}, \varepsilon_{n+1}) \subset \bar{\beta}(x_n, \varepsilon_n)$$

for all $n \in \boldsymbol{N}$, then

$$\bigcap_{n \in \boldsymbol{N}} \bar{\beta}(x_n, \varepsilon_n)$$

contains exactly one point.

Proof: We will show that the sequence $\{x_n\}$ of center points of the closed balls is a Cauchy sequence in M; since M is complete, this will imply that the sequence $\{x_n\}$ converges to a point x of M, and the point x will be the unique element of $\bigcap_{n \in N} \bar{\beta}(x_n, \varepsilon_n)$.

Let ε be a positive real number. Since $\mathrm{Lim}_{n \to \infty} \varepsilon_n = 0$ in $\boldsymbol{R}$, there is some $m \in N$ such that $0 < \varepsilon_n < \varepsilon/2$ for all $n \geqslant m$. But if $n \geqslant m$ and $k \geqslant m$ then $\bar{\beta}(x_n, \varepsilon_n) \subset \bar{\beta}(x_m, \varepsilon_m)$ and $\bar{\beta}(x_k, \varepsilon_k) \subset \bar{\beta}(x_m, \varepsilon_m)$, and thus

$$\rho(x_n, x_k) \leqslant \rho(x_n, x_m) + \rho(x_k, x_m) < \varepsilon_m + \varepsilon_m < \frac{\varepsilon}{2} + \frac{\varepsilon}{2} = \varepsilon$$

Hence $\{x_n\}$ is a Cauchy sequence in M and therefore converges to some $x \in M$.

We claim that $x \in \bigcap_{n \in N} \bar{\beta}(x_n, \varepsilon_n)$. If x is not an element of this intersection, then there is some $m \in N$ such that $x \in M - \bar{\beta}(x_m, \varepsilon_m)$, which is an open set. Therefore there is some $\varepsilon > 0$ such that

$$\beta(x, \varepsilon) \subset M - \bar{\beta}(x_m, \varepsilon_m)$$

But then $\beta(x, \varepsilon) \cap \beta(x_n, \varepsilon_n) = \emptyset$ for all $n \geqslant m$, and this in turn implies that $\rho(x, x_n) \geqslant \varepsilon$ for all $n \geqslant m$, which contradicts the fact that $\{x_n\}$ converges to x. Hence $x \in \bigcap_{n \in N} \bar{\beta}(x_n, \varepsilon_n)$.

Finally we show that x is the only element in $\bigcap_{n \in N} \bar{\beta}(x_n, \varepsilon_n)$. To see this, suppose that y is also a point in this intersection; then

$$0 \leqslant \rho(x, y) \leqslant \rho(x, x_n) + \rho(y, x_n) < 2\varepsilon_n$$

for all $n \in N$. But $0 \leqslant \rho(x, y) < 2\varepsilon_n$ for all $n \in N$ and $\mathrm{Lim}_{n \to \infty} \varepsilon_n = 0$ together imply that $\rho(x, y) = 0$ and hence that $y = x$. □

Corollary 6.3.15 Let (M, ρ) be a complete metric space. If

$$\{\bar{\beta}(x_n, \varepsilon_n) \mid n \in N\}$$

is a collection of closed balls in M such that

$$\bar{\beta}(x_{n+1}, \varepsilon_{n+1}) \subset \bar{\beta}(x_n, \varepsilon_n)$$

for all $n \in N$, then $\bigcap_{n \in N} \bar{\beta}(x_n, \varepsilon_n)$ is nonempty.

Corollary 6.3.16 (Nested Intervals Theorem) If $\{J_n \mid n \in N\}$ is a collection of closed intervals in $\boldsymbol{R}$ such that $J_{n+1} \subset J_n$ for all $n \in N$, then $\bigcap_{n \in N} J_n$ is nonempty.

The proofs of Corollaries 6.3.15 and 6.3.16 will be left to the reader. (See Exercise 7 at the end of this section.)

Example 6.3.17 As was previously noted, the conclusion of Theorem 6.3.14 need not hold if (M, ρ) is not a complete metric space. As an example, consider the metric space $(0, 1)$ with the absolute value metric. The sequence $\{1/(n+1)\}$ is a Cauchy sequence in $(0, 1)$ which does not converge to a point of the space, so the space is not complete. Note that the collection of closed balls $\{\bar{\beta}(0, 1/(n+1)) \mid n \in \boldsymbol{N}\}$ satisfies the hypotheses of Theorem 6.3.14, but that the intersection of these closed balls is empty.

Theorem 6.3.14 allows us to prove the important Bolzano–Weierstrass theorem.

Theorem 6.3.18 (Bolzano–Weierstrass) If S is a bounded infinite subset of $\boldsymbol{R}$ then S has a limit point.

Proof: Since S is bounded, there exists a closed interval $[a, b]$ such that $S \subset [a, b]$. Let x_1 be the midpoint of $[a, b]$ and let $\varepsilon_1 = (b-a)/2$; then $S \subset [a, b] = \bar{\beta}(x_1, \varepsilon_1)$. Because S is infinite, at least one of the intervals $[a, x_1]$, $[x_1, b]$ must contain an infinite subset of S. Choose one of the intervals $[a, x_1]$, $[x_1, b]$ which does contain an infinite subset of S; if x_2 is the midpoint of the chosen interval and

$$\varepsilon_2 = |x_1 - x_2| = \frac{b-a}{2^2}$$

then the interval may be written as $\bar{\beta}(x_2, \varepsilon_2)$. Note that

$$\bar{\beta}(x_2, \varepsilon_2) \subset \bar{\beta}(x_1, \varepsilon_1)$$

Now bisect the closed interval $\bar{\beta}(x_2, \varepsilon_2)$ and argue as before. In this manner we obtain for each $n \in \boldsymbol{N}$ a closed interval $\bar{\beta}(x_n, \varepsilon_n)$ which contains an infinite subset of S and such that $\varepsilon_n = (b-a)/2^n$ and $\bar{\beta}(x_{n+1}, \varepsilon_{n+1}) \subset \bar{\beta}(x_n, \varepsilon_n)$. Therefore by Theorem 6.3.14 there is a unique point $x \in \bigcap_{n \in \boldsymbol{N}} \bar{\beta}(x_n, \varepsilon_n)$. The point x is a limit point of S, because if G is any open set containing x then there is some $n \in \boldsymbol{N}$ such that $\bar{\beta}(x_n, \varepsilon_n) \subset G$ and $\bar{\beta}(x_n, \varepsilon_n)$ contains an infinite subset of S. □

Note that the conclusion of the Bolzano–Weierstrass theorem fails to hold if either of the hypotheses is contravened: thus, a finite subset of $\boldsymbol{R}$

will have no limit point, while N provides an example of an infinite unbounded subset of R which has no limit point.

EXERCISES

1. For each of the following sequences in R use Corollary 6.3.4 to establish convergence or divergence.

   (a) $\left\{\frac{1}{n}\right\}$

   (b) $\left\{\frac{n}{n^2+1}\right\}$

   (c) $\left\{\frac{n^2+1}{n}\right\}$

   (d) $\{a^n\}$, $a \geqslant 1$

   (e) $\{a^n\}$, $0 \leqslant a < 1$

2. Let M be a set and ρ the discrete metric on M. Prove that (M, ρ) is a complete metric space.
3. Prove that if (M, ρ) is complete metric space and S is a closed subset of M, then (S, ρ) is a complete metric space. Give an example to show that the conclusion fails to hold if S is not closed in M.
4. Let ρ be the absolute value metric on R. Give an example of an uncountable subset S of R such that (S, ρ) is a complete metric space which contains no open interval.
5. Let ρ be the product metric on R^2. Prove that (R^2, ρ) is a complete metric space.
6. Use Corollary 6.3.13 to establish convergence or divergence for the following sequences in R.

   (a) $\{x_n\}$, where $x_1 = 1$, $x_2 = 2$, and $x_n = (1/2)(x_{n-1} + x_{n-2})$ for all $n > 2$;

   (b) $\{x_n\}$, where $x_n = \Sigma_{k=1}^{n} (1/k)$ for all $n \in N$.
7. Prove Corollaries 6.3.15 and 6.3.16.
8. Let (M, ρ) be a metric space and let S be a subset of M. Define S to be *bounded* in M if there is some $x \in M$ and some positive real number ε such that $S \subset \bar{\beta}(x, \varepsilon)$. Define S to be *totally bounded* in M if for every positive real number ε there is a finite set of points $x_1, \ldots, x_n$ in M such that

   $$S \subset \bigcup_{1 \leqslant k \leqslant n} \bar{\beta}(x_k, \varepsilon)$$

   (a) Prove that if S is totally bounded in M, then S is bounded in M.

(b) Give an example of a metric space (M, ρ) and a subset S of M such that S is bounded in M but not totally bounded in M.

9. Prove that in $\boldsymbol{R}$, boundedness is equivalent to total boundedness; that is, prove that a subset S of $\boldsymbol{R}$ is bounded if and only if it is totally bounded.
10. Prove the generalized Bolzano–Weierstrass Theorem: if S is a totally bounded infinite subset of a complete metric space, then S has a limit point in the space.
11. Give an example of a metric space (M, ρ) and a subset S of M such that
    (a) S is infinite and totally bounded in M but has no limit point;
    (b) (M, ρ) is complete, S is totally bounded in M, but S has no limit point;
    (c) (M, ρ) is complete, S is infinite and bounded in M, but S has no limit point.

    Thus if any one of the three hypotheses of the generalized Bolzano–Weierstrass Theorem is removed, the theorem fails.
12. Prove that the property of Theorem 6.3.14 is equivalent to completeness; that is, prove that a metric space (M, ρ) is complete if and only if

$$\bigcap_{n \in N} \bar{\beta}(x_n, \varepsilon_n) \qquad \text{contains exactly one point}$$

whenever

$$\bar{\beta}(x_{n+1}, \varepsilon_{n+1}) \subset \bar{\beta}(x_n, \varepsilon_n) \text{ for all } n \in N \text{ and } \qquad \lim_{n \to \infty} \varepsilon_n = 0$$

13. Let M be the set of all sequences in $\boldsymbol{R}$ which have a finite number of nonzero terms. Thus

$$M = \{\{x_n\} \text{ in } \boldsymbol{R} \mid \text{for some } m \in \boldsymbol{N},\ x_n = 0 \text{ for all } n \geqslant m\}$$

If $\rho(\{x_n\}, \{y_n\}) = \sup\{|x_n - y_n| \mid n \in \boldsymbol{N}\}$ for all $\{x_n\}$ and $\{y_n\}$ in M, then (M, ρ) is a subspace of the metric space l^∞, and hence is itself a metric space. Show that (M, ρ) is not complete. (Hint: consider the sequence $\{s_n\}$ in M, where for each $n \in \boldsymbol{N}$, s_n is the sequence $\{1, 1/2, \ldots, 1/n, 0, 0, \ldots\}$.)

14. Let S be a subset of $\boldsymbol{R}$ and let $f\colon S \to \boldsymbol{R}$. A point $s \in S$ is a *fixed point* of f if $f(s) = s$. Suppose f satisfies a Lipschitz condition with positive

constant $c < 1$. (See Exercise 7 of Section 6.2.) Show that f has a unique fixed point. (Hint: let $s_1 \in S$, define $s_{n+1} = f(s_n)$ for all $n \in N$, and consider the limit of $\{s_n\}$.)

6.4 COMPLETION OF METRIC SPACES*

In this section we will show that every metric space can be identified with a dense subset of a complete metric space. Our proof of this fact will be constructive: that is, we will start with an arbitrary metric space (M, ρ) and demonstrate how to construct from it a complete metric space (M^*, ρ^*) which contains as a dense subset a "copy" of (M, ρ). Our proof will also suggest an alternative method of constructing the real numbers $\boldsymbol{R}$ from the rational numbers $\boldsymbol{Q}$.

It will be helpful to us to have available the notion of an equivalence relation on a set.

Definition 6.4.1 Let S be a set. A subset R of $S \times S$ is called a *relation* on S. If $(x, y) \in R$, we write xRy; if $(x, y) \notin R$, we write $x \not R y$. A relation R on S is an *equivalence relation* if it is

1. Reflexive: xRx for all $x \in S$;
2. Symmetric: if $x \in S$, $y \in S$, and xRy, then yRx; and
3. Transitive: if $x \in S$, $y \in S$, and $z \in S$, and xRy and yRz, then xRz.

If R is an equivalence relation on S and $x \in S$, then $[x] = \{y \in S \mid xRy\}$ is called the *equivalence class of* x under R, and x is referred to as the *representative* of the equivalence class.

Examples 6.4.2 1. Let $S = \{a, b, c\}$ and $R = \{(a, a), (b, b), (a, b)\}$, so that aRa, bRb, and aRb. The relation R is not an equivalence relation on S because, for instance, $c \in S$ but $c \not R c$. (Also, aRb but $b \not R a$.)

2. Let $S = \{a, b, c\}$ and $R = \{(a, a), (b, b), (c, c), (a, b), (b, a)\}$. The relation R is an equivalence relation on S. Note that $[a] = \{a, b\}$, $[b] = \{a, b\}$, and $[c] = \{c\}$; thus the equivalence class $\{a, b\}$ has two different representatives, namely a and b. Note also that distinct equivalence classes are mutually disjoint.

3. If x and y are real numbers, define xRy if $x = y$. This relation is an equivalence relation on $\boldsymbol{R}$. Note that $[x] = \{x\}$, for all $x \in \boldsymbol{R}$.

---

*This section is optional.

4. If x and y are real numbers, define xRy if $x \leqslant y$. This relation is reflexive and transitive, but not symmetric; hence it is not an equivalence relation on $\boldsymbol{R}$.

5. If $\{x_n\}$ and $\{y_n\}$ are Cauchy sequences in $\boldsymbol{Q}$, define $\{x_n\}R\{y_n\}$ if the sequence $\{|x_n - y_n|\}$ converges to zero. Let us show that this is an equivalence relation on the set of all Cauchy sequences in $\boldsymbol{Q}$:

Reflexivity: If $\{x_n\}$ is a Cauchy sequence in $\boldsymbol{Q}$, then surely the sequence $\{|x_n - x_n|\} = \{0\}$ converges to zero; hence $\{x_n\}R\{x_n\}$.

Symmetry: If $\{x_n\}$, $\{y_n\}$ are Cauchy sequences in $\boldsymbol{Q}$ with $\{x_n\}R\{y_n\}$, then $\{|x_n - y_n|\}$ converges to zero; but $|x_n - y_n| = |y_n - x_n|$, so $\{|y_n - x_n|\}$ converges to zero, and hence $\{y_n\}R\{x_n\}$.

Transitivity: If $\{x_n\}$, $\{y_n\}$, and $\{z_n\}$ are Cauchy sequences in $\boldsymbol{Q}$ with $\{|x_n - y_n|\}$ and $\{|y_n - z_n|\}$ converging to zero, then

$$|x_n - z_n| \leqslant |x_n - y_n| + |y_n - z_n|$$

implies that $\{|x_n - z_n|\}$ converges to zero (why?) and hence that $\{x_n\}R\{z_n\}$.

Note that $[\{0\}] = [\{1/n\}] = [\{a_n\}]$, where $\{a_n\}$ is any Cauchy sequence in $\boldsymbol{Q}$ which converges to zero.

As the preceding examples suggest, if R is an equivalence relation on a set S, then either xRy, in which case $[x] = [y]$ and the elements x and y are representatives of the same class, or $x \not R y$, in which case $[x] \cap [y] = \emptyset$. Thus an equivalence relation on a set S *partitions* the set; that is, it defines a collection of mutually disjoint subsets of S whose union is S. (The mutually disjoint subsets are the distinct equivalence classes of S under R.)

Proposition 6.4.3 Let S be a set and let R be an equivalence relation on S. If $x \in S$ and $y \in S$, then

1. xRy implies that $[x] = [y]$;
2. $x \not R y$ implies that $[x] \cap [y] = \emptyset$.

Hence the distinct equivalence classes of S under R partition the set S.

The proof of Proposition 6.4.3 is left to the reader. (See Exercise 1 at the end of this section.)

At the beginning of this section we spoke of a "copy" of a metric space (M, ρ); let us formalize this notion.

Definition 6.4.4 Let (M_1, ρ_1) and (M_2, ρ_2) be metric spaces and let f be a function from M_1 onto M_2. If $\rho_1(x, y) = \rho(f(x), f(y))$ for all $x \in M_1$, $y \in M_1$, then f is called an *isometry* and M_1 and M_2 are said to be *isometric spaces*.

It is easy to show that every isometry is a homeomorphism. (See Exercise 2 at the end of this section.) Indeed, since an isometry preserves distances, isometric spaces are identical as metric spaces; hence, if $f: M_1 \to M_2$ is an isometry, then we may think of M_2 as a "copy" of M_1.

Now we are ready to begin the process of completing a metric space. We wish to prove that every metric space is isometric to a dense subset of a complete metric space. Thus if (M, ρ) is any metric space, we must construct from it a complete metric space (M^*, ρ^*) and then produce a function $f: M \to M^*$ such that f is an isometry from (M, ρ) to $(f(M), \rho^*)$ and $f(M)$ is dense in M^*. The basic idea of the proof is to construct M^* as a set of equivalence classes of Cauchy sequences in M.

Definition 6.4.5 Let (M, ρ) be a metric space and let C be the set of all Cauchy sequences in M. Define a relation R on C by setting $\{x_n\}R\{y_n\}$ if the sequence of nonnegative real numbers $\{\rho(x_n, y_n)\}$ converges to zero.

Proposition 6.4.6 The relation R of Definition 6.4.5 is an equivalence relation on C.

The proof of the Proposition is similar to that of Example 5 of Examples 6.4.2, and is left to the reader. (See Exercise 3 at the end of this section.)

Definition 6.4.7 Let (M, ρ) be a metric space and let R be the equivalence relation of Definition 6.4.5. If C is the set of all Cauchy sequences in M and $\{x_n\} \in C$, denote the equivalence class of $\{x_n\}$ under R by $[x_n]$. Let M^* denote the set of distinct equivalence classes of C under R, and define $\rho^*: M^* \times M^* \to \mathbf{R}$ as follows: if $[x_n] \in M^*$ and $[y_n] \in M^*$, then

$$\rho^*([x_n], [y_n]) = \lim_{n \to \infty} \rho(x_n, y_n)$$

in $\mathbf{R}$.

Our first task is to show that $\rho^*([x_n], [y_n])$ exists and is well-defined.

Proposition 6.4.8 For all $[x_n] \in M^*$ and $[y_n] \in M^*$, $\rho^*([x_n], [y_n])$ exists and its value does not depend on the Cauchy sequences $\{x_n\}$, $\{y_n\}$ which are chosen as representatives for the equivalence classes $[x_n]$, $[y_n]$.

Proof: To show that $\rho^*([x_n], [y_n])$ exists for all $[x_n] \in M^*$ and $[y_n] \in M^*$, it will suffice to show that $\{\rho(x_n, y_n)\}$ is a Cauchy sequence in $\boldsymbol{R}$. (Why?)

By the triangle inequality in (M, ρ), we have

$$\rho(x_n, y_n) \leqslant \rho(x_n, x_m) + \rho(x_m, y_m) + \rho(y_m, y_n)$$

and

$$\rho(x_m, y_m) \leqslant (x_m, x_n) + \rho(x_n, y_n) + \rho(y_n, y_m)$$

for all $n \in \boldsymbol{N}$ and $m \in \boldsymbol{N}$. Therefore

$$(1) \qquad \rho(x_n, y_n) - \rho(x_m, y_m) \leqslant \rho(x_n, x_m) + \rho(y_m, y_n)$$

and

$$(2) \qquad \rho(x_m, y_m) - \rho(x_n, y_n) \leqslant \rho(x_n, x_m) + \rho(y_m, y_n)$$

for all $n \in \boldsymbol{N}$ and $m \in \boldsymbol{N}$. But because $\{x_n\}$ and $\{y_n\}$ are Cauchy sequences in (M, ρ), for every positive real number ε there is some $k \in \boldsymbol{N}$ such that $n \geqslant k$ and $m \geqslant k$ imply that

$$\rho(x_n, x_m) < \frac{\varepsilon}{2} \qquad \text{and} \qquad \rho(y_n, y_m) < \frac{\varepsilon}{2}$$

Hence the inequalities (1) and (2) together show that

$$|\rho(x_n, y_n) - \rho(x_m, y_m)| < \varepsilon$$

and therefore $\{\rho(x_n, y_n)\}$ is a Cauchy sequence in $\boldsymbol{R}$.

To show that the value of $\rho^*([x_n], [y_n])$ does not depend on the choice of the representatives used for the equivalence classes, suppose that $\{x_n\}$, $\{y_n\}$, $\{s_n\}$, and $\{t_n\}$ are Cauchy sequences in (M, ρ) with $\{x_n\}R\{s_n\}$ and $\{y_n\}R\{t_n\}$ It will suffice to show that the sequences $\{\rho(x_n, y_n)\}$ and $\{\rho(s_n, t_n)\}$ converge to the same limit in $\boldsymbol{R}$. (Why?)

Since $\{x_n\}R\{s_n\}$ and $\{y_n\}R\{t_n\}$, we have

$$\operatorname*{Lim}_{n \to \infty} \rho(x_n, s_n) = 0 \qquad \text{and} \qquad \operatorname*{Lim}_{n \to \infty} \rho(y_n, t_n) = 0$$

Also, since $\rho^*([x_n], [y_n])$ and $\rho^*([s_n], [t_n])$ exist by the first part of this proof, the sequences $\{\rho(x_n, y_n)\}$ and $\{\rho(s_n, t_n)\}$ converge in $\boldsymbol{R}$. By the triangle inequality,

$$\rho(s_n, t_n) \leqslant \rho(s_n, x_n) + \rho(x_n, y_n) + \rho(y_n, t_n)$$

for all $n \in \boldsymbol{N}$, and therefore

$$\begin{aligned}\lim_{n\to\infty} \rho(s_n, t_n) &\leqslant \lim_{n\to\infty} \rho(s_n, x_n) + \lim_{n\to\infty} \rho(x_n, y_n) + \lim_{n\to\infty} \rho(y_n, t_n)\\ &= \lim_{n\to\infty} \rho(x_n, y_n)\end{aligned}$$

Thus

$$\lim_{n\to\infty} \rho(s_n, t_n) \leqslant \lim_{n\to\infty} \rho(x_n, y_n)$$

and a similar argument shows that

$$\lim_{n\to\infty} \rho(x_n, y_n) \leqslant \lim_{n\to\infty} \rho(s_n, t_n)$$

Hence

$$\lim_{n\to\infty} \rho(x_n, y_n) = \lim_{n\to\infty} \rho(s_n, t_n)$$

and we are done. □

Now we know that ρ^* is a well-defined function from $M^* \times M^*$ to $\boldsymbol{R}$, it is easy to show that it is a metric on M^*.

Proposition 6.4.9 The function ρ^* of Definition 6.4.7 is a metric on M^*.

Proof: We will show that the triangle inequality holds, leaving the remainder of the proof to the reader. (See Exercise 4 at the end of this section.)

Let $[x_n]$, $[y_n]$, and $[z_n]$ be elements of M^*. By the triangle inequality for (M, ρ),

$$\rho(x_n, y_n) \leqslant \rho(x_n, z_n) + \rho(y_n, z_n)$$

for all $n \in \boldsymbol{N}$, and hence

$$\lim_{n\to\infty} \rho(x_n, y_n) \leqslant \lim_{n\to\infty} \rho(x_n, z_n) + \lim_{n\to\infty} \rho(y_n, z_n)$$

Therefore

$$\rho^*([x_n], [y_n]) \leqslant \rho^*([x_n], [z_n]) + \rho^*([y_n], [z_n]) \quad \square$$

We have now constructed from the metric space (M, ρ) a new metric space (M^*, ρ^*) whose points are equivalence classes of Cauchy sequences in M. We still must show that (M, ρ) is isometric to a dense subset of (M^*, ρ^*) and that (M^*, ρ^*) is a complete metric space. Note that if $x \in M$ then $\{x\}$ is the constant sequence all of whose terms are equal to x, and $[x] \in M^*$ is the equivalence class of this sequence.

Proposition 6.4.10 The function $f: M \to M^*$ defined by $f(x) = [x]$ for all $x \in M$ is an isometry from (M, ρ) onto $(f(M), \rho^*)$, and $f(M)$ is a dense subset of (M^*, ρ^*).

Proof: For all $x \in M$, $y \in M$, we have $[x] \in M^*$, $[y] \in M^*$, and

$$\rho^*([x], [y]) = \operatorname*{Lim}_{n \to \infty} \rho(x, y) = \rho(x, y)$$

thus f is an isometry. To show that $f(M)$ is dense in (M^*, ρ^*) it will suffice to show that for any $[x_n] \in M^*$ and any positive real number ε, there is some $[x] \in f(M)$ such that $\rho^*([x_n], [x]) \leqslant \varepsilon$. (Why?)

Since $\{x_n\}$ is a Cauchy sequence, there is some $k \in N$ such that $\rho(x_n, x_m) < \varepsilon$ for all $n \geqslant k$, $m \geqslant k$. Let $x = x_k$, and consider $\rho^*([x_n], [x]) = \operatorname{Lim}_{n \to \infty} \rho(x_n, x)$. Since $\rho(x_n, x) = \rho(x_n, x_k) < \varepsilon$ for all $n \geqslant k$, it follows that $\operatorname{Lim}_{n \to \infty} \rho(x_n, x) \leqslant \varepsilon$, and therefore that

$$\rho^*([x_n], [x]) \leqslant \varepsilon \quad \square$$

Now we are ready to begin the proof that the metric space (M^*, ρ^*) is complete.

Proposition 6.4.11 If $\{x_m{}^*\}$ is a sequence in (M^*, ρ^*), then for each $m \in N$ there is a Cauchy sequence $\{x_{nm} \mid n \in N\}$ in (M, ρ) such that $x_m{}^* = [x_{nm}]$ and $\rho(x_{nm}, x_{km}) < 1/m$ for all $n \in N$ and $k \in N$.

Proof: For each $m \in N$, $x_m{}^*$ is a point of M^*, and hence is an equivalence class of Cauchy sequences in (M, ρ); let $\{x_{nm}\}$ denote a representative of the class which determines $x_m{}^*$, so that $x_m{}^* = [x_{nm}]$. Since $\{x_{nm}\}$ is a Cauchy sequence, there is some $t_m \in N$ such that $\rho(x_{nm}, x_{km}) < 1/m$ for all $n \geqslant t_m$ and $k \geqslant t_m$. Consider the Cauchy sequences $\{x_{nm} \mid n \in N\}$ and $\{x_{nm} \mid n \geqslant t_m\}$: it is easy to show that these two sequences in (M, ρ) are equivalent under the equivalence relation R of Definition 6.4.5 (see Exercise 5 at the end of this section), and hence they define the same point, namely $x_m{}^*$, in M^*. If we now renumber the terms of the sequence

$\{x_{nm} \mid n \geqslant t_m\}$ so that its first term is x_{1n} rather than $x_{t_m m}$, we obtain a sequence which satisfies the requirements of the proposition. □

Proposition 6.4.12 Let $\{x_m{}^*\}$ be a Cauchy sequence in (M^*, ρ^*), and for each $m \in \boldsymbol{N}$ let $\{x_{nm}\}$ be a Cauchy sequence in (M, ρ) such that $x_m{}^* = [x_{nm}]$ and $\rho(x_{nm}, x_{km}) < 1/m$ for all $n \in \boldsymbol{N}$ and $k \in \boldsymbol{N}$. Then the sequence $\{x_{1m} \mid m \in \boldsymbol{N}\}$ is a Cauchy sequence in (M, ρ).

Proof: The existence of the sequences $\{x_{nm}\}$ is guaranteed by the previous proposition; the sequence $\{x_{1m} \mid m \in \boldsymbol{N}\}$ consists of the first terms of all these sequences. We must prove that $\{x_{1m}\}$ is a Cauchy sequence in (M, ρ).

If ε is a positive real number, then, since $\{x_m{}^*\}$ is a Cauchy sequence, there is some $p_1 \in \boldsymbol{N}$ such that $\rho^*(x_m{}^*, x_t{}^*) < \varepsilon/3$ for all $m \geqslant p_1$ and $t \geqslant p_1$. Choose $p \in \boldsymbol{N}$ such that $p > \max\{p_1, 3/\varepsilon\}$: then $1/p < \varepsilon/3$ and $\rho^*(x_m{}^*, x_t{}^*) < \varepsilon/3$ for all $m \geqslant p$ and $t \geqslant p$. Now,

$$\rho^*(x_m{}^*, x_t{}^*) = \rho^*([x_{nm}], [x_{nt}]) = \operatorname*{Lim}_{n \to \infty} \rho(x_{nm}, x_{nt})$$

in $\boldsymbol{R}$. Since $\rho^*(x_m, x_t) < \varepsilon/3$ for $m \geqslant p$ and $t \geqslant p$, it follows that there must be $q \in \boldsymbol{N}$ such that $\rho(x_{nm}, x_{nt}) < \varepsilon/3$ for all $n \geqslant q$.

Suppose $m \geqslant p$ and $t \geqslant p$ and consider $\rho(x_{1m}, x_{1t})$. By the triangle inequality,

$$\rho(x_{1m}, x_{1t}) \leqslant \rho(x_{1m}, x_{qm}) + \rho(x_{qm}, x_{qt}) + \rho(x_{qt}, x_{1t})$$

But

$$\rho(x_{1m}, x_{qm}) < \frac{1}{m} < \frac{1}{p} < \frac{\varepsilon}{3}$$

$$\rho(x_{qt}, x_{1t}) < \frac{1}{t} < \frac{1}{p} < \frac{\varepsilon}{3}$$

and

$$\rho(x_{qm}, x_{qt}) < \frac{\varepsilon}{3}$$

Therefore $\rho(x_{1m}, x_{1t}) < \varepsilon$ for all $m \geqslant p$ and $t \geqslant p$, and thus $\{x_{1m}\}$ is a Cauchy sequence in (M, ρ). □

Theorem 6.4.13 The metric space (M^*, ρ^*) is complete.

Proof: By virtue of the previous two propositions we may assume that if $\{x_m^*\}$ is a Cauchy sequence in (M^*, ρ^*), then for each $m \in N$ there is a Cauchy sequence $\{x_{nm} \mid n \in N\}$ in (M, ρ) such that $x_m^* = [x_{nm}]$ and $\rho(x_{nm}, x_{km}) < 1/m$. Furthermore, we may also assume that the sequence $(x_{1m} \mid m \in N\}$ is a Cauchy sequence in (M, ρ).

Let ε be a positive real number. Since $\{x_{1m}\}$ is a Cauchy sequence, there is some $k_1 \in N$ such that $\rho(x_{1m}, x_{1t}) < \varepsilon/2$ for all $m \geqslant k_1$ and $t \geqslant k_1$. Choose $k \in N$ such that $k > \max\{k_1, 2/\varepsilon\}$: then $1/k < \varepsilon/2$ and $\rho(x_{1m}, x_{1t}) < \varepsilon/2$ for all $m \geqslant k$ and $t \geqslant k$. But then

$$\rho(x_{1m}, x_{mt}) \leqslant \rho(x_{1m}, x_{1t}) + \rho(x_{1t}, x_{mt}) < \frac{\varepsilon}{2} + \frac{1}{t} \leqslant \frac{\varepsilon}{2} + \frac{1}{k} < \varepsilon$$

We have thus shown that for any positive real number ε there is some $k \in N$ such that $m \geqslant k$ and $t \geqslant k$ imply $\rho(x_{1m}, x_{mt}) < \varepsilon$.

The Cauchy sequence $\{x_{1m}\}$ in (M, ρ) defines a point $x^* = [x_{1m}]$ in M^*, and we claim that the Cauchy sequence $\{x_m^*\}$ in (M^*, ρ^*) converges to x^*. In order to prove this, we must show that if ε is a positive real number, then there is some $k \in N$ such that $\rho^*(x_t^*, x^*) < \varepsilon$ for $t \geqslant k$. However, by the result proved in the preceding paragraph, there is some $k \in N$ such that $\rho(x_{1m}, x_{mt}) < \varepsilon/2$ for $m \geqslant k$ and $t \geqslant k$. Therefore for $t \geqslant k$ we have

$$\rho^*(x_t^*, x^*) = \rho^*([x_{mt}], [x_{1m}]) = \lim_{n \to \infty} \rho(x_{mt}, x_{1m}) < \frac{\varepsilon}{2} < \varepsilon$$

and we are finished. □

We have thus shown that every metric space (M, ρ) is isometric to a dense subset of a complete metric space (M^*, ρ^*). The metric space (M^*, ρ^*), which was explicitly constructed from (M, ρ), is called the *completion* of (M, ρ). The completion (M^*, ρ^*) is essentially unique, in the sense that if (M, ρ) is isometric to a dense subset of another metric space (M_1, ρ_1), then (M_1, ρ_1) must be isometric to (M^*, ρ^*). (See Exercise 7 at the end of this section.)

Since the metric space $\boldsymbol{Q}$ of rational numbers is not complete and is a dense subset of the complete metric space $\boldsymbol{R}$, it is reasonable to ask if in fact $\boldsymbol{R}$ is the completion of $\boldsymbol{Q}$. The answer is yes: the procedure we have used to construct the completion of a metric space can, with some modification, be used to construct $\boldsymbol{R}$ from $\boldsymbol{Q}$. (Modification is required because several of our definitions and proofs made use of real numbers, and of course if we intend to construct $\boldsymbol{R}$ from $\boldsymbol{Q}$, we cannot use real numbers in so doing.) Thus, we begin with the number system $\boldsymbol{Q}$, and assume that the operations of arithmetic, the order relation $<$, and absolute value have been defined in $\boldsymbol{Q}$. We then define the concept of an open interval in $\boldsymbol{Q}$ and allow the open intervals to generate the natural topology on $\boldsymbol{Q}$ in the usual manner. Imitating the proof of Proposition 6.3.3, we show that a sequence $\{x_n\}$ in the topological space $\boldsymbol{Q}$ converges to $x \in \boldsymbol{Q}$ if and only if for every $q \in \boldsymbol{Q}$, $q > 0$, there is some $m \in \boldsymbol{N}$ such that $|x_n - x| < q$ for all $n \geqslant m$. Next we define $\{x_n\}$ to be a Cauchy sequence in $\boldsymbol{Q}$ if for every $q \in \boldsymbol{Q}$, $q > 0$, there is some $m \in \boldsymbol{N}$ such that $|x_n - x_k| < q$ for all $n \geqslant m, k \geqslant m$. Now it is possible to define a relation R on the set of all Cauchy sequences in $\boldsymbol{Q}$ by setting $\{x_n\} R \{y_n\}$ if the sequence $\{|x_n - y_n|\}$ converges to 0 in $\boldsymbol{Q}$; as Example 5 of Examples 6.4.2 shows, this relation is an equivalence relation. Finally, we define the set of real numbers to be the set of all equivalence classes of rational Cauchy sequences under the relation R.

Once the construction of $\boldsymbol{R}$ outlined above has been accomplished, it is possible to define the usual operations of arithmetic, the order relation $<$, and the concepts of convergent sequence and Cauchy sequence in $\boldsymbol{R}$. It then requires only minor modifications of the relevant proofs of this section to show that $\boldsymbol{Q}$ is isometric to a dense subset of $\boldsymbol{R}$ and that every Cauchy sequence in $\boldsymbol{R}$ converges. We do not intend to pursue this topic any further, but the interested reader may consult any one of a number of texts on advanced real analysis. Incidentally, if $\boldsymbol{R}$ has been constructed both by means of Dedekind cuts and as equivalence classes of rational Cauchy sequences, then it becomes necessary to show that the two constructions yield the same number system. This can be done by defining a one-to-one function from the set of all equivalence classes of rational Cauchy sequences onto the set of all Dedekind cuts, which preserves the arithmetic operations and ordering. (An excellent source for the complete details of the construction of the reals by both methods is C. Goffman, *Real Analysis*, Holt, Rinehart and Winston.)

EXERCISES

1. Prove Proposition 6.4.3.
2. Prove that every isometry from one metric space to another is a homeomorphism.

3. Prove Proposition 6.4.6.
4. Prove Proposition 6.4.9.
5. Prove that the Cauchy sequences $\{x_{nm} \mid n \in N\}$ and $\{x_{nm} \mid n \geqslant t_m\}$ in the proof of Proposition 6.4.11 are equivalent.
6. Let M be a dense subset of the metric space (M_1, ρ_1) and M' be a dense subset of the metric space (M_2, ρ_2). Prove that if (M, ρ_1) is isometric to (M', ρ_2), then (M_1, ρ_1) is isometric to (M_2, ρ_2). (Hint: if f is an isometry from M to M', extend f to M_1 by defining

$$f(x) = \operatorname*{Lim}_{n \to \infty} f(x_n)$$

whenever $\{x_n\}$ is a sequence in M which converges to $x \in M_1$.)
7. Prove that if the metric space (M, ρ) is isometric to dense subsets of both the metric space (M_1, ρ_1) and the metric space (M_2, ρ_2), then (M_1, ρ_1) is isometric to (M_2, ρ_2).
8. Suppose that $\boldsymbol{R}$ has been constructed from $\boldsymbol{Q}$ in the manner outlined in Section 6.4, so that a real number is an equivalence class of Cauchy sequences in $\boldsymbol{Q}$, where by definition, $\{x_n\}$ is a Cauchy sequence in $\boldsymbol{Q}$ if for every $q \in \boldsymbol{Q}$, $q > 0$, there is some $m \in N$ such that $n \geqslant m$, $k \geqslant m$ imply $|x_n - x_k| < q$.
   (a) Show that if $\{x_n\}$ and $\{y_n\}$ are Cauchy sequences in $\boldsymbol{Q}$, then so is $\{x_n + y_n\}$.
   (b) Define addition of real numbers (equivalence classes of rational Cauchy sequences) as follows: if $\{x_n\}$ and $\{y_n\}$ are Cauchy sequences in $\boldsymbol{Q}$, with equivalence classes $[x_n]$ and $[y_n]$, then

$$[x_n] + [y_n] = [x_n + y_n]$$

   Prove that addition of real numbers is well-defined by showing that if $[x_n] = [r_n]$ and $[y_n] = [s_n]$, then

$$[x_n + y_n] = [r_n + s_n]$$

   (c) Show that if $\{x_n\}$ and $\{y_n\}$ are Cauchy sequences in $\boldsymbol{Q}$, then so is $\{x_n y_n\}$.
   (d) Define multiplication of real numbers (equivalence classes of rational Cauchy sequences) as follows: if $\{x_n\}$ and $\{y_n\}$ are Cauchy sequences in $\boldsymbol{Q}$, with equivalence classes $[x_n]$ and $[y_n]$,

then

$$[x_n][y_n] = [x_n y_n]$$

Prove that multiplication of real numbers is well-defined by showing that if $[x_n] = [r_n]$ and $[y_n] = [s_n]$, then

$$[x_n y_n] = [r_n s_n]$$

7
Sequences of Functions

In this chapter we examine the behavior of sequences of functions in metric spaces. We define and illustrate the concepts of pointwise and uniform convergence of a sequence of functions, show that certain properties of functions (e.g., continuity) are preserved by uniform convergence, introduce and briefly study several metric spaces whose points are functions, and finally prove two important approximation theorems for functions. For simplicity, we will only consider functions from a metric space to the real numbers, although many of our definitions and results can be extended without difficulty to the more general setting of functions from one metric space to another.

7.1 POINTWISE AND UNIFORM CONVERGENCE

In this section we define two types of convergence for sequences of functions and show how these preserve or fail to preserve certain properties. We begin with the definition of pointwise convergence for a sequence of functions.

Definition 7.1.1 Let (M, ρ) be a metric space, let f be a function from M to $\boldsymbol{R}$, and for each $n \in \boldsymbol{N}$ let $f_n\colon M \to \boldsymbol{R}$. We say that *the sequence of functions* $\{f_n\}$ *converges pointwise to the function* f if for each $x \in M$ the sequence of real numbers $\{f_n(x)\}$ converges to the real number $f(x)$.

Thus $\{f_n\}$ converges pointwise to f if $\text{Lim}_{n\to\infty} f(x_n) = f(x)$ for all $x \in M$.

Examples 7.1.2 1. For each $n \in \boldsymbol{N}$, define $f_n: \boldsymbol{R} \to \boldsymbol{R}$ by $f_n(x) = x/n$ for all $x \in \boldsymbol{R}$ (see Figure 7.1). Since $\text{Lim}_{n\to\infty} f(x_n) = \text{Lim}_{n\to\infty} x/n = 0$ for all $x \in \boldsymbol{R}$, the sequence of functions $\{x/n\}$ converges pointwise to the function f defined by $f(x) = 0$ for all $x \in \boldsymbol{R}$.

2. For each $n \in \boldsymbol{N}$, define $f_n: \boldsymbol{R} \to \boldsymbol{R}$ by $f_n(x) = x^n$ for all $x \in \boldsymbol{R}$ (see Figure 7.2). If $x > 1$, $\text{Lim}_{n\to\infty} f_n(x) = \text{Lim}_{n\to\infty} x^n$ does not exist, so the sequence $\{x^n\}$ does not converge pointwise.

3. Pointwise convergence of a sequence of functions depends on the metric space as well as on the functions. To see this, let $f_n: [0, 1] \to \boldsymbol{R}$ be given by $f_n(x) = x^n$ for all $n \in \boldsymbol{N}$ and $x \in [0, 1]$ (see Figure 7.3). These are the same functions as in the previous example, but here

$$\lim_{n\to\infty} f_n(x) = \lim_{n\to\infty} x^n = \begin{cases} 0 & \text{if } x \in [0, 1) \\ 1 & \text{if } x = 1 \end{cases}$$

hence the sequence $\{x^n\}$ converges pointwise to the function $f: [0, 1] \to \boldsymbol{R}$

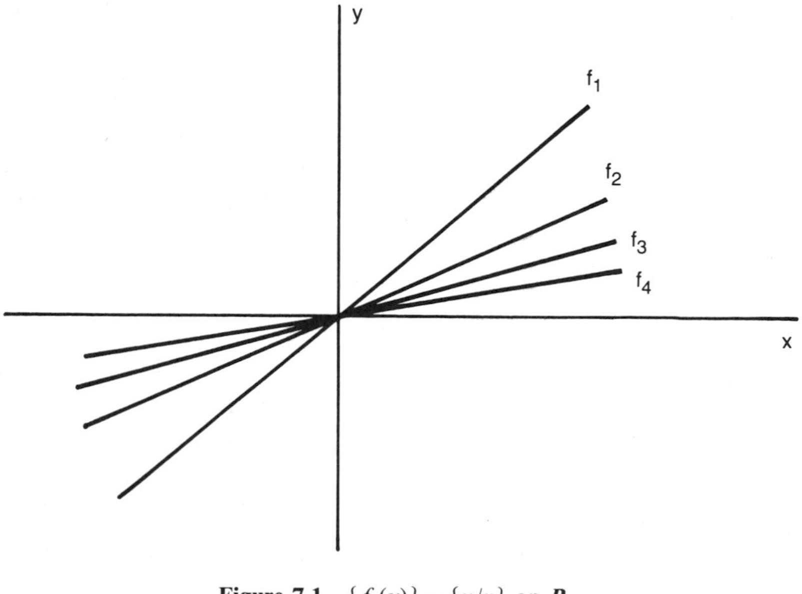

Figure 7.1 $\{f_n(x)\} = \{x/n\}$ on $\boldsymbol{R}$.

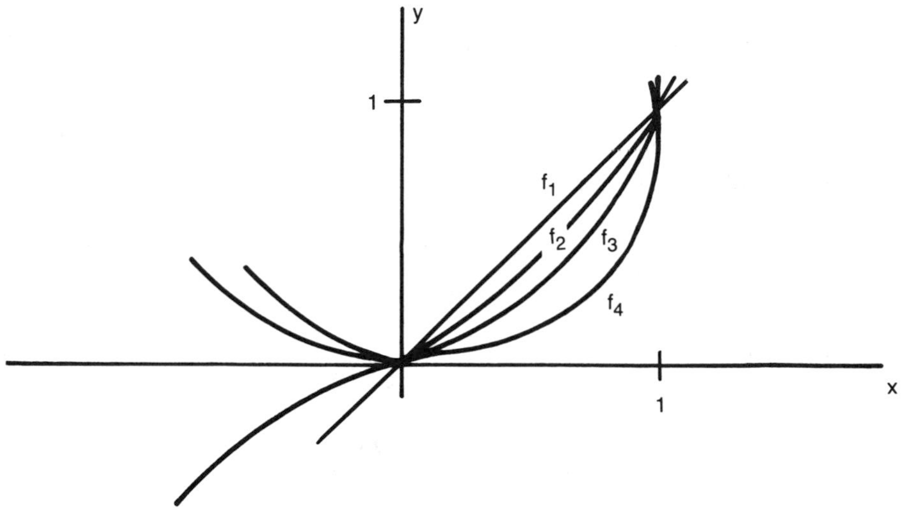

Figure 7.2 $\{f_n(x)\} = \{x^n\}$ on $\boldsymbol{R}$.

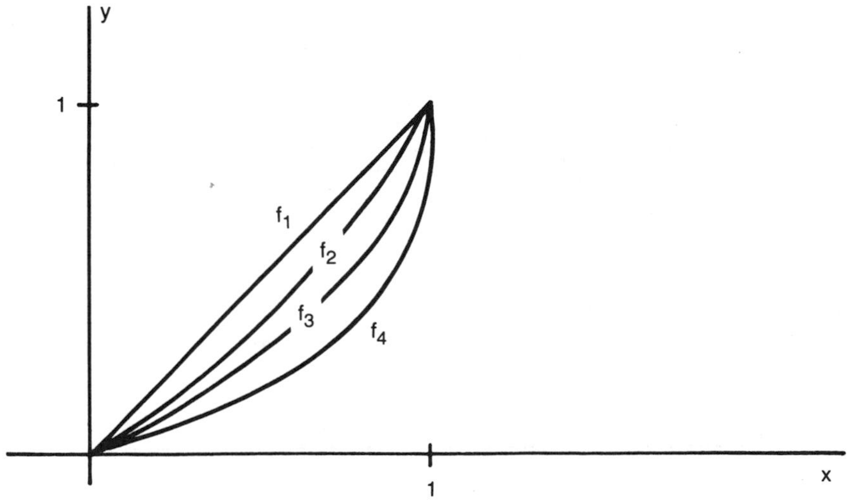

Figure 7.3 $\{f_n(x)\} = \{x^n\}$ on $[0, 1]$.

defined by

$$f(x) = \begin{cases} 0 & \text{if } x \in [0, 1) \\ 1 & \text{if } x = 1 \end{cases}$$

4. Let $f_n: (0, +\infty) \to \boldsymbol{R}$ be defined by $f_n(x) = 1 + x/(1 + nx)$ for all $n \in \boldsymbol{N}$ and all $x > 0$ (see Figure 7.4). Since

$$1 < 1 + \frac{x}{1+nx} < 1 + \frac{x}{nx} = 1 + \frac{1}{n}$$

we have

$$1 \leqslant \operatorname*{Lim}_{n \to \infty} f_n(x) \leqslant \operatorname*{Lim}_{n \to \infty} \left(1 + \frac{1}{n}\right) = 1$$

Therefore the sequence $\{f_n\}$ converges pointwise to the function f defined by $f(x) = 1$ for all $x \in (0, +\infty)$.

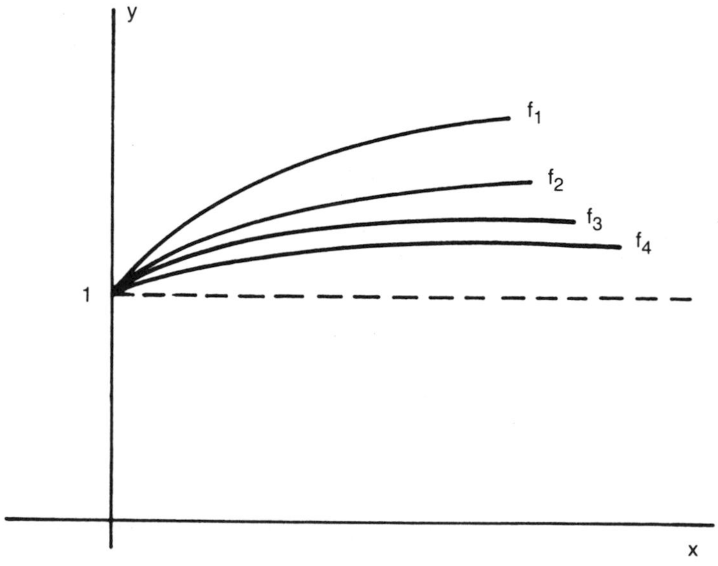

Figure 7.4 $\{f_n(x)\} = \{1 + [x/(1 + nx)]\}$ on $(0, +\infty)$.

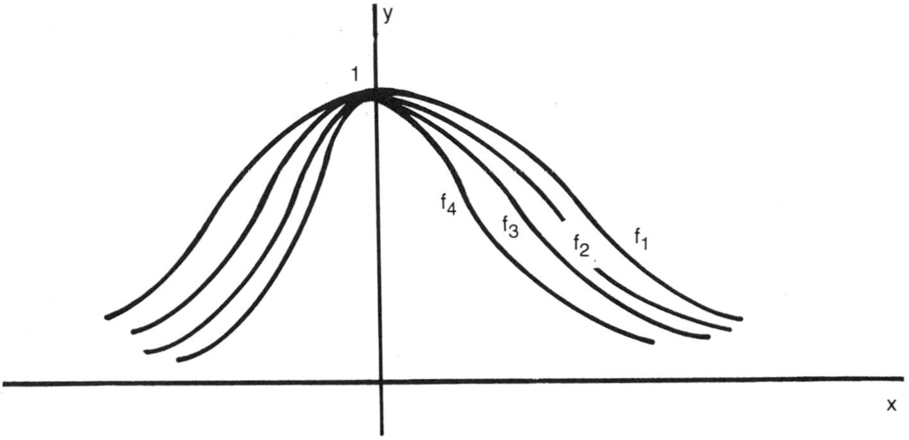

Figure 7.5 $\{f_n(x)\} = \{1/(1+nx^2)\}$.

5. Let f_n: $\boldsymbol{R} \to \boldsymbol{R}$ be defined by $f_n(x) = 1/(1+nx^2)$ for all $n \in \boldsymbol{N}$ and all $x \in \boldsymbol{R}$ (see Figure 7.5). It is easy to check that if $x = 0$, then $\text{Lim}_{n\to\infty} f_n(x) = 1$, while if $x \neq 0$, then $\text{Lim}_{n\to\infty} f_n(x) = 0$. Thus the sequence $\{f_n\}$ converges pointwise to the function f: $\boldsymbol{R} \to \boldsymbol{R}$ given by

$$f(x) = \begin{cases} 0 & \text{if } x \neq 0 \\ 1 & \text{if } x = 0 \end{cases}$$

Our first result in this section characterizes pointwise convergence in terms of the absolute value metric on $\boldsymbol{R}$.

Proposition 7.1.3 Let (M, ρ) be a metric space, f be a function from M to $\boldsymbol{R}$, and f_n: $M \to \boldsymbol{R}$ for all $n \in \boldsymbol{N}$. The sequence of functions $\{f_n\}$ converges pointwise to f if and only if for each $x \in M$ and for each positive real number ε there is some $m \in \boldsymbol{N}$ such that $|f_n(x) - f(x)| < \varepsilon$ whenever $n \geqslant m$.

Proof: The sequence $\{f_n\}$ converges pointwise to f if and only if for each $x \in M$, $\text{Lim}_{n\to\infty} f_n(x) = f(x)$. But $\text{Lim}_{n\to\infty} f_n(x) = f(x)$ if and only if for every positive real number ε there is some $m \in \boldsymbol{N}$ such that $n \geqslant m$ implies $|f_n(x) - f(x)| < \varepsilon$. □

Note that the proposition states that $\{f_n\}$ converges pointwise to f if for *each* $x \in M$ and $\varepsilon \in \boldsymbol{R}$, $\varepsilon > 0$, there is some $m \in \boldsymbol{N}$ such that $n \geqslant m$ implies $|f_n(x) - f(x)| < \varepsilon$; in general, then, m will depend on x as well as on ε. In other words, for a particular $\varepsilon > 0$ it may be necessary to choose different values of m for each $x \in M$; on the other hand, it may be possible to select a single value of m which will serve for all $x \in M$. To illustrate, let us examine two of our previous examples using Proposition 7.1.3.

Examples 7.1.4 1. Let $\{f_n\}$ be the sequence $\{x/n\}$ of Example 1, Examples 7.1.2. If we wish to use Proposition 7.1.3 to show that this sequence converges pointwise to the zero function, then we must show that given any $\varepsilon > 0$ we can find $m \in \boldsymbol{N}$ such that $|x/n - 0| = |x|/n < \varepsilon$ whenever $n \geqslant m$. Clearly, it suffices to choose $m > |x|/\varepsilon$. Note that m depends on both ε and x.

2. Let $\{f_n\}$ be the sequence $\{1 + x/(1 + nx)\}$ of Example 4, Examples 7.1.2. If $x \in M$ and $\varepsilon > 0$, we seek $m \in \boldsymbol{N}$ such that $n \geqslant m$ implies

$$|f_n(x) - f(x)| = \left|1 + \frac{x}{1+nx} - 1\right| = \frac{x}{1+nx} < \varepsilon$$

But since

$$\frac{x}{1+nx} < \frac{1}{n}$$

for all $x \in (0, +\infty)$, if we choose $m \in \boldsymbol{N}$ such that $m > 1/\varepsilon$, then $n \geqslant m$ implies

$$|f_n(x) - f(x)| = \frac{x}{1+nx} < \frac{1}{n} \leqslant \frac{1}{m} < \varepsilon$$

Note that here m does not depend on x but only on ε: for a given $\varepsilon > 0$, the same value of m will work for all $x \in (0, +\infty)$.

Suppose the sequence $\{f_n\}$ converges pointwise to the function f. As we have just seen, the value of m guaranteed by Proposition 7.1.3 may depend on ε alone or on both ε and x. This is reminiscent of the situation encountered in Section 6.2 of Chapter 6, where we discussed continuity in metric spaces. Just as we there defined a stronger form of continuity called uniform continuity, so here we may define a stronger form of convergence known as *uniform convergence*.

Definition 7.1.5 Let (M, ρ) be a metric space, let f be a function from M to $\boldsymbol{R}$, and for each $n \in \boldsymbol{N}$ let $f_n\colon M \to \boldsymbol{R}$. The sequence of functions $\{f_n\}$

converges uniformly to f on M if for every positive real number ε there is some $m \in \boldsymbol{N}$ such that $n \geqslant m$ implies $|f_n(x) - f(x)| < \varepsilon$ for all $x \in M$.

Thus $\{f_n\}$ converges uniformly to f if, given $\varepsilon > 0$, there is some $m \in \boldsymbol{N}$, m depending only on ε, such that $|f_n(x) - f(x)| < \varepsilon$ whenever $n \geqslant m$, for all $x \in M$.

Examples 7.1.6 1. Example 1 of Examples 7.1.2 shows that the sequence $\{x/n\}$ converges pointwise to the function defined by $f(x) = 0$ for all $x \in \boldsymbol{R}$, while Example 1 of Examples 7.1.4 suggests (but does not prove) that this sequence does not converge uniformly to this function. Here we will prove that the convergence is not uniform. To see this, suppose the sequence $\{x/n\}$ does converge uniformly to the zero function on $\boldsymbol{R}$; then there is some $m \in \boldsymbol{N}$, m depending only on $\varepsilon = 1$, such that $n \geqslant m$ implies

$$|f_n(x) - f(x)| = |x|/n < 1$$

for all $x \in \boldsymbol{R}$. But this cannot be true for all $x \in \boldsymbol{R}$, for if $n = m$ and $x = m$, then $|x|/m = 1$.

2. Example 2 of Examples 7.1.4 shows that the sequence $\{1 + x/(1 + nx)\}$ converges uniformly on $(0, +\infty)$ to the function f defined by $f(x) = 1$ for all $x \in (0, +\infty)$.

3. Uniform convergence depends on the metric space as well as on the functions. For instance, let $f_n\colon [0, 1] \to \boldsymbol{R}$ be given by $f_n(x) = x/n$ for all $n \in \boldsymbol{N}$ and all $x \in [0, 1]$. These are the same functions as in Example 1 above, but here the sequence $\{x/n\}$ converges uniformly to the zero function on $[0, 1]$. To see this, suppose $\varepsilon > 0$ is given and choose $m > 1/\varepsilon$. Since $0 \leqslant x \leqslant 1$, $n \geqslant m$ implies that

$$|f_n(x) - f(x)| = \left|\frac{x}{n} - 0\right| = \frac{x}{n} \leqslant \frac{1}{n} \leqslant \frac{1}{m} < \varepsilon$$

for all $x \in [0, 1]$, and hence $\{x/n\}$ converges uniformly on $[0, 1]$.

4. The sequence of functions $\{1/(1 + nx^2)\}$ of Example 5, Examples 7.1.2, does not converge uniformly on $\boldsymbol{R}$ to the function f defined by

$$f(x) = \begin{cases} 0 & \text{if } x \neq 0 \\ 1 & \text{if } x = 0 \end{cases}$$

To see this, suppose the convergence is uniform; then there is some $m \in \boldsymbol{N}$ such that $n \geqslant m$ implies $|f_n(x) - f(x)| < \frac{1}{2}$ for all $x \in \boldsymbol{R}$. But it is easy to check that this is impossible when $n = m$ and $0 \leqslant x < 1/\sqrt{m}$.

5. Let $a \in \boldsymbol{R}$, $a > 0$, and for each $n \in \boldsymbol{N}$ define $f_n(x) = 1/(1 + nx^2)$ for all $x \in [a, +\infty)$. The sequence $\{f_n\}$ converges uniformly to the zero function on $[a, +\infty)$, because if $m \in \boldsymbol{N}$, $m > (1 - \varepsilon)/a^2$, then $n \geqslant m$ implies

$$|f_n(x) - 0| = \frac{1}{1 + nx^2} \leqslant \frac{1}{1 + mx^2} \leqslant \frac{1}{1 + ma^2} < \varepsilon$$

for all $x \in [a, +\infty)$.

It is clear that if a sequence of functions $\{f_n\}$ converges uniformly to f on a metric space, then the sequence also converges pointwise to f; on the other hand, Examples 1 and 4 of Examples 7.1.6 show that pointwise convergence does not imply uniform convergence. Thus uniform convergence is a stronger condition than pointwise convergence.

Application of the definition of uniform convergence requires knowledge of the function f which is the limit of the sequence $\{f_n\}$. It would be convenient to have a characterization of uniformly convergent sequences which does not utilize the limit function f. The next result, known as the Cauchy criterion for uniform convergence, provides such a characterization.

Proposition 7.1.7 (Cauchy Criterion) Let (M, ρ) be a metric space and let $f_n: M \to \boldsymbol{R}$ for all $n \in \boldsymbol{N}$. The sequence $\{f_n\}$ converges uniformly to a function from M to $\boldsymbol{R}$ if and only if for every positive real number ε there is some $m \in \boldsymbol{N}$ such that $n \geqslant m$ and $k \geqslant m$ imply $|f_n(x) - f_k(x)| < \varepsilon$ for all $x \in M$.

Proof: If the sequence of functions $\{f_n\}$ converges uniformly to f on M, then for every positive real number ε there is some $m \in \boldsymbol{N}$ such that $n \geqslant m$ implies $|f_n(x) - f(x)| < \varepsilon/2$ for all $x \in M$. Therefore $n \geqslant m$ and $k \geqslant m$ imply that

$$|f_n(x) - f_k(x)| \leqslant |f_n(x) - f(x)| + |f_k(x) - f(x)| < \frac{\varepsilon}{2} + \frac{\varepsilon}{2} = \varepsilon$$

for all $x \in M$.

Conversely, suppose that $\{f_n\}$ is a sequence of functions from M to $\boldsymbol{R}$ such that for every positive real number ε there is some $m \in \boldsymbol{N}$ such that $n \geqslant m$ and $k \geqslant m$ imply $|f_n(x) - f_k(x)| < \varepsilon$ for all $x \in M$. For fixed $x \in M$, this implies that the sequence of real numbers $\{f_n(x)\}$ is a Cauchy sequence, and hence that $\operatorname{Lim}_{n\to\infty} f_n(x)$ exists. Define $f: M \to \boldsymbol{R}$ by

$$f(x) = \operatorname*{Lim}_{n\to\infty} f_n(x)$$

for all $x \in M$. We claim that $\{f_n\}$ converges uniformly to this function f.

If $\varepsilon \in \boldsymbol{R}$, $\varepsilon > 0$, then there is some $m \in \boldsymbol{N}$ such that $n \geqslant m$ and $k \geqslant m$ imply $|f_n(x) - f_k(x)| < \varepsilon/2$ for all $x \in M$. For fixed k, $k \geqslant m$, and fixed $x \in M$, consider the sequence $\{|f_n(x) - f_k(x)| \mid n \in \boldsymbol{N}\}$: since

$$\operatorname*{Lim}_{n\to\infty} f_n(x) = f(x) \qquad \text{and} \qquad |f_n(x) - f_k(x)| < \frac{\varepsilon}{2} \quad \text{for } n \geqslant m$$

we have

$$\operatorname*{Lim}_{n\to\infty} |f_n(x) - f_k(x)| = |f(x) - f_k(x)| \leqslant \frac{\varepsilon}{2}$$

Thus if $k \geqslant m$, then $|f(x) - f_k(x)| < \varepsilon$ for all $x \in M$, which shows that $\{f_n\}$ converges uniformly to f on M. □

Suppose $\{f_n\}$ is a sequence of functions which converges to f and f_n has property P for each $n \in \boldsymbol{N}$. Will the limit function f necessarily have property P? The answer is often no if the convergence is pointwise, but yes if it is uniform. Thus uniformly convergent sequences of functions are important because properties shared by the functions of the sequence tend to be preserved in the limit, whereas this is frequently not the case for pointwise convergence. We will illustrate by showing that the properties of boundedness and continuity are preserved by uniform convergence, but not by pointwise convergence.

First we must define what is meant by a bounded function from a metric space to $\boldsymbol{R}$.

Definition 7.1.8 If (M, ρ) is a metric space and f is a function from M to $\boldsymbol{R}$, then f is *bounded on M* if $f(M)$ is a bounded subset of $\boldsymbol{R}$.

Now we prove that uniform convergence preserves boundedness.

Proposition 7.1.9 Let (M, ρ) be a metric space and let $\{f_n\}$ be a sequence of bounded functions from M to $\boldsymbol{R}$. If $\{f_n\}$ converges uniformly on M to the function f, then f is bounded on M.

Proof: Since $\{f_n\}$ converges uniformly to f on M, there is some $m \in \boldsymbol{N}$

such that $n \geq m$ implies $|f(x) - f_n(x)| < 1$ for all $x \in M$. Since f_m is bounded on M, there is some $r \in \boldsymbol{R}$ such that $|f_m(x)| < r$ for all $x \in M$. Therefore

$$|f(x)| \leq |f(x) - f_m(x)| + |f_m(x)| < 1 + r$$

for all $x \in M$, and hence f is bounded on M. □

Example 7.1.10 This example shows that the conclusion of Proposition 7.1.9 does not hold if the convergence is pointwise but not uniform.

For each $n \in \boldsymbol{N}$, define f_n: $[0, +\infty) \to \boldsymbol{R}$ as follows:

$$f_n(x) = \begin{cases} 0 & \text{if } x \geq n \\ k & \text{if } k \in \boldsymbol{N} \text{ and } k - 1 \leq x < k \leq n \end{cases}$$

If $x \in [0, +\infty)$, then there is some $m \in \boldsymbol{N}$ such that $m - 1 \leq x < m$, and it follows that $f_n(x) = m$ for all $n \geq m$. Therefore if $m - 1 \leq x < m$, $m \in \boldsymbol{N}$, then $\operatorname{Lim}_{n \to \infty} f_n(x) = m$, and hence $\{f_n\}$ converges pointwise to the function f defined by $f(x) = m$ if $m - 1 \leq x < m$, $m \in \boldsymbol{N}$. Note that each of the functions f_n is bounded on $[0, +\infty)$ because $0 \leq f_n(x) \leq n$ for all $x \in [0, +\infty)$, but that the limit function f is not bounded on $[0, +\infty)$. The reader should check that $\{f_n\}$ does not converge uniformly to f.

We conclude this section by showing that uniform convergence preserves continuity while pointwise convergence need not do so.

Proposition 7.1.11 Let (M, ρ) be a metric space and let $\{f_n\}$ be a sequence of continuous functions from M to $\boldsymbol{R}$. If $\{f_n\}$ converges uniformly on M to the function f, then f is continuous on M.

Proof: Since $\{f_n\}$ converges uniformly to f on M, for every $\varepsilon \in \boldsymbol{R}$, $\varepsilon > 0$, there is some $m \in \boldsymbol{N}$ such that $n \geq m$ implies $|f_n(y) - f(y)| < \varepsilon/3$ for all $y \in M$. If $x \in M$, the continuity of f_m at x implies that there is a real number $\delta > 0$ such that

$$\rho(x, y) < \delta \qquad \text{implies} \qquad |f_m(x) - f_m(y)| < \frac{\varepsilon}{3}$$

Thus if $\rho(x, y) < \delta$, then

$$\begin{aligned} |f(x) - f(y)| &\leq |f(x) - f_m(x)| + |f_m(x) - f_m(y)| + |f_m(y) - f(y)| \\ &< \frac{\varepsilon}{3} + \frac{\varepsilon}{3} + \frac{\varepsilon}{3} = \varepsilon \end{aligned}$$

which proves that the limit function f is continuous at x. □

Example 7.1.12 This example shows that the conclusion of Proposition 7.1.11 does not hold if the convergence is pointwise but not uniform.

The sequence $\{f_n\}$ of Example 5, Examples 7.1.2 and Example 4, Examples 7.1.6 converges pointwise but not uniformly on $\boldsymbol{R}$ to the function f defined by

$$f(x) = \begin{cases} 0 & \text{if } x \neq 0 \\ 1 & \text{if } x = 0 \end{cases}$$

It is clear that f_n is continuous on $\boldsymbol{R}$ for all $n \in \boldsymbol{N}$ but that f is not continuous on $\boldsymbol{R}$.

EXERCISES

1. For each of the following, consider the sequence of functions $\{f_n\}$ defined on the given subset of $\boldsymbol{R}$. If the sequence converges uniformly to a function f, find f and show that the convergence is uniform; if the convergence is pointwise but not uniform, show this; if the sequence does not converge, show this.

   (a) $f_n(x) = \dfrac{x}{2^n}$, on $\boldsymbol{R}$

   (b) $f_n(x) = \dfrac{x}{2^n}$, on $[0, 1]$

   (c) $f_n(x) = nx$, on $[a, b]$

   (d) $f_n(x) = \dfrac{nx}{1 + nx}$, on $(0, +\infty)$

   (e) $f_n(x) = \dfrac{nx}{1 + nx}$, on $[a, +\infty)$, $a > 0$

   (f) $f_n(x) = \dfrac{x^n}{1 + x^n}$, on $[0, 1]$

   (g) $f_n(x) = \dfrac{x^n}{1 + x^n}$, on $[0, b]$, $0 < b < 1$

2. Let $f: \boldsymbol{R} \to \boldsymbol{R}$ be uniformly continuous. For each $n \in \boldsymbol{N}$, define $f_n(x) = f(x + 1/n)$, for all $x \in \boldsymbol{R}$. Prove that the sequence $\{f_n\}$ converges uniformly to f on $\boldsymbol{R}$.
3. Let (M, ρ) be a metric space and for each $n \in \boldsymbol{N}$ let f_n be a function from M to $\boldsymbol{R}$. Prove that if $\{f_n\}$ converges uniformly on a dense subset of M, then it converges uniformly on M.

4. Let χ be the characteristic function of the rationals, defined by

$$\chi(x) = \begin{cases} 1 & \text{if } x \in \boldsymbol{Q} \\ 0 & \text{if } x \notin \boldsymbol{Q} \end{cases}$$

Prove that there is no sequence of continuous functions on $\boldsymbol{R}$ which converges uniformly to χ.

5. Give an example of a sequence of functions $\{f_n\}$ which converges pointwise to a bounded function f but does not converge uniformly to f. This shows that the converse of Proposition 7.1.9 does not hold.

6. Give an example of a sequence of functions $\{f_n\}$ which converges pointwise to a continuous function f but does not converge uniformly to f. This shows that the converse of Proposition 7.1.11 does not hold.

7. Let (M, ρ) be a metric space and for all $n \in \boldsymbol{N}$ let f_n be a function from M to $\boldsymbol{R}$. Show that if $\{f_n\}$ converges uniformly on each of the subsets $S_1, S_2, \ldots, S_k$ of M, then it converges uniformly on $\bigcup_{1 \leqslant i \leqslant k} S_i$. Give an example to show that the conclusion does not hold for denumerable unions.

8. Let (M, ρ) be a metric space and for each $n \in \boldsymbol{N}$ let f_n be a function from M to $\boldsymbol{R}$. The sequence $\{f_n\}$ is *uniformly bounded on M* if there is some $r \in \boldsymbol{R}$ such that $|f_n(x)| \leqslant r$ for all $n \in \boldsymbol{N}$ and all $x \in M$.
   (a) Prove that if $\{f_n\}$ is uniformly bounded on M, then f_n is bounded on M for all $n \in \boldsymbol{N}$.
   (b) Give an example of a metric space (M, ρ) and a sequence $\{f_n\}$ which is not uniformly bounded on M even though f_n is bounded on M for all $n \in \boldsymbol{N}$.

9. Let (M, ρ) be a metric space and for each $n \in \boldsymbol{N}$ let f_n be a function from M to $\boldsymbol{R}$. The sequence $\{f_n\}$ is *equicontinuous on M* if for every positive real number ε there is a positive real number δ such that if $x \in M$ and $y \in M$ with $\rho(x, y) < \delta$, then $|f_n(x) - f_n(y)| < \varepsilon$ for all $n \in \boldsymbol{N}$.
   (a) Prove that if $\{f_n\}$ is equicontinuous on M, then f_n is a continuous function from M to $\boldsymbol{R}$ for all $n \in \boldsymbol{N}$.
   (b) For each $n \in \boldsymbol{N}$, define $f_n\colon [0, 1] \to \boldsymbol{R}$ by $f_n(x) = x/n$ for all $x \in [0, 1]$. Show that $\{f_n\}$ is equicontinuous on $[0, 1]$.
   (c) For each $n \in \boldsymbol{N}$, define $f_n\colon [0, 1] \to \boldsymbol{R}$ by $f_n(x) = 1/(1 + nx^2)$ for all $x \in [0, 1]$. Show that the sequence of continuous functions $\{f_n\}$ is uniformly bounded on $[0, 1]$ but not equicontinuous there.

10. Let (M, ρ) be a metric space and let $\{f_n\}$ be a sequence of continuous function from M to $\boldsymbol{R}$ which converges uniformly on M. Prove that if M is compact, then
    (a) $\{f_n\}$ is uniformly bounded on M;
    (b) $\{f_n\}$ is equicontinuous on M.
11. We utilize this exercise to prove the Arzela–Ascoli Theorem, which is a partial converse to the result of Exercise 10. The reader is asked to fill in the details.

    Arzela–Ascoli Theorem. Let (M, ρ) be a compact metric space. If $\{f_n\}$ is a sequence of functions from M to $\boldsymbol{R}$ which is uniformly bounded and equicontinuous on M, then some subsequence of $\{f_n\}$ converges uniformly on M.

    (a) Since (M, ρ) is compact, it contains a countable dense subset S. (See Exercise 15 of Section 6.1, Chapter 6.)
    (b) Let $S = \{x_1, x_2, \ldots\}$. Since $\{f_n\}$ is uniformly bounded, the sequence $\{f_n(x_1)\}$ is bounded in $\boldsymbol{R}$ and hence has a convergent subsequence

    $$\{f_{1m}(x_1)\} = \{f_{11}(x_1), f_{12}(x_1), \ldots\}$$

    (See Exercise 10, Section 5.2, Chapter 5.) Similarly, the sequence $\{f_{1m}(x_2)\}$ is bounded and therefore has a convergent subsequence

    $$\{f_{2m}(x_2)\} = \{f_{21}(x_2), f_{22}(x_2), \ldots\}$$

    Thus we construct a family of subsequences of $\{f_n\}$, namely

    $$\{f_{1m}\}, \{f_{2m}\}, \ldots, \{f_{nm}\}, \ldots$$

    each of which is a subsequence of the preceding one, and such that $\{f_{nm}(x_n)\}$ converges in $\boldsymbol{R}$, for all $n \in N$.
    (c) Consider the subsequence $\{f_{mm}\}$ of $\{f_n\}$. If $x_n \in S$, then for $m > n$ we have $\{f_{mm}(x_n)\}$ a subsequence of the convergent sequence $\{f_{nm}(x_n)\}$, and therefore $\{f_{mm}(x_n)\}$ converges for $m > n$. But this implies that the subsequence $\{f_{mm}\}$ converges at every point of S.
    (d) We now use the Cauchy Criterion for uniform convergence, the equicontinuity of $\{f_n\}$, and the fact that S is dense in M to show that $\{f_{mm}\}$ converges uniformly on S. But Exercise 3 above then shows that $\{f_{mm}\}$ converges uniformly on M.

7.2 FUNCTION SPACES AND UNIFORM APPROXIMATION†

In this section we introduce several important metric spaces whose elements are functions. We also consider the approximation of a given function f by a sequence of functions which converges uniformly to f; our major result in this regard will be the famous Weierstrass Approximation Theorem.

It is possible to make the set of all bounded functions from a metric space to the reals into a metric space. In order to do this, we need the concept of the supremum norm of a bounded real-valued function.

Definition 7.2.1 Let (M, ρ) be a metric space and let

$$B(M) = \{f\colon M \to \boldsymbol{R} \mid f \text{ is bounded on } M\}$$

If $f \in B(M)$, define the *supremum norm of* f, denoted by $\|f\|$, as follows:

$$\|f\| = \sup \{|f(x)| \mid x \in M\}$$

Proposition 7.2.2 Let (M, ρ) be a metric space. If $f \in B(M)$ and $g \in B(M)$, then

1. $\|f\| > 0$;
2. $\|f\| = 0$ if and only if f is the zero function from M to $\boldsymbol{R}$;
3. $\|rf\| = |r| \, \|f\|$ for all $r \in \boldsymbol{R}$;
4. $\|f+g\| \leqslant \|f\| + \|g\|$.

The proof of Proposition 7.2.2 will be left to the reader. (See Exercise 1 at the end of this section.)

Now suppose that for all $f \in B(M)$ and $g \in B(M)$, we define

$$\rho^*(f, g) = \|f - g\|$$

then ρ^* will be a metric on $B(M)$. In fact, not only is $(B(M), \rho^*)$ a metric space, it is a complete metric space.

Proposition 7.2.3 Let (M, ρ) be a metric space. If $\rho^*\colon B(M) \times B(M) \to \boldsymbol{R}$ is defined by $\rho^*(f, g) = \|f - g\|$ for all $f \in B(M)$ and $g \in B(M)$, then $(B(M), \rho^*)$ is a complete metric space.

---

†This section is optional.

Proof: As indicated above, the fact that $(B(M), \rho^*)$ is a metric space follows from Proposition 7.2.2. We leave the details to the reader. (See Exercise 2 at the end of this section.) To prove that $(B(M), \rho^*)$ is complete, we must show that every Cauchy sequence in $(B(M), \rho^*)$ converges to an element of $B(M)$.

If $\{f_n\}$ is a Cauchy sequence in $B(M)$, then for every positive real number ε there is some $m \in \boldsymbol{N}$ such that $n \geqslant m$ and $k \geqslant m$ imply

$$\rho^*(f_n, f_k) = \|f_n - f_k\| < \varepsilon$$

But

$$\|f_n - f_k\| = \sup \{|f_n(x) - f_k(x)| \mid x \in M\}$$

and thus $\|f_n - f_k\| < \varepsilon$ surely implies that $|f_n(x) - f_k(x)| < \varepsilon$ for all $x \in M$. Therefore if $\{f_n\}$ is a Cauchy sequence in $(B(M), \rho^*)$, then for every $\varepsilon \in \boldsymbol{R}$, $\varepsilon > 0$, there is some $m \in \boldsymbol{N}$ such that $n \geqslant m$ and $k \geqslant m$ imply $|f_n(x) - f_k(x)| < \varepsilon$ for all $x \in M$. But this is just the Cauchy Criterion for uniform convergence, and hence we conclude that if $\{f_n\}$ is a Cauchy sequence in $(B(M), \rho^*)$, then it converges uniformly to some function f from M to $\boldsymbol{R}$. Since f_n is bounded on M for all $n \in \boldsymbol{N}$, so is f, by Proposition 7.1.9. Therefore $f \in B(M)$, and we are done. □

Corollary 7.2.4 A sequence $\{f_n\}$ in the metric space $B(M)$ converges if and only if it converges uniformly as a sequence of functions from M to $\boldsymbol{R}$.

The proof of the corollary is left as an exercise. (See Exercise 3 at the end of this section.)

Since uniform convergence preserves continuity in the limit, it follows from Corollary 7.2.4 that the set consisting of all continuous functions in $B(M)$ is also a complete metric space under ρ^*.

Proposition 7.2.5 Let (M, ρ) be a metric space. If $C(M)$ denotes the set of all continuous bounded functions from M to $\boldsymbol{R}$, then $(C(M), \rho^*)$ is a complete metric space.

The proof of the proposition is left to the reader. (See Exercise 4 at the end of this section.)

The metric spaces $(B(M), \rho^*)$ and $(C(M), \rho^*)$ are referred to as *function spaces*, and the metric ρ^* is the *supremum metric*. Study of these spaces yields many interesting results concerning real-valued functions on metric spaces. We will not pursue this topic any further, but merely remark that function spaces are of considerable importance in higher analysis.

Now we turn our attention to the uniform approximation of functions. Suppose $\{f_n\}$ is a sequence of functions which converges uniformly to f on M. Then for any $\varepsilon \in \boldsymbol{R}$, $\varepsilon > 0$, there is some $m \in \boldsymbol{N}$ such that $n \geqslant m$ implies $|f(x) - f_n(x)| < \varepsilon$ for all $x \in M$. Therefore by making ε small enough we can find a function f_m which is arbitrarily close to f on M. For this reason we say that the limit function f is *uniformly approximated* by the functions $\{f_n\}$. There are many approximation theorems of the following form: let f be a function having property P; then f can be uniformly approximated by functions having property P'. This is the same as saying that if f has property P, then there is a sequence of functions having property P' which converges uniformly to f. We will prove two important approximation theorems of this type for continuous functions on $\boldsymbol{R}$.

Definition 7.2.6 Let S be a subset of $\boldsymbol{R}$ and let f be a function from S to $\boldsymbol{R}$. We say that f is a *step function on* S if S is a finite union of mutually disjoint intervals or rays and f is constant on each of the intervals or rays.

Example 7.2.7 Let $f\colon \boldsymbol{R} \to \boldsymbol{R}$ be defined by

$$f(x) = \begin{cases} 2 & \text{if } x \leqslant -1 \\ -3 & \text{if } -1 < x < \frac{1}{2} \\ 1 & \text{if } \frac{1}{2} \leqslant x \leqslant 1 \\ 0 & \text{if } x > 1 \end{cases}$$

Then f is a step function on $\boldsymbol{R}$. See Figure 7.6.

Our first approximation theorem says that a function which is continuous on a closed interval can be uniformly approximated there by step functions.

Proposition 7.2.8 If $f\colon [a, b] \to \boldsymbol{R}$ is continuous, then f can be uniformly approximated on $[a, b]$ by step functions.

Proof: We must produce a sequence $\{f_n\}$ of step functions which converges uniformly to f on $[a, b]$.

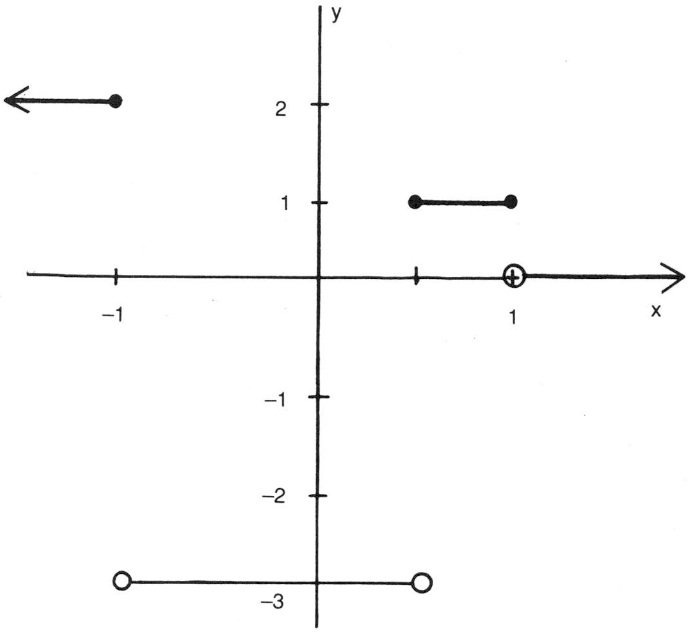

Figure 7.6 A step function.

For each $n \in \boldsymbol{N}$, set

$$I_{n1} = \left[a, a + \frac{b-a}{n}\right)$$

$$I_{n2} = \left[a + \frac{b-a}{n}, a + \frac{2(b-a)}{n}\right)$$

$$\vdots$$

$$I_{nn} = \left[a + \frac{(n-1)(b-a)}{n}, b\right]$$

Then for each $n \in \boldsymbol{N}$, $[a, b] = \bigcup_{1 \leqslant k \leqslant n} I_{nk}$ and the intervals I_{nk} are mutually disjoint. For each k, $1 \leqslant k \leqslant n$, let x_{nk} denote the midpoint of the interval I_{nk}. For each $n \in \boldsymbol{N}$, define f_n: $[a, b] \to \boldsymbol{R}$ by

$$f_n(x) = f(x_{nk}) \qquad \text{if } x \in I_{nk},\ 1 \leqslant k \leqslant n$$

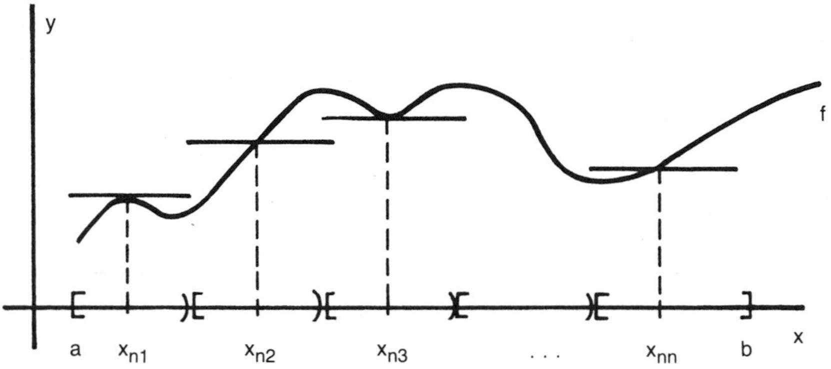

Figure 7.7 $f_n(x) = f(x_{nk})$ on I_{nk}.

See Figure 7.7. Clearly each f_n is a step function on $[a, b]$. We will show that the sequence $\{f_n\}$ converges uniformly to f on $[a, b]$.

Suppose ε is a positive real number. Since f is continuous on the compact set $[a, b]$ it is uniformly continuous there, and hence there exists a positive real number δ such that whenever x and y are in $[a, b]$ with

$$|x - y| < \delta$$

then

$$|f(x) - f(y)| < \varepsilon$$

Choose $m \in \boldsymbol{N}$ such that $(b - a)/m < \delta$. For $n \geqslant m$, $x \in I_{nk}$ and $y \in I_{nk}$ imply

$$|x - y| < \frac{b - a}{n} \leqslant \frac{b - a}{m} < \delta$$

In particular, this is so if $y = x_{nk}$, the midpoint of I_{nk}. Therefore if $n \geqslant m$, then

$$|f(x) - f_n(x)| = |f(x) - f(x_{nk})| < \varepsilon$$

for all $x \in [a, b]$, and we are done. □

Our final result of this chapter is the important Weierstrass Approximation Theorem, which says that a function continuous on a closed interval can be uniformly approximated there by polynomials. Before we can prove this theorem, we need several definitions and a proposition.

Recall the $n!$ (n factorial) is defined by

$$0! = 1 \qquad n! = n(n-1)(n-2)\cdots 1 \qquad \text{for all } n \in \boldsymbol{N}$$

Also, the binomial coefficient

$$\binom{n}{k} = \frac{n!}{k!(n-k)!}$$

for $0 \leqslant k \leqslant n$, and the Binomial Theorem states that

$$(x+y)^n = \sum_{k=0}^{n} \binom{n}{k} x^k y^{n-k}$$

for all $n \in \boldsymbol{N}$.

We will need the following inequality in the proof of the Weierstrass Approximation Theorem.

Proposition 7.2.9 For all $x \in [0, 1]$ and all $n \in \boldsymbol{N}$,

$$\sum_{k=0}^{n} \binom{n}{k} x^k (1-x)^{n-k} \left(x - \frac{k}{n}\right)^2 \leqslant \frac{1}{4n}$$

The proof is left to the reader. (See Exercise 9 at the end of this section.)

Next we define the polynomial functions which we will use to approximate a given continuous function.

Definition 7.2.10 Let f: $[0, 1] \to \boldsymbol{R}$. For all $n \in \boldsymbol{N}$, the polynomial

$$B_n = \sum_{k=0}^{n} \binom{n}{k} x^k (1-x)^{n-k} f\left(\frac{k}{n}\right)$$

is the nth *Bernstein polynomial for* f.

Example 7.2.11 The nth Bernstein polynomial for f is calculated from the values of f at the rational points $0, 1/n, 2/n, \ldots, 1$. Let us find the first two Bernstein polynomials for $f(x) = x^2$:

$$n = 1: \quad B_n = \binom{1}{0} x^0(1-x)^1 f(0) + \binom{1}{1} x^1(1-x)^0 f(1)$$
$$= (1-x)0 + x1 = x$$
$$n = 2: \quad B_n = \binom{2}{0} x^0(1-x)^2 f(0) + \binom{2}{1} x^1(1-x)^1 f\left(\frac{1}{2}\right)$$
$$+ \binom{2}{2} x^2(1-x)^0 f(1)$$
$$= (1-x)^2 0 + 2x(1-x)\tfrac{1}{4} + x^2 1 = \tfrac{1}{2}x + \tfrac{1}{2}x^2$$

In fact, it is easy to check that

$$B_n = \frac{1}{n} x + \left(\frac{n-1}{n}\right) x^2$$

for all $n \in \mathbf{N}$. Note that the sequence of Bernstein polynomials converges uniformly to $f(x) = x^2$ on $[0, 1]$.

Now we are ready to prove the Weierstrass Approximation Theorem.

Theorem 7.2.12 (Weierstrass Approximation Theorem) If $f: [a, b] \to \mathbf{R}$ is continuous, then f can be uniformly approximated on $[a, b]$ by polynomials.

Proof: We first prove the theorem for the case where $[a, b] = [0, 1]$.

Let $f: [0, 1] \to \mathbf{R}$ be continuous. We will show that the sequence of Bernstein polynomials for f converges uniformly to f on $[0, 1]$. Since

$$|f(x) - B_n(x)| \leqslant \sum_{k=0}^{n} \binom{n}{k} x^k(1-x)^{n-k} \left| f(x) - f\left(\frac{k}{n}\right) \right|$$

it will suffice to show that if ε is a positive real number, then there is some $m \in \mathbf{N}$ such that $n \geqslant m$ implies

$$\sum_{k=0}^{n} \binom{n}{k} x^k(1-x)^{n-k} \left| f(x) - f\left(\frac{k}{n}\right) \right| < \varepsilon$$

for all $x \in [0, 1]$.

Since f is continuous on the compact set $[0, 1]$, it is both uniformly continuous and bounded there. Hence for every $\varepsilon \in \boldsymbol{R}$, $\varepsilon > 0$, there is some $\delta \in \boldsymbol{R}$, $\delta > 0$, such that $|x - y| < \delta$ implies $|f(x) - f(y)| < \varepsilon/2$, and also there is some $r \in \boldsymbol{R}$ such that $|f(x)| < r$ for all $x \in [0, 1]$. Choose $n \in \boldsymbol{N}$ such that $1/n < \delta$, and write

$$\begin{aligned}\sum_{k=0}^{n} \binom{n}{k} x^k(1-x)^{n-k}\left|f(x) - f\left(\frac{k}{n}\right)\right| &= \sum_{p} \binom{n}{p} x^p(1-x)^{n-p}\left|f(x) - f\left(\frac{p}{n}\right)\right| \\ &\quad + \sum_{q} \binom{n}{q} x^q(1-x)^{n-q}\left|f(x) - f\left(\frac{q}{n}\right)\right|\end{aligned}$$

where p ranges over all values of k such that $|x - (k/n)| < \delta$, and q ranges over all values of k such that $|x - (k/n)| \geqslant \delta$. Because $|x - (p/n)| < \delta$ for each p, we have $|f(x) - f(p/n)| < \varepsilon/2$ for each p, and thus

$$\begin{aligned}\sum_{p} \binom{n}{p} x^p(1-x)^{n-p}\left|f(x) - f\left(\frac{p}{n}\right)\right| &< \frac{\varepsilon}{2}\sum_{p} \binom{n}{p} x^p(1-x)^{n-p} \\ &< \frac{\varepsilon}{2}\sum_{k=0}^{n} \binom{n}{k} x^k(1-x)^{n-k}\end{aligned}$$

But by the Binomial Theorem,

$$\sum_{k=0}^{n} \binom{n}{k} x^k(1-x)^{n-k} = (x + (1-x))^n = 1$$

Therefore we have shown that if $1/n < \delta$, then

$$\sum_{p} \binom{n}{p} x^p(1-x)^{n-p}\left|f(x) - f\left(\frac{p}{n}\right)\right| < \frac{\varepsilon}{2}$$

for all $x \in [0, 1]$.

Now consider

$$\sum_{q} \binom{n}{q} x^q(1-x)^{n-q}\left|f(x) - f\left(\frac{q}{n}\right)\right|$$

Since $|f(x)| < r$ for all $x \in [0, 1]$, we have $|f(x) - f(q/n)| < 2r$, and hence

$$\sum_{q} \binom{n}{q} x^q(1-x)^{n-q}\left|f(x) - f\left(\frac{q}{n}\right)\right| < 2r \sum_{q} \binom{n}{q} x^q(1-x)^{n-q} \frac{(x - q/n)^2}{(x - q/n)^2}$$

But $|x - q/n| \geqslant \delta$, so $(x - q/n)^2 \geqslant \delta^2$, and thus

$$\sum_q \binom{n}{q} x^q(1-x)^{n-q}\left|f(x) - f\left(\frac{q}{n}\right)\right| < \left(\frac{2r}{\delta^2}\right)\sum_q \binom{n}{q} x^q(1-x)^{n-q}\left(x - \frac{q}{n}\right)^2$$

But by the inequality of Proposition 7.2.9,

$$\left(\frac{2r}{\delta^2}\right)\sum_q \binom{n}{q} x^q(1-x)^{n-q}\left(x - \frac{q}{n}\right)^2 \leqslant \left(\frac{2r}{\delta^2}\right)\frac{1}{4n} = \frac{r}{2n\delta^2}$$

and $r/2n\delta^2 < \varepsilon/2$ if $1/n < \delta^2\varepsilon/r$. Thus if we choose $m \in \boldsymbol{N}$ such that $1/m < \min\{\delta, \delta^2\varepsilon/r\}$, what we have done shows that $n \in \boldsymbol{N}$, $n \geqslant m$ implies

$$\sum_{k=0}^{n} \binom{n}{k} x^k(1-x)^{n-k}\left|f(x) - f\left(\frac{k}{n}\right)\right| < \frac{\varepsilon}{2} + \frac{\varepsilon}{2} = \varepsilon$$

Therefore the theorem is proved for the case where $[a, b] = [0, 1]$.

Now suppose $f\colon [a, b] \to \boldsymbol{R}$ is continuous. Define

$$h(x) = f(a + (b - a)x)$$

for all $x \in [0, 1]$. Then h is a continuous function on $[0, 1]$, since it is the composite of the continuous function $g(x) = a + (b - a)x$ with the continuous function f. By what we have already shown, for every positive real number ε there is some polynomial $P(x)$ such that $|h(x) - P(x)| < \varepsilon$ for all $x \in [0, 1]$. But then

$$P\left(\frac{x - a}{b - a}\right)$$

is a polynomial and

$$\left|f(x) - P\left(\frac{x - a}{b - a}\right)\right| < \varepsilon$$

for all $x \in [a, b]$, so we are done. □

EXERCISES

1. Prove Proposition 7.2.2.
2. Prove that, as claimed in Proposition 7.2.3, $(B(M), \rho^*)$ is a metric space.
3. Prove Corollary 7.2.4.

4. Prove Proposition 7.2.5.
5. Let (M, ρ) be a metric space and let a, b be real numbers with $a < b$. Consider the subset $C(a, b)$ of $C(M)$ defined by

$$C(a, b) = \{f \in C(M) \mid a \leqslant f(x) \leqslant b \text{ for all } x \in M\}$$

Prove that $(C(a, b), \rho^*)$ is a complete metric space. (Hint: use the result of Exercise 3 of Section 6.3.)
6. Find a sequence of step functions which converges uniformly to f on $[0, 1]$ if
(a) $f(x) = x$ for all $x \in [0, 1]$;
(b) $f(x) = x^2$ for all $x \in [0, 1]$.
7. A function g: $[a, b] \to \boldsymbol{R}$ is *piecewise linear* on $[a, b]$ if there exist points $t_0, t_1, \ldots, t_n$ and real numbers $r_0, r_1, \ldots, r_n, s_0, s_1, \ldots, s_n$ such that

$$a = t_0 < t_1 < \cdots < t_n = b$$

and

$$g(x) = r_k x + s_k, \qquad \text{for all } x \in [t_k, t_{k+1}], 0 \leqslant k \leqslant n - 1$$

Prove that if f: $[a, b] \to \boldsymbol{R}$ is continuous, then f can be uniformly approximated on $[a, b]$ by piecewise linear functions.
8. Find a sequence of polynomials which converges uniformly to f on the indicated interval:
(a) $f(x) = |x|$ for all $x \in [-1, 1]$;
(b) $f(x) = 1/x$ for all $x \in [1, 2]$.
9. Prove Proposition 7.2.9. (Hint: write

$$\sum_{k=0}^{n} \binom{n}{k} x^k (1-x)^{n-k} \left(x - \frac{k}{n}\right)^2$$
$$= \sum_{k=0}^{n} x^2 \binom{n}{k} x^k (1-x)^{n-k} + \sum_{k=0}^{n} -2x \frac{k}{n} \binom{n}{k} x^k (1-x)^{n-k}$$
$$+ \sum_{k=0}^{n} \left(\frac{k^2}{n^2}\right) \binom{n}{k} x^k (1-x)^{n-k}$$

and show that the three terms on the right-hand side reduce to x^2, $-2x^2$, and $x/n + [(n-1)x^2]/n$, respectively.)

8
Calculus

In this chapter we will study the basic concepts of calculus, the definite integral and the derivative. We will not be concerned with techniques of integration and differentiation, but rather with the theory of integration and differentiation. We begin with a section devoted to the definition of the Riemann integral and integrable functions, proceed to consider the properties of the Riemann integral, and conclude with an examination of the derivative and its properties.

8.1 THE RIEMANN INTEGRAL

In this section we will define the integral of calculus, which is known as the Riemann integral, and discuss briefly the question of the Riemann integrability of certain functions. Before we can define the integral, we need some preliminary results about partitions of intervals and Riemann sums.

Definition 8.1.1 Let $[a, b]$ be a closed interval in $\boldsymbol{R}$. A *partition* of $[a, b]$ is a finite set of real numbers

$$P = \{x_0, x_1, \ldots, x_n\}$$

such that

$$a = x_0 < x_1 < \cdots < x_n = b$$

The closed intervals

$$[x_{k-1}, x_k] \qquad 1 \leqslant k \leqslant n$$

are called *subintervals* of the partition P. If P and P' are both partitions of $[a, b]$ and P is a proper subset of P', we say that P' is a *refinement* of P.

Example 8.1.2 Let $[a, b] = [0, 1]$ and let

$$P = \left\{0, \frac{1}{4}, \frac{1}{2}, 1\right\}$$

$$P' = \left\{0, \frac{1}{4}, \frac{1}{2}, \frac{3}{4}, 1\right\}$$

and

$$P'' = \left\{0, \frac{1}{4}, \frac{1}{2}, \frac{1}{\sqrt{2}}, 1\right\}$$

The sets P, P', and P'' are all partitions of $[0, 1]$. The partition P' is a refinement of P and P'' is also a refinement of P; however, P' is not a refinement of P'', nor is P'' a refinement of P'.

Now we define the upper and lower Riemann sums of a bounded function over a partition.

Definition 8.1.3 Let a, b be real numbers, with $a < b$, and let

$$P = \{a = x_0, x_1, \ldots, x_n = b\}$$

be a partition of $[a, b]$. If $f\colon [a, b] \to \boldsymbol{R}$ is bounded, define the *upper Riemann sum $U(f, P)$ of f over P* as follows:

$$U(f, P) = \sum_{k=1}^{n} \sup \{f(x) \mid x \in [x_{k-1}, x_k]\}(x_k - x_{k-1})$$

Similarly, define the *lower Riemann sum $L(f, P)$ of f over P* as follows:

$$L(f, P) = \sum_{k=1}^{n} \inf \{f(x) \mid x \in [x_{k-1}, x_k]\}(x_k - x_{k-1})$$

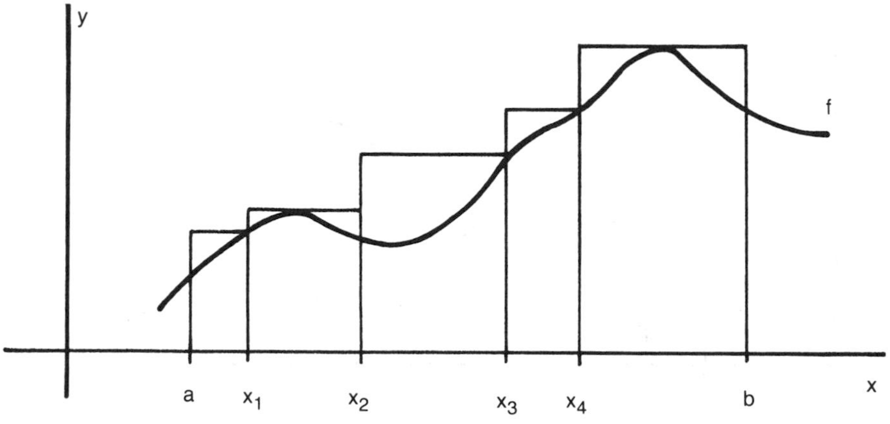

Figure 8.1 An upper Riemann sum for f.

Note that since f is bounded on $[a, b]$ both $U(f, P)$ and $L(f, P)$ exist as finite real numbers for any partition P of $[a, b]$. Furthermore, it is clear from the definition that $L(f, P) \leqslant U(f, P)$ for any partition P of $[a, b]$.

If f is a continuous function on $[a, b]$, then $U(f, P)$ is the sum of the areas of the rectangles whose bases are the subintervals $[x_{k-1}, x_k]$ and whose heights are equal to the maximum values which f attains on the subintervals (see Figure 8.1). Similarly, if f is continuous on $[a, b]$, then $L(f, P)$ is the sum of the areas of the rectangles whose bases are the subintervals $[x_{k-1}, x_k]$ and whose heights are equal to the minimum values which f attains on the subintervals (see Figure 8.2). Notice, however, that Definition 8.1.3 does not require that f be continuous on $[a, b]$, but only that it be bounded there.

Examples 8.1.4 1. Let $c \in \boldsymbol{R}$ and let f be the constant function defined by $f(x) = c$ for all $x \in [a, b]$. If $P = \{a = x_0, x_1, \ldots, x_n = b\}$ is any partition of $[a, b]$, then

$$U(f, P) = \sum_{k=1}^{n} c(x_k - x_{k-1}) = c(x_n - x_0) = c(b - a)$$

and

$$L(f, P) = \sum_{k=1}^{n} c(x_k - x_{k-1}) = c(x_n - x_0) = c(b - a)$$

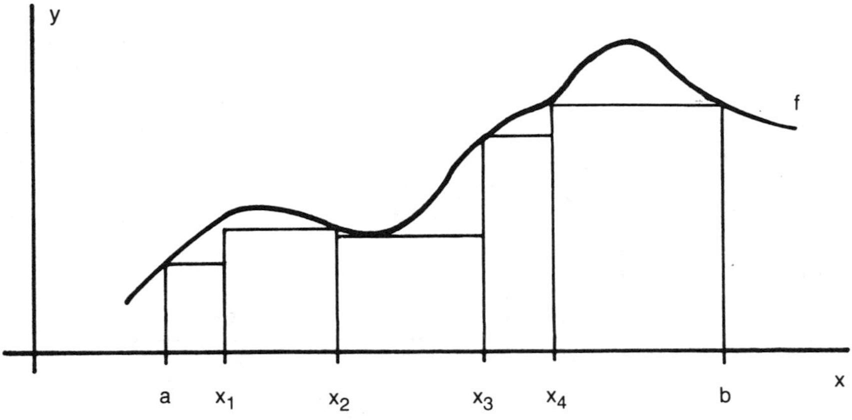

L(f,P) is the sum of the areas of the rectangles

Figure 8.2 A lower Riemann sum for f.

2. Let $f\colon [a, b] \to \boldsymbol{R}$ be given by $f(x) = x$ for all $x \in [a, b]$. For each $n \in \boldsymbol{N}$ let P_n be the partition

$$P_n = \left\{a, a + \frac{b-a}{n}, a + \frac{2(b-a)}{n}, \ldots, a + \frac{(n-1)(b-a)}{n}, b\right\}$$

Then

$$U(f, P_n) = \sum_{k=1}^{n} \left(a + \frac{k(b-a)}{n}\right)\left(\frac{b-a}{n}\right) = \frac{b-a}{n} \sum_{k=1}^{n} a + \frac{(b-a)^2}{n^2} \sum_{k=1}^{n} k$$

$$= \frac{b-a}{n} na + \frac{(b-a)^2}{n^2} \sum_{k=1}^{n} k$$

Using the well-known identity

$$1 + 2 + \cdots + n = \frac{n(n+1)}{2}$$

for the sum of the first n natural numbers, we see that

$$U(f, P_n) = a(b-a) + \frac{(b-a)^2}{n^2} \frac{n(n+1)}{2}$$

$$= \frac{1}{2}(b^2 - a^2) + \frac{1}{2n}(b-a)^2$$

Similarly,

$$L(f, P_n) = \frac{1}{2}(b^2 - a^2) - \frac{1}{2n}(b - a)^2$$

3. Let f be defined on $[0, 1]$ by

$$f(x) = \begin{cases} 0 & \text{if } 0 \leqslant x < \frac{1}{2} \\ 1 & \text{if } \frac{1}{2} \leqslant x \leqslant 1 \end{cases}$$

If $P = \{0 = x_0, x_1, \ldots, x_n = 1\}$ is any partition of $[0, 1]$, there is some $i \in \boldsymbol{N}$, $1 \leqslant i \leqslant n$, such that $x_{i-1} \leqslant \frac{1}{2} < x_i$. Thus

$$\begin{aligned} U(f, P) &= \sum_{k=1}^{i-1} 0(x_k - x_{k-1}) + \sum_{k=i}^{n} 1(x_k - x_{k-1}) \\ &= x_n - x_{i-1} = 1 - x_{i-1} \end{aligned}$$

Similarly, there is some $j \in \boldsymbol{N}$, $1 \leqslant j \leqslant n$, such that $x_{j-1} < \frac{1}{2} \leqslant x_j$, and thus

$$\begin{aligned} L(f, P) &= \sum_{k=1}^{j} 0(x_k - x_{k-1}) + \sum_{k=j+1}^{n} 1(x_k - x_{k-1}) \\ &= x_n - x_j = 1 - x_j \end{aligned}$$

4. Let f be the function defined on $[0, 1]$ by

$$f(x) = \begin{cases} 0 & \text{if } x \neq \dfrac{1}{n} \text{ for any } n \in \boldsymbol{N} \\ \dfrac{1}{n} & \text{if } x = \dfrac{1}{n} \text{ for some } n \in \boldsymbol{N} \end{cases}$$

For any partition P of $[0, 1]$ we will have $L(f, P) = 0$, because every subinterval of P must contain a point x such that $f(x) = 0$. Now let us consider upper Riemann sums. We claim that for every positive real number ε we can find a partition P_ε of $[0, 1]$ such that $0 < U(f, P_\varepsilon) < \varepsilon$.

We may assume that $\varepsilon < 1$, so that there is some $m \in \boldsymbol{N}$ such that

$$0 < \frac{1}{m+1} < \frac{\varepsilon}{2} < \frac{1}{m}$$

It is possible to choose $\delta \in \boldsymbol{R}$, $\delta > 0$, such that $\delta < \varepsilon/(4m)$ and also such that

$$\frac{1}{n} \notin \left[\frac{1}{k} - \delta, \frac{1}{k} + \delta\right] \qquad \text{for } 1 \leqslant k \leqslant m \text{ unless } n = k$$

(Why?) Let

$$P = \left\{0, \frac{\varepsilon}{2}, \frac{1}{m} - \delta, \frac{1}{m} + \delta, \frac{1}{m-1} - \delta, \frac{1}{m-1} + \delta, \ldots, 1 - \delta, 1\right\}$$

We have

$$\sup\left\{f(x) \,\middle|\, x \in \left[0, \frac{\varepsilon}{2}\right]\right\} = \frac{1}{m+1}$$

$$\sup\left\{f(x) \,\middle|\, x \in \left[\frac{1}{k} - \delta, \frac{1}{k} + \delta\right]\right\} = \frac{1}{k}, \quad 2 \leqslant k \leqslant m$$

$$\sup \{f(x) \mid x \in [1 - \delta, 1] = 1$$

and $\sup \{f(x)\} = 0$ on the remaining subintervals of P. Therefore

$$\begin{aligned} 0 < U(f, P_\varepsilon) &= \frac{1}{m+1}\frac{\varepsilon}{2} + \left(\sum_{k=2}^{m} \frac{1}{k} 2\delta\right) + 1\delta \\ &\leqslant \frac{\varepsilon}{2} + \sum_{k=1}^{m} 2\delta \\ &< \frac{\varepsilon}{2} + \sum_{k=1}^{m} \frac{\varepsilon}{2m} = \frac{\varepsilon}{2} + \frac{\varepsilon}{2} = \varepsilon \end{aligned}$$

5. Let χ denote the characteristic function of the rationals restricted to $[0, 1]$, so that

$$\chi(x) = \begin{cases} 1 & \text{if } x \in [0, 1] \text{ is rational} \\ 0 & \text{if } x \in [0, 1] \text{ is irrational} \end{cases}$$

If P is any partition of $[0, 1]$, then every subinterval of P contains both rational and irrational numbers, and hence

$$\sup \{\chi(x)\} = 1 \qquad \text{and} \qquad \inf \{\chi(x)\} = 0$$

on every subinterval of P. Therefore $U(\chi, P) = 1$ and $L(\chi, P) = 0$ for every partition P of $[0, 1]$.

We need to show that for a given bounded function, every lower Riemann sum is less than or equal to every upper Riemann sum. This result will appear as a corollary to the following proposition.

Proposition 8.1.5 Let f be a bounded real-valued function on $[a, b]$, where $a < b$. If P and P' are partitions of $[a, b]$ with P' a refinement of P, then

$$L(f, P) \leqslant L(f, P') \leqslant U(f, P') \leqslant U(f, P)$$

Proof: We have already observed that, as a consequence of the definition of upper and lower Riemann sums, the lower Riemann sum over a given partition is less than or equal to the upper Riemann sum over the same partition. Therefore

$$L(f, P') \leqslant U(f, P')$$

and thus it will suffice to show that

$$L(f, P) \leqslant L(f, P') \qquad \text{and} \qquad U(f, P') \leqslant U(f, P)$$

Let us first consider the special case in which P' has been obtained by adding one point to P; say

$$P = \{a = x_0, x_1, \ldots, x_n = b\}$$

and

$$P' = \{a = x_0, x_1, \ldots, x_{i-1}, y, x_i, \ldots, x_n = b\}$$

It is clear that

$$\sup \{f(x) \mid x \in [x_{i-1}, y]\} \leqslant \sup \{f(x) \mid x \in [x_{i-1}, x_i]\}$$

and

$$\sup \{f(x) \mid x \in [y, x_i]\} \leqslant \sup \{f(x) \mid x \in [x_{i-1}, x_i]\}$$

Hence

$$\begin{aligned}
U(f, P') &= \sum_{\substack{k=1 \\ k \neq i}}^{n} \sup\{f(x) \mid x \in [x_{k-1}, x_k]\}(x_k - x_{k-1}) \\
&\quad + \sup\{f(x) \mid x \in [x_{i-1}, y]\}(y - x_{i-1}) \\
&\quad + \sup\{f(x) \mid x \in [y, x_i]\}(x_i - y) \\
&\leqslant \sum_{\substack{k=1 \\ k \neq i}}^{n} \sup\{f(x) \mid x \in [x_{k-1}, x_k]\}(x_k - x_{k-1}) \\
&\quad + \sup\{f(x) \mid x \in [x_{i-1}, x_i]\}(y - x_{i-1}) \\
&\quad + \sup\{f(x) \mid x \in [x_{i-1}, x_i]\}(x_i - y) \\
&= \sum_{k=1}^{n} \sup\{f(x) \mid x \in [x_{k-1}, x_k]\}(x_k - x_{k-1}) \\
&= U(f, P)
\end{aligned}$$

We have thus shown that if P' is a refinement of P obtained by adding one point to P, then $U(f, P') \leqslant U(f, P)$, and a similar calculation establishes that $L(f, P) \leqslant L(f, P')$. (See Exercise 2 at the end of this section.) But every refinement of P can be obtained by adding finitely many points to P, so it follows by induction that if P' is any refinement of P, then

$$U(f, P') \leqslant U(f, P) \qquad \text{and} \qquad L(f, P) \leqslant L(f, P')$$

We leave the details of the induction argument to the reader. (See Exercise 3 at the end of this section.) □

Corollary 8.1.6 Let f be a bounded real-valued function on $[a, b]$, where $a < b$. If P and P' are any partitions of $[a, b]$, then $L(f, P) \leqslant U(f, P')$.

Proof: If P and P' are the same partition, the conclusion is clearly true. If P and P' are distinct partitions, then $P \cup P'$ is a refinement of P and also a refinement of P', and hence by the proposition

$$L(f, P) \leqslant L(f, P \cup P') \leqslant U(f, P \cup P') \leqslant U(f, P') \ \square$$

Thus for a given bounded function f, every lower Riemann sum is less than or equal to every upper Riemann sum, and it follows that

$$\{U(f, P) \mid P \text{ is a partition of } [a, b]\}$$

is bounded below and therefore has a greatest lower bound, and that

$$\{L(f, P) \mid P \text{ is a partition of } [a, b]\}$$

is bounded above and therefore has a least upper bound.

Definition 8.1.7 Let f be a bounded real-valued function defined on $[a, b]$, where $a < b$. Set

$$\sup_P L(f, P) = \sup \{L(f, P) \mid P \text{ is a partition of } [a, b]\}$$

and

$$\inf_P U(f, P) = \inf \{U(f, P) \mid P \text{ is a partition of } [a, b]\}$$

It is a very important fact about Riemann sums that

$$\sup_P L(f, P) \leqslant \inf_P U(f, P)$$

Proposition 8.1.8 If f is a bounded real-valued function defined on $[a, b]$, where $a < b$, then

$$\sup_P L(f, P) \leqslant \inf_P U(f, P)$$

Proof: If

$$\sup_P L(f, P) > \inf_P U(f, P)$$

then by virtue of Definition 8.1.7 there is some partition P of $[a, b]$ such that

$$L(f, P) > \inf_P U(f, P)$$

But then, again by virtue of Definition 8.1.7, there is some partition P' such that

$$L(f, P) > U(f, P')$$

However, this last inequality contradicts Corollary 8.1.6. □

Since

$$\sup_P L(f, P) \leqslant \inf_P U(f, P)$$

always holds, we may single out those bounded functions f for which

$$\sup_P L(f, P) = \inf_P U(f, P)$$

Such functions are said to be Riemann integrable.

Definition 8.1.9 Let f be a bounded real-valued function defined on the closed interval $[a, b]$, where $a < b$. If

$$\sup_P L(f, P) = \inf_P U(f, P)$$

on $[a, b]$, then f is *Riemann integrable on* $[a, b]$, and the *Riemann integral of f on* $[a, b]$, denoted by

$$\int_a^b f\,dx$$

is defined by

$$\int_a^b f\,dx = \sup_P L(f, P) = \inf_P U(f, P)$$

The Riemann integral is the definite integral studied in calculus. There are more general definite integrals, one of which, the Riemann–Stieltjes integral, is introduced in the exercises at the end of this section. We also remark here that as far as we are concerned the dx which appears in

$$\int_a^b f\,dx$$

does not stand for any specific quantity, but is merely part of the notation for the Riemann integral. When dealing with a more general integral than the Riemann integral, the dx is replaced by another, more appropriate, symbol. Thus the function of the dx is to tell us that we are dealing with the Riemann integral of f rather than one of the more general integrals of f.

The reader should note that the Riemann integral as defined above has no connection with derivatives. As we shall see in Section 8.3, it is indeed true that for *some* Riemann integrable functions f, it is possible to calculate the value of

$$\int_a^b f\,dx$$

by finding a function whose derivative is f and evaluating this function at a and b. However, this is not a *definition* of the Riemann integral, but a *theorem* about it. Furthermore, there are many Riemann integrable functions for which this theorem does not hold.

We have not yet demonstrated that there are any Riemann integrable functions. The first example below shows that every constant function from a closed interval to $\boldsymbol{R}$ is Riemann integrable.

Examples 8.1.10 1. Let $c \in \boldsymbol{R}$ and f be defined on $[a, b]$ by $f(x) = c$ for all $x \in [a, b]$. From Example 1 of Examples 8.1.4,

$$U(f, P) = L(f, P) = c(b - a)$$

for every partition P of $[a, b]$, and therefore

$$\sup_P L(f, P) = \inf_P U(f, P) = c(b - a)$$

Hence f is Riemann integrable on $[a, b]$ and

$$\int_a^b f\,dx = \int_a^b c\,dx = c(b - a)$$

2. Let f be defined on $[a, b]$ by $f(x) = x$ for all $x \in [a, b]$. From Example 2 of Examples 8.1.4, we know that for each $n \in \boldsymbol{N}$ we can find a partition P_n such that

$$U(f, P_n) = \frac{1}{2}(b^2 - a^2) + \frac{1}{2n}(b - a)^2$$

and

$$L(f, P_n) = \frac{1}{2}(b^2 - a^2) - \frac{1}{2n}(b - a)^2$$

Therefore

$$\inf_P U(f, P) \leqslant \frac{1}{2}(b^2 - a^2) + \frac{1}{2n}(b - a)^2$$

for all $n \in N$ and

$$\sup_P L(f, P) \geqslant \frac{1}{2}(b^2 - a^2) - \frac{1}{2n}(b - a)^2$$

for all $n \in N$. But these inequalities imply that

$$\inf_P U(f, P) \leqslant \frac{1}{2}(b^2 - a^2) \leqslant \sup_P L(f, P)$$

and since

$$\sup_P L(f, P) \leqslant \inf_P U(f, P)$$

always holds by Proposition 8.1.8, we conclude that

$$\inf_P U(f, P) = \sup_P U(f, P) = \frac{1}{2}(b^2 - a^2)$$

Hence f is Riemann integrable on $[a, b]$ and

$$\int_a^b f\,dx = \int_a^b x\,dx = \frac{1}{2}(b^2 - a^2)$$

3. Let f be defined on $[0, 1]$ by

$$f(x) = \begin{cases} 0 & \text{if } 0 \leqslant x < \frac{1}{2} \\ 1 & \text{if } \frac{1}{2} \leqslant x \leqslant 1 \end{cases}$$

Let P be any partition of $[0, 1]$ such that $x_k = \frac{1}{2}$ for some k. Example 3 of Examples 8.1.4 shows that

$$U(f, P) = 1 - x_k = 1 - \frac{1}{2} = \frac{1}{2}$$

and

$$L(f, P) = 1 - x_k = \frac{1}{2}$$

Therefore

$$\inf_P U(f, P) \leqslant \frac{1}{2} \qquad \text{and} \qquad \sup_P L(f, P) \geqslant \frac{1}{2}$$

Again, since

$$\sup_P L(f, P) \leqslant \inf_P U(f, P)$$

always holds, it follows that

$$\sup_P L(f, P) = \inf_P U(f, P) = \frac{1}{2}$$

Hence f is Riemann integrable on [0, 1] and

$$\int_0^1 f\,dx = \frac{1}{2}$$

4. Let f be the function defined on [0, 1] by

$$f(x) = \begin{cases} 0 & \text{if } x \neq \dfrac{1}{n} \text{ for any } n \in \boldsymbol{N} \\ \dfrac{1}{n} & \text{if } x = \dfrac{1}{n} \text{ for some } n \in \boldsymbol{N} \end{cases}$$

From Example 4 of Examples 8.1.4, $L(f, P) = 0$ for every partition P of [0, 1], and also for every positive real number ε there is some partition P_ε such that

$$0 < U(f, P_\varepsilon) < \varepsilon$$

It follows that

$$\sup_P L(f, P) = 0 = \inf_P U(f, P)$$

Hence f is Riemann integrable on [0, 1] and

$$\int_0^1 f\,dx = 0$$

5. Let χ denote the characteristic function of the rationals, restricted

to [0, 1]. From Example 5 of Examples 8.1.4, for every partition P of [0, 1] we have $U(\chi, P) = 1$ and $L(\chi, P) = 0$. Thus

$$\inf_P U(\chi, P) = 1 \qquad \text{and} \qquad \sup_P L(\chi, P) = 0$$

so the function χ is not Riemann integrable on [0, 1].

Let us consider the question of the integrability of a real-valued function f defined on a closed interval $[a, b]$. What must we do to show that f is Riemann integrable on $[a, b]$? We must first prove that it is bounded on $[a, b]$, for if it is not, we cannot form its Riemann sums and it certainly is not Riemann integrable. Having established that f is bounded on $[a, b]$, it then suffices to show that

$$\inf_P U(f, P) \leqslant \sup_P L(f, P)$$

since this, together with the result of Proposition 8.1.8, implies that

$$\inf_P U(f, P) = \sup_P L(f, P)$$

and hence that f is Riemann integrable on $[a, b]$. The task of showing that

$$\inf_P U(f, P) \leqslant \sup_P L(f, P)$$

can be simplified somewhat, as our next result demonstrates.

Proposition 8.1.11 If f is a bounded real-valued function defined on $[a, b]$, where $a < b$, then f is Riemann integrable on $[a, b]$ if and only if for every positive real number ε there is some partition P_ε of $[a, b]$ such that

$$U(f, P_\varepsilon) - L(f, P_\varepsilon) < \varepsilon$$

Proof: Let ε be a positive real number. Because

$$\inf_P U(f, P) = \inf \{U(f, P) \mid P \text{ a partition of } [a, b]\}$$

there must be a partition P' of $[a, b]$ such that

$$U(f, P') < \inf_P U(f, P) + \frac{\varepsilon}{2}$$

Similarly, there must be a partition P'' of $[a, b]$ such that

$$L(f, P'') > \sup_P L(f, P) - \frac{\varepsilon}{2}$$

Let $P_\varepsilon = P' \cup P''$. Since P_ε is a refinement of both P' and P'', it follows that

$$U(f, P_\varepsilon) \leqslant U(f, P') \qquad \text{and} \qquad L(f, P_\varepsilon) \geqslant L(f, P'')$$

Hence

$$U(f, P_\varepsilon) < \inf_P U(f, P) + \frac{\varepsilon}{2} \qquad \text{and} \qquad L(f, P_\varepsilon) > \sup_P L(f, P) - \frac{\varepsilon}{2}$$

But if f is Riemann integrable on $[a, b]$ then $\inf_P U(f, P) = \sup_P L(f, P)$, so

$$U(f, P) - L(f, P) < \frac{\varepsilon}{2} + \frac{\varepsilon}{2} = \varepsilon$$

To prove the converse, suppose that for every $\varepsilon \in \boldsymbol{R}$, $\varepsilon > 0$, there is some partition P_ε such that

$$U(f, P) - L(f, P) < \varepsilon$$

then

$$\inf_P U(f, P) \leqslant U(f, P_\varepsilon) < L(f, P_\varepsilon) + \varepsilon \leqslant \sup_P L(f, P) + \varepsilon$$

Thus

$$\inf_P U(f, P) < \sup_P L(f, P) + \varepsilon$$

for every positive real number ε, and it follows that

$$\inf_P U(f, P) \leqslant \sup_P L(f, P)$$

and hence that f is Riemann integrable on $[a, b]$. □

Every function continuous on a closed interval $[a, b]$ should be Riemann integrable there: its upper Riemann sums are approximations from above the graph of the function to the area between the graph and the x-axis from $x = a$ to $x = b$ (see Figure 8.1), its lower Riemann sums are approximations from below the graph to this same area (see Figure 8.2), and these approximations can be made to come arbitrarily close to the true value of the area, and hence to each other, by taking the partition fine enough. Therefore for a continuous function f, it should be the case that

$$\inf_P U(f, P) = \sup_P L(f, P)$$

and thus that f is Riemann integrable on $[a, b]$. We now prove this.

Proposition 8.1.12 Every continuous real-valued function defined on a closed interval $[a, b]$, where $a < b$, is Riemann integrable on $[a, b]$.

Proof: Let f be a continuous real-valued function defined on $[a, b]$. Since f is continuous on the compact set $[a, b]$, it is both bounded and uniformly continuous on $[a, b]$. Since f is uniformly continuous on $[a, b]$, for every positive real number ε there is some positive real number δ such that

$$x \in [a, b],\ y \in [a, b],\ |x - y| < \delta \qquad \text{implies } |f(x) - f(y)| < \frac{\varepsilon}{b - a}$$

Let $P_\varepsilon = \{a = x_0, x_1, \ldots, x_n = b\}$ be a partition of $[a, b]$ such that $x_k - x_{k-1} < \delta$ for all k, $1 \leqslant k \leqslant n$. Since f is continuous on $[x_{k-1}, x_k]$, it attains its maximum and minimum there, and hence for each k, $1 \leqslant k \leqslant n$, there are points α_k, β_k in $[x_{k-1}, x_k]$ such that

$$f(\alpha_k) = \inf \{f(x) \mid x \in [x_{k-1}, x_k]\}$$

and

$$f(\beta_k) = \sup \{f(x) \mid x \in [x_{k-1}, x_k]\}$$

Thus

$$U(f, P_\varepsilon) - L(f, P_\varepsilon) = \sum_{k=1}^{n} (f(\beta_k) - f(\alpha_k))(x_k - x_{k-1})$$

But $|\beta_k - \alpha_k| < \delta$, so

$$f(\beta_k) - f(\alpha_k) = |f(\beta_k) - f(\alpha_k)| < \frac{\varepsilon}{b - a}$$

for $1 \leqslant k \leqslant n$, and therefore

$$U(f, P_\varepsilon) - L(f, P_\varepsilon) < \sum_{k=1}^{n} \frac{\varepsilon}{b-a}(x_k - x_{k-1}) = \frac{\varepsilon}{b-a}(b-a) = \varepsilon$$

Hence f is Riemann integrable by Proposition 8.1.11. □

We have thus established that continuity of a function on a closed interval is a sufficient condition for its Riemann integrability on the interval. Continuity is not a necessary condition for Riemann integrability, however; Examples 3 and 4 of Examples 8.1.10 show that a function may be discontinuous at a finite or even a denumerable number of points and still be Riemann integrable. On the other hand, the mere fact of boundedness is not sufficient for Riemann integrability, as Example 5 of Examples 8.1.10 shows. These examples and Proposition 8.1.12 suggest that necessary and sufficient conditions for the Riemann integrability of a function will involve its boundedness together with some other condition concerning the set of its discontinuities. In fact, it turns out that a function is Riemann integrable on a closed interval if and only if it is bounded there and its set of discontinuities in the interval has measure zero. Here the concept of the *measure* of a set is a generalization of the idea of length: the measure of an interval is its length, and all finite and denumerable sets have measure zero, as do some uncountable sets (such as the Cantor set, for instance). We will not prove this characterization of Riemann integrable functions here, but instead refer the reader to Exercises 13–16 at the end of this section, where the relevant definitions are given and a proof is outlined.

EXERCISES

1. Let f, g, and h be defined on $[0, 1]$ as follows:

$$f(x) = x^2 \text{ for all } x \in [0, 1]$$

$$g(x) = \begin{cases} -1 & \text{if } 0 \leqslant x \leqslant \frac{1}{3} \\ 1 & \text{if } \frac{1}{3} < x \leqslant 1 \end{cases}$$

$$h(x) = \begin{cases} 0 & \text{if } x \text{ is irrational, } x \in [0, 1] \\ x & \text{if } x \text{ is rational, } x \in [0, 1] \end{cases}$$

Let $n \in \boldsymbol{N}$ and let $P_n = \{0, 1/n, 2/n, \ldots, (n-1)/n, 1\}$. Write out the upper and lower Riemann sums $U(f, P_n)$, $L(f, P_n)$, $U(g, P_n)$, $L(g, P_n)$, $U(h, P_n)$, and $L(h, P_n)$.

2. In Proposition 8.1.5, show that $L(f, P) \leqslant L(f, P')$.
3. Carry out the induction argument which completes the proof of Proposition 8.1.5.
4. Using only the definition of the Riemann integral and the identity

$$1^2 + 2^2 + \cdots + m^2 = \frac{m(m+1)(2m+1)}{6}$$

show that $f(x) = x^2$ is Riemann integrable on $[a, b]$ and evaluate

$$\int_a^b x^2\, dx$$

5. Let f be defined on $[0, 1]$ by

$$f(x) = \begin{cases} 0 & \text{if } 0 \leqslant x < \frac{1}{3} \\ 2 & \text{if } \frac{1}{3} \leqslant x < \frac{3}{5} \\ -1 & \text{if } \frac{3}{5} \leqslant x \leqslant 1 \end{cases}$$

Show that f is Riemann integrable on $[0, 1]$ and evaluate

$$\int_0^1 f\, dx$$

6. Let $\boldsymbol{Q} \cap [0, 1] = \{q_1, q_2, \ldots\}$. For each $n \in \boldsymbol{N}$, let

$$\chi_n(x) = \begin{cases} 1 & \text{if } x \in \{q_1, q_2, \ldots, q_n\} \\ 0 & \text{if } x \in [0, 1] - \{q_1, q_2, \ldots, q_n\} \end{cases}$$

(Thus χ_n is the characteristic function of the set $\{q_1, \ldots, q_n\}$.) Show that χ_n is Riemann integrable on $[0, 1]$ and evaluate

$$\int_0^1 \chi_n\, dx$$

Now show that the sequence of functions $\{\chi_n \mid n \in \boldsymbol{N}\}$ converges pointwise on $[0, 1]$ to χ, the characteristic function of the rationals.

Since χ is not Riemann integrable (Example 5 of Examples 8.1.10), this demonstrates that the pointwise limit of a sequence of Riemann integrable functions need not be Riemann integrable.

7. Let f be defined on $[a, b]$ by

$$f(x) = \begin{cases} 0 & \text{if } x \text{ is irrational, } x \in [a, b] \\ x & \text{if } x \text{ is rational, } x \in [a, b] \end{cases}$$

Prove that f is not Riemann integrable on $[a, b]$.

8. Let f be defined on $[0, 1]$ by $f(0) = 1$, $f(x) = 0$ if x is irrational, and $f(x) = 1/n$ if $x = m/n$, where m, n are natural numbers which have no common factor. Prove that f is Riemann integrable on $[0, 1]$ and evaluate

$$\int_0^1 f\,dx$$

9. Prove that if f is Riemann integrable on $[a, b]$, and g is obtained from f by changing the values of f at finitely many points of $[a, b]$, then g is Riemann integrable on $[a, b]$ and

$$\int_a^b f\,dx = \int_a^b g\,dx$$

(Hint: prove for the case where g differs from f at only one point, then use induction.)

10. Let f be a step function defined on $[a, b]$, where $a < b$. (Thus $[a, b]$ is a union of mutually disjoint intervals $I_1, \ldots, I_n$, and $f(x) = c_k$ for all $x \in I_k$.) Prove that f is Riemann integrable on $[a, b]$ and evaluate

$$\int_a^b f\,dx$$

11. Prove that a real-valued function f defined on $[a, b]$, where $a < b$, is Riemann integrable on $[a, b]$ if and only if for every $\varepsilon > 0$ there are step functions g_ε and h_ε defined on $[a, b]$ such that

$$g_\varepsilon(x) \leqslant f(x) \leqslant h_\varepsilon(x) \qquad \text{for all } x \in [a, b]$$

and

$$\int_a^b (h_\varepsilon - g_\varepsilon)\, dx < \varepsilon$$

This result is sometimes used as the definition of Riemann integrability for a function f.

12. Let $P = \{a = x_0, x_1, \ldots, x_n = b\}$ be a partition of $[a, b]$, where $a < b$, and let $f\colon [a, b] \to \boldsymbol{R}$ be bounded. Define

$$R(f, P) = \sum_{k=1}^{n} f(y_k)(x_k - x_{k-1})$$

where for each k, $1 \leqslant k \leqslant n$, y_k may be any point in the subinterval $[x_{k-1}, x_k]$. Such a sum is called a *Riemann sum*, and its value depends on the choice of both P and the points y_k (see Figure 8.3, and contrast it with Figures 8.1 and 8.2). Show that f is Riemann integrable on $[a, b]$ with Riemann integral equal to α if and only if for every positive real number ε there is a partition P_ε of $[a, b]$ such that if P is any refinement of P_ε, then

$$|R(f, P) - \alpha| < \varepsilon$$

(Hint: $P = P_\varepsilon \cup \{y_1, \ldots, y_n\}$ is a refinement of P_ε.)

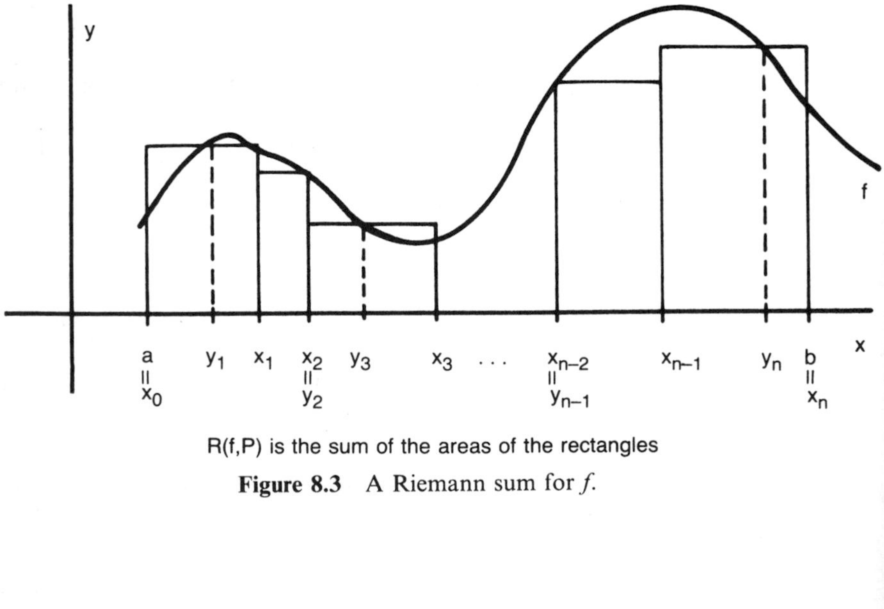

R(f,P) is the sum of the areas of the rectangles

Figure 8.3 A Riemann sum for f.

This result is sometimes used as the definition of Riemann integrability for a function f.

We utilize Exercises 13–16 to prove that a function is Riemann integrable on a closed interval if and only if it is bounded there and its set of discontinuities within the interval is a set of measure zero.

13. If I is an interval in $\boldsymbol{R}$, then $I = (a, b)$, $I = [a, b]$, $I = (a, b]$, or $I = [a, b)$, and the length of I is defined to be $m(I) = b - a$. A subset S of $\boldsymbol{R}$ is said to be a *set of measure zero* if for every positive real number ε there exists a collection of open intervals $\{I_\alpha \mid \alpha \in A\}$ such that (i) A is a countable set; (ii) $S \subset \bigcup_{\alpha \in A} I_\alpha$; and (iii) $\Sigma_{\alpha \in A}\, m(I_\alpha) < \varepsilon$. Note that the collection of open intervals $\{I_\alpha \mid \alpha \in A\}$ can be either finite or denumerable. Prove:
    (a) If S is a finite subset of $\boldsymbol{R}$, then S is a set of measure zero;
    (b) If S is the union of countably many sets of measure zero, then S is a set of measure zero. (Hint: if $S = \bigcup_n S_n$, then $S_n \subset \bigcup_k I_{nk}$, where $\Sigma_k\, m(I_{nk}) < \varepsilon/2^n$.)
    (c) If S is a countable subset of $\boldsymbol{R}$, then S is a set of measure zero;
    (d) If T is a subset of S and S is a set of measure zero, then T is a set of measure zero;
    (e) If S contains an open interval, then S is not a set of measure zero.
14. Let f be a bounded function on $[a, b]$, where $a < b$. If I is any interval contained in $[a, b]$, define $\omega_f(I)$, the *oscillation of f over I*, by

$$\omega_f(I) = \sup \{f(x) \mid x \in I\} - \inf \{f(x) \mid x \in I\}$$

    If $t \in [a, b]$, define $\omega_f(t)$, the *oscillation of f at t*, by

$$\omega_f(t) = \inf \{\omega_f(I) \mid I \text{ an open interval containing } t\}$$

    Prove that
    (a) If f is continuous at t, then $\omega_f(t) = 0$;
    (b) If f is discontinuous at t, then $\omega_f(t) > 0$;
    (c) If $S_n = \{t \in [a, b] \mid \omega_f(t) \geqslant 1/n\}$, then S_n is closed. (Hint: if r is a limit point of S_n, any open interval I containing r must contain a point s of S_n, and then $\omega_f(I) \geqslant \omega_f(s) \geqslant 1/n$.)
15. Let f be a real-valued function defined on $[a, b]$, where $a < b$. Prove that if f is Riemann integrable on $[a, b]$, then the set of discontinuities of f within $[a, b]$ is a set of measure zero.

(a) Let $S_n = \{t \in [a, b] \mid \omega_f(t) \geqslant 1/n\}$. Show that S_n is a set of measure zero:

(i) There is a partition P of $[a, b]$ such that

$$U(f, P) - L(f, P) < \frac{\varepsilon}{2n}$$

(Why?) Show that

$$\sum_{k=1}^{n} \omega_f([x_{k-1}, x_k])(x_k - x_{k-1}) = U(f, P) - L(f, P)$$

(ii) Show that $S_n = T_1 \cup T_2$, where

$$T_1 = \{x \in S_n \mid x \in P\}$$

and

$$T_2 = \{x \in S_n \mid x \text{ is in the interior of a subinterval of } P\}$$

(iii) Show that T_1 can be covered by a union of open intervals the sum of whose lengths is less than $\varepsilon/2$.

(iv) Now consider T_2: let $J_{k_1}, \ldots, J_{k_r}$ be the subintervals of P whose interiors cover T_2. If $t \in T_2$, show that

$$\omega_f(J_{k_i}) \geqslant \omega_f(t) \geqslant \frac{1}{n}$$

and then use this and the result of (i) above to show that

$$\frac{1}{n} \sum_{i=1}^{r} m(J_{k_i}) \leqslant \sum_{i=1}^{r} \omega_f(J_{k_i}) m(J_{k_i})$$

and thus that

$$\sum_{i=1}^{r} m(J_{k_i}) < \frac{\varepsilon}{2}$$

(v) Use the results of (iii) and (iv) above to prove that S_n is a set of measure zero.

(b) Show that the set of discontinuities of f within $[a, b]$ is the union of the sets S_n, $n \in \mathbf{N}$, and then conclude that the set of discontinuities of f is a set of measure zero.

16. Let f be a real-valued function defined on $[a, b]$, where $a < b$. Prove that if f is bounded on $[a, b]$ and its set of discontinuities on $[a, b]$ is a set of measure zero, then f is Riemann integrable on $[a, b]$.

(a) Show that we may assume that $\omega_f([a, b]) > 0$.

(b) Choose $n \in \mathbf{N}$ such that $(b - a)/n < \varepsilon/2$, and define

$$S_n = \left\{ t \in [a, b] \;\middle|\; \omega_f(t) \geqslant \frac{1}{n} \right\}$$

Show that there exists a countable collection of open intervals $\{J_\alpha \mid \alpha \in A\}$ such that

$$\sum_{\alpha \in A} m(J_\alpha) < \frac{\varepsilon}{2\omega_f([a, b])}$$

(c) Show that a finite number of the open intervals $\{J_\alpha\}$ of (b) above must cover S_n. Let $J_{k_1}, \ldots, J_{k_r}$ be these open intervals, and show that

$$[a, b] = \left(\bigcup_{1 \leqslant i \leqslant r} J_{k_i} \right) \cup \left(\bigcup_{1 \leqslant j \leqslant s} I_j \right)$$

where I_j is a closed interval for $1 \leqslant j \leqslant s$.

(d) Show that for each $x \in I_j$ there is an open interval K_x which contains x and such that $\omega_f(\bar{K}_x) < 1/n$, and then show that a finite number of the open intervals K_x must cover I_j.

(e) Let P_j be a partition of I_j whose points are the endpoints of the finite number of open intervals K_x which cover I_j. Show that

$$U(f, P_j) - L(f, P_j) < \frac{m(I_j)}{n}$$

(f) Define a partition P of $[a, b]$ by setting

$$P = \bigcup_{1 \leqslant j \leqslant s} P_j$$

Show that the subintervals of P are the subintervals of P_j, for

$1 \leqslant j \leqslant s$, together with the intervals $J_{k_1}, \ldots, J_{k_r}$, and then show that

$$U(f, P) - L(f, P) < \frac{1}{n} \sum_{j=1}^{s} m(I_j) + \sum_{i=1}^{r} \omega_f(J_{k_i}) m(J_{k_i}) < \varepsilon$$

17. We utilize this exercise to introduce the Riemann–Stieltjes integral. Let f, g be bounded real-valued functions defined on $[a, b]$, where $a < b$. Let

$$P = \{a = x_0, x_1, \ldots, x_n = b\}$$

be a partition of $[a, b]$. Define the *Riemann–Stieltjes sum for f with respect to g over P*, denoted $R(f, g, P)$, as follows:

$$R(f, g, P) = \sum_{k=1}^{n} f(y_k)(g(x_k) - g(x_{k-1}))$$

where for each k, $1 \leqslant k \leqslant n$, y_k may be any point of the subinterval $[x_{k-1}, x_k]$ (compare Exercise 12 above). Define f to be *Riemann–Stieltjes integrable with respect to g* on $[a, b]$, with *Riemann–Stieltjes integral* α, if there is a real number α such that for every $\varepsilon \in \boldsymbol{R}$, $\varepsilon > 0$, there is some partition P_ε of $[a, b]$ such that if P is any refinement of P_ε, then $|R(f, g, P) - \alpha| < \varepsilon$. If f is Riemann–Stieltjes integrable with respect to g on $[a, b]$, then the *Riemann–Stieltjes integral of f with respect to g over* $[a, b]$ is

$$\int_a^b f\, dg = \alpha$$

The Riemann–Stieltjes integral is a generalization of the Riemann integral which reduces to the Riemann integral when $g(x)$ is the identity function on $[a, b]$ (compare Exercise 12 above).

(a) Let

$$f(x) = g(x) = \begin{cases} 0 & \text{if } 0 \leqslant x < \frac{1}{2} \\ 1 & \text{if } \frac{1}{2} \leqslant x \leqslant 1 \end{cases}$$

Prove that f is not Riemann–Stieltjes integrable with respect to g on $[0, 1]$.

(b) Let f be continuous on $[0, 1]$ and let

$$g(x) = \begin{cases} 0 & \text{if } 0 \leqslant x < \frac{1}{2} \\ 1 & \text{if } \frac{1}{2} \leqslant x \leqslant 1 \end{cases}$$

Prove that f is Riemann–Stieltjes integrable with respect to g on $[0, 1]$ and show that

$$\int_0^1 f\,dg = f\left(\frac{1}{2}\right)$$

(c) Let g be a constant function on $[a, b]$. Prove that if f is any bounded function on $[a, b]$, then f is Riemann–Stieltjes integrable with respect to g on $[a, b]$ and

$$\int_a^b f\,dg = 0$$

(d) Let $f(x) = 1$ for all $x \in [a, b]$ and let $g(x)$ be bounded on $[a, b]$. Prove that f is Riemann–Stieltjes integrable with respect to g on $[a, b]$ and that

$$\int_a^b f\,dg = g(b) - g(a)$$

(e) Evaluate the integrals

$$\int_0^1 x\,dx^2 \qquad \text{and} \qquad \int_{-1}^1 x\,d|x|$$

if they exist.

18. (a) If f and g are bounded functions on $[a, b]$, prove that f is Riemann–Stieltjes integrable with respect to g on $[a, b]$ if and only if for every positive real number ε there is a partition P_ε of $[a, b]$ such that whenever P', P'' are refinements of P_ε, then

$$|R(f, g, P') - R(f, g, P'')| < \varepsilon$$

(Hint: if the property holds, let P_n be the partition such that $R(f, g, P') - R(f, g, P'') < 1/n$ for all refinements P', P'' of P_n.

Show that the sequence of real numbers $\{R(f, g, P_n)\}$ converges and that its limit is the Riemann–Stieltjes integral of f with respect to g.)

(b) A function g defined on $[a, b]$ is *monotonically increasing* if

$$x \in [a, b],\ y \in [a, b],\ x \leqslant y \qquad \text{imply } g(x) \leqslant g(y)$$

Prove that if f is bounded and continuous on $[a, b]$ and g is monotonically increasing on $[a, b]$, then f is Riemann–Stieltjes integrable with respect to g on $[a, b]$.
(Hint: use the uniform continuity of f to find $\delta > 0$ such that

$$|f(x) - f(y)| < \frac{\varepsilon}{2(g(b) - g(a))} \qquad \text{whenever } |x - y| < \delta$$

Let P be a partition such that $\max\{x_k - x_{k-1}\} < \delta$, and show that if P' is any refinement of P, then $|R(f, g, P) - R(f, g, P')| < \varepsilon/2$. Finally, show that if P' and P'' are refinements of P, then $|R(f, g, P') - R(f, g, P'')| < \varepsilon$.)

8.2 PROPERTIES OF THE INTEGRAL

In this section we prove some of the elementary properties of the Riemann integral and also discuss the question of whether integrability is or is not preserved under pointwise and uniform convergence.

Our first result is a technical one which will be useful to us as we develop the properties of the integral.

Proposition 8.2.1 Let f be a real-valued function defined on $[a, b]$, where $a < b$. For every partition P of $[a, b]$,

1. If $r \in \boldsymbol{R}$, $r \geqslant 0$, then

$$U(rf, P) = r \cdot U(f, P) \qquad \text{and} \qquad L(rf, P) = r \cdot L(f, P)$$

2. If $r \in \boldsymbol{R}$, $r < 0$, then

$$U(rf, P) = r \cdot L(f, P) \qquad \text{and} \qquad L(rf, P) = r \cdot U(f, P)$$

Proof: We prove the first part of (1), leaving the remainder of the proof as an exercise (See Exercise 1 at the end of this section.)

If $r \in \boldsymbol{R}$, $r \geqslant 0$, and $P = \{a = x_0, x_1, \ldots, x_n = b\}$ is a partition of $[a, b]$, then

$$\begin{aligned} U(rf, P) &= \sum_{k=1}^{n} \sup\{(rf)(x) \mid x \in [x_{k-1}, x_k]\}(x_k - x_{k-1}) \\ &= \sum_{k=1}^{n} \sup\{rf(x) \mid x \in [x_{k-1}, x_k]\}(x_k - x_{k-1}) \\ &= \sum_{k=1}^{n} r \sup\{f(x) \mid x \in [x_{k-1}, x_k]\}(x_k - x_{k-1}) \\ &= r \sum_{k=1}^{n} \sup\{f(x) \mid x \in [x_{k-1}, x_k]\}(x_k - x_{k-1}) \\ &= r \cdot U(f, P) \quad \square \end{aligned}$$

Corollary 8.2.2 If f is a real-valued function defined on $[a, b]$, where $a < b$, then

1. $r \in \boldsymbol{R}$, $r \geqslant 0$ implies

$$\inf_P U(rf, P) = r \cdot \inf_P U(f, P) \qquad \text{and} \qquad \sup_P L(rf, P) = r \cdot \sup_P L(f, P)$$

2. $r \in \boldsymbol{R}$, $r < 0$ implies

$$\inf_P U(rf, P) = r \cdot \sup_P L(f, P) \qquad \text{and} \qquad \sup_P L(rf, P) = r \cdot \inf_P U(f, P)$$

Proof: If $r \in \boldsymbol{R}$, $r \geqslant 0$, then $U(rf, P) = r \cdot U(f, P)$ for every partition P of $[a, b]$, so

$$\inf U(rf, P) = \inf\{r \cdot U(f, P)\} = r \cdot \inf U(f, P)$$

We leave the remainder of the proof as an exercise. (See Exercise 2 at the end of this section.) □

Now we are ready to prove the familiar properties of the Riemann integral. We begin by showing that the integral of a constant times a function is the constant times the integral of the function.

Proposition 8.2.3 If f is a function which is Riemann integrable on $[a, b]$, then rf is Riemann integrable on $[a, b]$ for all $r \in \boldsymbol{R}$, and

$$\int_a^b rf\,dx = r\int_a^b f\,dx$$

Proof: Let $r \in \boldsymbol{R}$. Since f is Riemann integrable on $[a, b]$ it is bounded there, and this surely implies that rf is bounded on $[a, b]$.

If $r \geqslant 0$, then

$$\sup_P L(rf, P) = r \cdot \sup_P L(f, P) \qquad \text{and} \qquad \inf_P U(rf, P) = r \cdot \inf_P U(f, P)$$

by Corollary 8.2.2. But since f is Riemann integrable on $[a, b]$,

$$\sup_P L(f, P) = \inf_P U(f, P) = \int_a^b f\,dx$$

Thus we have

$$\sup_P L(rf, P) = r\int_a^b f\,dx \qquad \text{and} \qquad \inf_P U(rf, P) = r\int_a^b f\,dx$$

and the proposition is proved when $r \geqslant 0$. The proof for the case $r < 0$ is similar. □

Next we prove that the integral of a sum (difference) is the sum (difference) of the integrals.

Proposition 8.2.4 If f and g are functions which are Riemann integrable on $[a, b]$, then $f+g$ and $f-g$ are Riemann integrable on $[a, b]$, and

$$\int_a^b (f+g)\,dx = \int_a^b f\,dx + \int_a^b g\,dx$$

$$\int_a^b (f-g)\,dx = \int_a^b f\,dx - \int_a^b g\,dx$$

Proof: It will suffice to prove the proposition for $f+g$, since the conclusion for $f-g$ will then follow from Proposition 8.2.3 and the fact that $f-g = f+(-1)g$.

Since f and g are Riemann integrable on $[a, b]$, they are bounded there, and thus $f+g$ is also bounded there. Because f is Riemann integrable on $[a, b]$, there is for every positive real number ε a partition P_ε of $[a, b]$ such that $U(f, P_\varepsilon) - L(f, P_\varepsilon) < \varepsilon/2$, and it follows that

$$U(f, P_\varepsilon) < \sup_P L(f, P) + \frac{\varepsilon}{2} = \int_a^b f\,dx + \frac{\varepsilon}{2}$$

Similarly, there is for every $\varepsilon > 0$ a partition P'_ε such that

$$U(g, P'_\varepsilon) < \int_a^b g\,dx + \frac{\varepsilon}{2}$$

For any closed interval $[\alpha, \beta]$ it is clear that

$$\sup\{f(x) + g(x) \mid x \in [\alpha, \beta]\} \leqslant \sup\{f(x) \mid x \in [\alpha, \beta]\} + \sup\{g(x) \mid x \in [\alpha, \beta]\}$$

and this implies that for every partition P of $[a, b]$,

$$U(f+g, P) \leqslant U(f, P) + U(g, P)$$

In particular, then,

$$U(f+g, P_\varepsilon \cup P'_\varepsilon) \leqslant U(f, P_\varepsilon \cup P'_\varepsilon) + U(g, P_\varepsilon \cup P'_\varepsilon)$$

But

$$U(f, P_\varepsilon \cup P'_\varepsilon) \leqslant U(f, P_\varepsilon) < \int_a^b f\,dx + \frac{\varepsilon}{2}$$

and

$$U(g, P_\varepsilon \cup P'_\varepsilon) \leqslant U(g, P'_\varepsilon) < \int_a^b g\,dx + \frac{\varepsilon}{2}$$

Therefore

$$U(f+g, P_\varepsilon \cup P'_\varepsilon) < \int_a^b f\,dx + \int_a^b g\,dx + \varepsilon$$

which implies that for every $\varepsilon > 0$,

$$\inf_P U(f+g, P) < \int_a^b f\,dx + \int_a^b g\,dx + \varepsilon$$

Hence

$$(*)\qquad \inf_P U(f+g, P) \leqslant \int_a^b f\,dx + \int_a^b g\,dx$$

Since the preceding inequality holds for every pair of functions which are Riemann integrable on $[a, b]$, it must hold for $-f$ and $-g$, which are Riemann integrable on $[a, b]$ by Proposition 8.2.3. Therefore

$$\inf_P U(-(f+g), P) = \inf_P U(-f-g, P) \leqslant \int_a^b -f\,dx + \int_a^b -g\,dx$$
$$= -\int_a^b f\,dx - \int_a^b g\,dx$$

But by Proposition 8.2.1 with $r = -1$,

$$\inf_P U(-(f+g), P) = -\sup_P L(f+g, P)$$

Thus

$$-\sup_P L(f+g, P) \leqslant -\int_a^b f\,dx - \int_a^b g\,dx$$

or

$$(**)\qquad \sup_P L(f+g, P) \geqslant \int_a^b f\,dx + \int_a^b g\,dx$$

Now, $\sup_P L(f+g, P) \leqslant \inf_P U(f+g, P)$ by Proposition 8.1.8, and this together with the inequalities $(*)$ and $(**)$ implies that

$$\inf_P U(f+g, P) = \sup_P (f+g, P) = \int_a^b f\,dx + \int_a^b g\,dx$$

and we are done. □

Now we are ready to show that the integral is additive on intervals.

Proposition 8.2.5 Let f be a function which is Riemann integrable on $[a, b]$. If $a < c < b$, then f is Riemann integrable on $[a, c]$ and on $[c, b]$, and

$$\int_a^b f\,dx = \int_a^c f\,dx + \int_c^b f\,dx$$

Proof: It is clear that f is bounded on $[a, c]$ and on $[c, b]$. Furthermore, for every positive real number ε there is a partition P_ε of $[a, b]$ such that

$$U(f, P_\varepsilon) - L(f, P_\varepsilon) < \varepsilon$$

We may assume that $c \in P_\varepsilon$. (If $c \notin P_\varepsilon$, then $P'_\varepsilon = P_\varepsilon \cup \{c\}$ is a partition of $[a, b]$ such that $c \in P'_\varepsilon$ and $U(f, P'_\varepsilon) - L(f, P'_\varepsilon) < \varepsilon$.)

Let P_a consist of those points of P_ε which belong to $[a, c]$ and let P_b consist of those points of P_ε which belong to $[c, b]$. Clearly,

$$U(f, P_a) - L(f, P_a) + U(f, P_b) - L(f, P_b) = U(f, P_\varepsilon) - L(f, P_\varepsilon) < \varepsilon$$

Since $U(f, P_a) - L(f, P_a)$ and $U(f, P_b) - L(f, P_b)$ are nonnegative, it follows that

$$U(f, P_a) - L(f, P_a) < \varepsilon \qquad \text{and} \qquad U(f, P_b) - L(f, P_b) < \varepsilon$$

Hence f is Riemann integrable on $[a, c]$ and on $[c, b]$ by virtue of Proposition 8.1.11.

Now we show that

$$\int_a^b f\,dx = \int_a^c f\,dx + \int_c^b f\,dx$$

If P is any partition of $[a, c]$ and P' any partition of $[c, b]$, then

$$\int_a^b f\,dx \leqslant U(f, P \cup P') = U(f, P) + U(f, P')$$

But this implies that

$$\int_a^b f\,dx \leqslant \inf_P U(f, P) + \inf_P U(f, P')$$

or

$$(*) \qquad \int_a^b f\,dx \leqslant \int_a^c f\,dx + \int_c^b f\,dx$$

The preceding inequality is valid for all functions which are Riemann integrable on $[a, b]$, hence is valid for $-f$, and thus

$$\int_a^b -f\,dx \leqslant \int_a^c -f\,dx + \int_c^b -f\,dx$$

or

$$(**) \qquad \int_a^b f\,dx \geqslant \int_a^c f\,dx + \int_c^b f\,dx$$

The inequalities $(*)$ and $(**)$ now imply that

$$\int_a^b f\,dx = \int_a^c f\,dx + \int_c^b f\,dx$$

and we are done. □

It is the usual practice to define the Riemann integral on a closed interval of the form $[a, a]$ to be zero; thus, by definition,

$$\int_a^a f\,dx = 0$$

(This makes sense, for the only "partition" of $[a, a]$ is $\{a = x_0, x_1 = a\}$, and certainly every upper and lower Riemann sum over this "partition" will be zero.) Also, if $a < b$, we define

$$\int_b^a f\,dx = -\int_a^b f\,dx$$

provided the integral on the right-hand side of the equation exists. With these definitions established, the equality

$$\int_a^b f\,dx = \int_a^c f\,dx + \int_c^b f\,dx$$

of Proposition 8.2.5 holds for all real numbers a, b, and c, as long as all three of the integrals exist. (See Exercise 2 at the end of this section.)

We return to our study of the properties of the Riemann integral. Our next result is often useful when it is necessary to bound an integral.

Proposition 8.2.6 Let f and g be functions which are Riemann integrable on $[a, b]$. If $f(x) \leqslant g(x)$ for all $x \in [a, b]$, then

$$\int_a^b f\,dx \leqslant \int_a^b g\,dx$$

Proof: Let $h = g - f$. Then h is Riemann integrable on $[a, b]$ and

$$\int_a^b h\,dx = \int_a^b g\,dx - \int_a^b f\,dx$$

But $h(x) \geqslant 0$ for all $x \in [a, b]$, so $L(h, P) \geqslant 0$ for every partition P of $[a, b]$, which implies that

$$\sup_p L(h, p) = \int_a^b h\,dx \geqslant 0 \quad \square$$

Corollary 8.2.7 If f is a function which is Riemann integrable on $[a, b]$, then $|f|$ is also Riemann integrable on $[a, b]$, and

$$\left| \int_a^b f\,dx \right| \leqslant \int_a^b |f|\,dx$$

Proof: If $\varepsilon \in \boldsymbol{R}$, $\varepsilon > 0$, there is a partition $P_\varepsilon = \{x_0, x_1, \ldots, x_n\}$ of $[a, b]$ such that $U(f, P_\varepsilon) - L(f, P_\varepsilon) < \varepsilon$. We may write

$$\begin{aligned} U(f, P_\varepsilon) - L(f, P_\varepsilon) = (&\sup \{f(x) \mid x \in [x_{k-1}, x_k]\} \\ &- \inf \{f(x) \mid x \in [x_{k-1}, x_k]\})(x_k - x_{k-1}) \end{aligned}$$

The integrability of $|f|$ now follows from this identity and the inequality

$$\big||f(x)| - |f(y)|\big| \leqslant |f(x) - f(y)|$$

which holds for all x and y in $[a, b]$. We leave the details to the reader. (See Exercise 5 at the end of this section.)

Since $-|f(x)| \leqslant f(x) \leqslant |f(x)|$ for all $x \in [a, b]$, it follows from Proposition 8.2.6 that

$$-\int_a^b |f|\,dx \leqslant \int_a^b f\,dx \leqslant \int_a^b |f|\,dx$$

and thus

$$\left|\int_a^b f\,dx\right| \leqslant \int_a^b |f|\,dx \;\Box$$

Next we state the First Mean Value Theorem for Integrals, which we will need in the next section to prove the Fundamental Theorem of Calculus. The proof of the First Mean Value Theorem is left to the reader. (See Exercise 7 at the end of this section.)

Proposition 8.2.8 (First Mean Value Theorem for Integrals) If f is a real-valued function defined and continuous on $[a, b]$, there is some $c \in (a, b)$ such that

$$\int_a^b f\,dx = f(c)(b-a)$$

This concludes our development of the basic properties of the Riemann integral. The reader is asked to prove additional properties of the integral in the exercises.

We end this section with a discussion of the realtionship between Riemann integrability and the convergence of a sequence of functions. The questions are these: If $\{f_n\}$ is a sequence of Riemann integrable functions which converges to a function f, is f Riemann integrable? If f is Riemann integrable, is it true that

$$\int_a^b f\,dx = \operatorname*{Lim}_{n\to\infty} \int_a^b f_n\,dx \;?$$

The answer to both these questions is no if the convergence is merely pointwise, but yes if it is uniform.

Examples 8.2.9 1. This example shows that if $\{f_n\}$ is a sequence of Riemann integrable functions which converges pointwise to f, then f need not be Riemann integrable. Let $\boldsymbol{Q} \cap [0, 1] = \{q_1, q_2, \ldots\}$ and for each $n \in \boldsymbol{N}$ let

$$\chi_n(x) = \begin{cases} 1 & \text{if } x \in \{q_1, q_2, \ldots, q_n\} \\ 0 & \text{if } x \in [0, 1] - \{q_1, q_2, \ldots, q_n\} \end{cases}$$

Then (see Exercise 6 of Exercises 8.1) χ_n is Riemann integrable on [0, 1] for each $n \in \boldsymbol{N}$, with

$$\int_a^b \chi_n \, dx = 0$$

and the sequence $\{\chi_n\}$ converges pointwise (but not uniformly) to χ, the characteristic function of the rationals restricted to [0, 1]. But as we showed in Example 5 of Examples 8.1.10, χ is not Riemann integrable on (0, 1].

2. This example will show that even if $\{f_n\}$ is a sequence of Riemann integrable functions which converges pointwise to a Riemann integrable function f, it need not be the case that

$$\int_a^b f \, dx = \operatorname*{Lim}_{n \to \infty} \int_a^b f_n \, dx$$

For each $n \in \boldsymbol{N}$, let f_n be the function whose graph over $(1/n, 1]$ is zero and whose graph over $[0, 1/n]$ is the isosceles triangle with altitude n (see Figure 8.4). The defining equations for f_n are

$$f_n(x) = \begin{cases} 2n^2 x & \text{if } 0 \leqslant x \leqslant 1/2n \\ 2n - 2n^2 x & \text{if } 1/2n < x \leqslant 1/n \\ 0 & \text{if } 1/n < x \leqslant 1 \end{cases}$$

Each function f_n is continuous on [0, 1], hence Riemann integrable there, and using the results of Propositions 8.2.3, 8.2.4, and 8.2.5 and of Examples 1 and 2 of Examples 8.1.10, we have

$$\begin{aligned} \int_0^1 f_n \, dx &= 2n^2 \int_0^{1/2n} x \, dx + 2n \int_{1/2n}^{1/n} 1 \, dx - 2n^2 \int_{1/2n}^{1/n} x \, dx + \int_{1/n}^1 0 \, dx \\ &= 2n^2 \frac{1}{2}\left(\frac{1}{4n^2} - 0\right) + 2n\left(\frac{1}{n} - \frac{1}{2n}\right) \\ &\quad - 2n^2 \frac{1}{2}\left(\frac{1}{n^2} - \frac{1}{4n^2}\right) + 0\left(1 - \frac{1}{n}\right) \\ &= \frac{1}{2} \end{aligned}$$

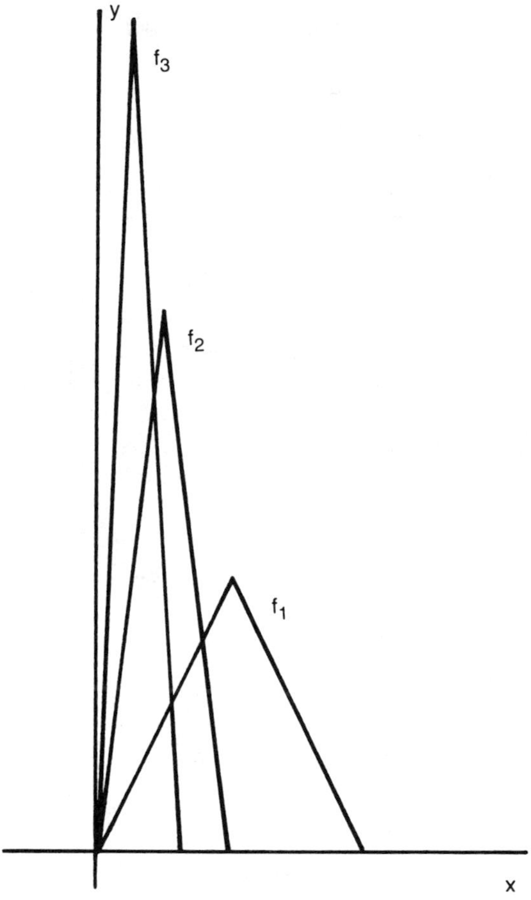

Figure 8.4 The sequence $\{f_n\}$ of Example 2, Examples 8.2.9.

The sequence of functions $\{f_n\}$ converges pointwise (but not uniformly) to the zero function on [0, 1]. (Check this.) Thus the limit function is certainly Riemann integrable, with integral equal to 0, whereas

$$\operatorname*{Lim}_{n\to\infty} \int_0^1 f_n\,dx = \operatorname*{Lim}_{n\to\infty} \frac{1}{2} = \frac{1}{2}$$

Now let us show that uniform convergence does preserve integrability, and that under uniform convergence the limit of the integrals is the integral of the limit.

Proposition 8.2.10 For each $n \in N$ let f_n be Riemann integrable on $[a, b]$. If $\{f_n\}$ converges uniformly to f on $[a, b]$, then f is Riemann integrable on $[a, b]$ and

$$\int_a^b f\,dx = \operatorname*{Lim}_{n \to \infty} \int_a^b f_n\,dx$$

Proof: Since f_n is Riemann integrable on $[a, b]$, it is bounded there, and hence by Proposition 7.1.9, the limit function f is bounded on $[a, b]$.

Let ε be a positive real number. Since $\{f_n\}$ converges uniformly to f, there is some $m \in N$ such that $n \geqslant m$ implies

$$|f(x) - f_n(x)| < \frac{\varepsilon}{3(b-a)}$$

for all $x \in [a, b]$. Because f_m is Riemann integrable on $[a, b]$, there is a partition $P = \{x_0, \ldots, x_n\}$ of $[a, b]$ such that

$$U(f_m, P) - L(f_m, P) < \frac{\varepsilon}{3}$$

Now,

$$|f(x) - f_m(x)| < \frac{\varepsilon}{3(b-a)}$$

for all $x \in [a, b]$ implies that

$$f_m(x) - \frac{\varepsilon}{3(b-a)} < f(x) < f_m(x) + \frac{\varepsilon}{3(b-a)}$$

for all $x \in [a, b]$. Since this last inequality must also hold on any subinterval of $[a, b]$, it follows that

$$U(f, P) \leqslant \sum_{k=1}^{n} \sup\{f_m(x) \mid x \in [x_{k-1}, x_k]\} + \frac{\varepsilon}{3(b-a)}(x_k - x_{k-1})$$

$$= U(f_m, P) + \frac{\varepsilon}{3(b-a)} \sum_{k=1}^{n} (x_k - x_{k-1})$$

$$= U(f_m, P) + \frac{\varepsilon}{3}$$

Similarly,

$$L(f, P) \geqslant L(f_m, P) - \frac{\varepsilon}{3}$$

Therefore

$$U(f, P) - L(f, P) \leqslant U(f_m, P) - L(f_m, P) + \frac{\varepsilon}{3} + \frac{\varepsilon}{3} < \frac{\varepsilon}{3} + \frac{\varepsilon}{3} + \frac{\varepsilon}{3} = \varepsilon$$

Hence f is Riemann integrable over $[a, b]$.

To complete the proof we must show that

$$\int_a^b f\,dx = \operatorname*{Lim}_{n\to\infty} \int_a^b f_n\,dx$$

Since $\{f_n\}$ converges uniformly to f on $[a, b]$ there is for every positive real number ε some $m \in N$ such that $n \geqslant m$ implies

$$|f(x) - f_n(x)| < \frac{\varepsilon}{b-a}$$

for all $n \in [a, b]$. By virtue of Corollary 8.2.7, Proposition 8.2.6, and Example 1 of Examples 8.1.10, we have that

$$\left|\int_a^b f\,dx - \int_a^b f_n\,dx\right| = \left|\int_a^b (f - f_n)\,dx\right| \leq \int_a^b |f - f_n|\,dx$$

$$\leqslant \int_a^b \frac{\varepsilon}{b-a}\,dx = \frac{\varepsilon}{b-a}(b-a) = \varepsilon$$

This proves that the sequence of real numbers

$$\left\{\int_a^b f_n\,dx\right\}$$

converges to the real number

$$\int_a^b f\,dx$$

and we are done. □

EXERCISES

1. (a) Complete the proof of Proposition 8.2.1.
   (b) Complete the proof of Corollary 8.2.2.
2. If $f(a)$ exists, define

$$\int_a^a f\,dx = 0$$

   If $a < b$ and

$$\int_a^b f\,dx$$

   exists, define

$$\int_b^a f\,dx = -\int_a^b f\,dx$$

   Prove that for any real numbers a, b, and c,

$$\int_a^b f\,dx = \int_a^c f\,dx + \int_c^b f\,dx$$

   provided all three of these integrals exist.
3. Let f be a real-valued function defined on $[a, b]$ and let $a < c < b$. If f is Riemann integrable on $[a, c]$ and on $[c, b]$, show that f is Riemann integrable on $[a, b]$ and that

$$\int_a^b f\,dx = \int_a^c f\,dx + \int_c^b f\,dx$$

4. Let f be a bounded real-valued function defined on $[a, b]$. If the set of discontinuities of f is a finite subset of $[a, b]$, prove that f is Riemann integrable on $[a, b]$.
5. Fill in the details in the proof of Corollary 8.2.7.
6. Give an example which shows that the converse of Corollary 8.2.7 does not hold.
7. (a) Let f be Riemann integrable on $[a, b]$ with

$$\alpha = \inf\{f(x) \mid x \in [a, b]\} \qquad \text{and} \qquad \beta = \sup\{f(x) \mid x \in [a, b]\}$$

Prove that

$$\alpha(b-a) \leqslant \int_a^b f\,dx \leqslant \beta(b-a)$$

(b) Prove the First Mean Value Theorem for Riemann integrals.

8. Let f be continuous on $[a, b]$ and let

$$F(t) = \int_a^t f\,dx$$

for all $t \in [a, b]$. (Note that $F(a) = 0$ by definition.) Prove that F is a continuous function on $[a, b]$.

9. (a) Let f be continuous on $[a, b]$ with $f(x) \geqslant 0$ for all $n \in [a, b]$. Prove that if $f(x_0) > 0$ for some $x_0 \in [a, b]$, then

$$\int_a^b f\,dx > 0$$

(b) Let f, g be continuous on $[a, b]$ with $f(x) \leqslant g(x)$ for all $x \in [a, b]$. Prove that if $f(x_0) < g(x_0)$ for some $x_0 \in [a, b]$, then

$$\int_a^b f\,dx < \int_a^b g\,dx$$

10. Give an example to show that the converse of Proposition 8.2.10 is false. That is, find a sequence of functions $\{f_n\}$ which converges pointwise but not uniformly to a function f, and such that

$$\operatorname*{Lim}_{n\to\infty} \int_a^b f_n\,dx = \int_a^b f\,dx$$

(Hint: modify Example 2 of Examples 8.2.9 by changing the heights of the triangles.)

11. If f_n is continuous on $[a, b]$ for all $n \in N$, the sequence $\{f_n\}$ converges uniformly on $[a, b]$ to f, and g is continuous on $[a, b]$, prove that fg is Riemann integrable on $[a, b]$ and that

$$\operatorname*{Lim}_{n\to\infty} \int_a^b f_n g\,dx = \int_a^b fg\,dx$$

12. This exercise continues the study of the Riemann–Stieltjes integral begun in Exercise 17 of Exercises 8.1. Most of the properties of the Riemann integral have Riemann–Stieltjes analogs. Prove that the following properties hold for Riemann–Stieltjes integrals, provided the integrals on the right-hand side of each equation exist.

(a) $$\int_a^b rf\,dg = r\int_a^b f\,dg \qquad \text{for all } r \in \boldsymbol{R}$$

(b) $$\int_a^b f\,d(rg) = r\int_a^b f\,dg \qquad \text{for all } r \in \boldsymbol{R}$$

(c) $$\int_a^b (f+g)\,dh = \int_a^b f\,dh + \int_a^b g\,dh$$

(d) $$\int_a^b f\,d(g+h) = \int_a^b f\,dg + \int_a^b f\,dh$$

(e) If $a < c < b$,

$$\int_a^b f\,dg = \int_a^c f\,dg + \int_c^b f\,dg$$

(f) $$f(b)g(b) - f(a)g(a) = \int_a^b f\,dg + \int_a^b g\,df$$

8.3 THE DERIVATIVE

In this section we define the derivative of a function on the reals, study the properties of the derivative, and examine the connection between the derivative and the Riemann integral. The derivative will be defined by means of a functional limit, and thus we will make repeated use of the results concerning functional limits developed in Section 5.4, Chapter 5; the reader is urged to review that section now. In particular, we will find it useful to have the definition of a functional limit (Definition 5.4.1) restated in δ-ε terms; we include this restatement as our first result of this section.

Proposition 8.3.1 Let f be a real-valued function defined on a subset S of $\boldsymbol{R}$. If c is a limit point of S and $x_0 \in \boldsymbol{R}$, then

$$\operatorname*{Lim}_{x \to c} f(x) = x_0$$

if and only if for every $\varepsilon \in \boldsymbol{R}$, $\varepsilon > 0$, there is some $\delta \in \boldsymbol{R}$, $\delta > 0$, such that

$$x \in S,\ 0 < |x - c| < \delta \qquad \text{imply} \qquad |f(x) - x_0| < \varepsilon$$

The proof of the proposition will be left to the reader. (See Exercise 1 at the end of this section.)

Now let us define the derivative of a function at a point.

Definition 8.3.2 Let f be a real-valued function whose domain is a subset S of $\boldsymbol{R}$, and let $c \in S$ be a limit point of S. The *derivative of f at c*, denoted by $f'(c)$, is defined by

$$f'(c) = \operatorname*{Lim}_{x \to c} \frac{f(x) - f(c)}{x - c}$$

Note that we consider the derivative of f at c only for those points c of the domain S which are also limit points of S. (If S is an interval or ray, as will frequently be the case, then every point of S is a limit point of S, and we may consider the derivative of f at c for all $c \in S$.) However, even if $c \in S$ is a limit point of S, the derivative of f at c need not exist, for the limit as x approaches c of the function

$$\frac{f(x) - f(c)}{x - c}$$

may not exist. If $f'(c)$ does exist for all $c \in S$, we say that f is *differentiable on S*. The function

$$\frac{f(x) - f(c)}{x - c}$$

called the *difference quotient of f at c*, is a real-valued function of x whose domain is $S - \{c\}$. Thus when working with the derivative we need only consider the values of the difference quotient when $x \in S$, $x \neq c$. Also, if

$$S' = \{c \in S \mid f'(c) \text{ exists}\}$$

then the *derivative function f'* is the function from S' to $\boldsymbol{R}$ which assigns to each $c \in S'$ the real number $f'(c)$.

Examples 8.3.3 1. Let S be any subset of $\boldsymbol{R}$, let $r \in \boldsymbol{R}$, and let f be the constant function defined by $f(x) = r$ for all $x \in S$. If $c \in S$ is a limit point of S, then

$$\operatorname*{Lim}_{x \to c} \frac{f(x) - f(c)}{x - c} = \operatorname*{Lim}_{x \to c} \frac{r - r}{x - c} = \operatorname*{Lim}_{x \to c} \frac{0}{x - c}$$

But since $x \neq c$ in the limit

$$\frac{0}{x - c} = 0$$

so

$$f'(c) = \operatorname*{Lim}_{x \to c} 0$$

provided this limit exists. But the constant function $g(x) = 0$ is certainly continuous at c, so by Proposition 5.4.4 of Chapter 5,

$$\operatorname*{Lim}_{x \to c} 0 = \operatorname*{Lim}_{x \to c} g(x) = g(c) = 0$$

Thus we have shown that if f is a constant function on S and $c \in S$ is a limit point of S, then $f'(c)$ exists and $f'(c) = 0$. Therefore we may say that if $f(x) = r$ on S, then $f'(x) = 0$ for all $x \in S$ such that x is a limit point of S.

2. Let S be any subset of $\boldsymbol{R}$, and let f be the identity function on S, defined by $f(x) = x$ for all $x \in S$. If $c \in S$ is a limit point of S, then

$$f'(c) = \operatorname*{Lim}_{x \to c} \frac{f(x) - f(c)}{x - c} = \operatorname*{Lim}_{x \to c} \frac{x - c}{x - c}$$

Again, since $x \neq c$ in the limit,

$$\frac{x - c}{x - c} = 1$$

and therefore the continuity of the constant function $g(x) = 1$ implies that

$$f'(c) = \operatorname*{Lim}_{x \to c} 1 = 1$$

Thus if $f(x) = x$ on S, then $f'(x) = 1$ for all $x \in S$ such that x is a limit point of S.

3. Let $f(x) = x^2$ for all $x \in \boldsymbol{R}$. If $c \in \boldsymbol{R}$, then

$$\operatorname*{Lim}_{x \to c} \frac{f(x) - f(c)}{x - c} = \operatorname*{Lim}_{x \to c} \frac{x^2 - c^2}{x - c} = \operatorname*{Lim}_{x \to c} \frac{(x + c)(x - c)}{x - c}$$

Since $x \neq c$ in the limit,

$$\operatorname*{Lim}_{x \to c} \frac{(x + c)(x - c)}{x - c} = \operatorname*{Lim}_{x \to c} (x + c)$$

and the continuity of the linear function $g(x) = x + c$ then implies that

$$f'(c) = \operatorname*{Lim}_{x \to c} (x + c) = \operatorname*{Lim}_{x \to c} g(x) = g(c) = 2c$$

Thus we have shown that if $f(x) = x^2$ on $\boldsymbol{R}$, then $f'(x) = 2x$ on $\boldsymbol{R}$.

4. Let f be the absolute value function on $\boldsymbol{R}$. Then $f(x) = x$ on the ray $(0, +\infty)$, so if $c \in \boldsymbol{R}$, $c > 0$, then by Example 2 above, $f'(c) = 1$. Similarly, since $f(x) = -x$ on the ray $(-\infty, 0)$, if $c < 0$ we have

$$f'(c) = \operatorname*{Lim}_{x \to c} \frac{f(x) - f(c)}{x - c} = \operatorname*{Lim}_{x \to c} \frac{-x + c}{x - c} = \operatorname*{Lim}_{x \to c} (-1) = -1$$

However, $f'(0)$ does not exist. To see this, suppose $f'(0)$ does exist; then

$$f'(0) = \operatorname*{Lim}_{x \to 0} \frac{f(x) - f(0)}{x - 0}$$

and by Proposition 8.3.1, there is a $\delta \in \boldsymbol{R}$, $\delta > 0$, such that $0 < |x| < \delta$ implies

$$\left| \frac{f(x) - f(0)}{x - 0} - f'(0) \right| = \left| \frac{|x|}{x} - f'(0) \right| < 1$$

But no matter what the value of δ, there are real numbers y and z such that $-\delta < y < 0 < z < \delta$, and hence such that

$$\left| \frac{|y|}{y} - f'(0) \right| = |-1 - f'(0)| < 1$$

and

$$\left|\frac{|z|}{z} - f'(0)\right| = |1 - f'(0)| < 1$$

Since there is no real number r such that $|-1-r| < 1$ and $|1-r| < 1$, we conclude that $f'(0)$ cannot exist. Thus the absolute value function does not have a derivative at 0.

5. Let $f(x) = \sqrt{x}$, for all $x \in [0, +\infty)$. If $c > 0$, then

$$\begin{aligned}\operatorname*{Lim}_{x\to c}\frac{f(x)-f(c)}{x-c} &= \operatorname*{Lim}_{x\to c}\frac{\sqrt{x}-\sqrt{c}}{x-c} = \operatorname*{Lim}_{x\to c}\frac{\sqrt{x}-\sqrt{c}}{x-c}\,\frac{\sqrt{x}+\sqrt{c}}{\sqrt{x}+\sqrt{c}} \\ &= \operatorname*{Lim}_{x\to c}\frac{x-c}{(x-c)(\sqrt{x}+\sqrt{c})} = \operatorname*{Lim}_{x\to c}\frac{1}{\sqrt{x}+\sqrt{c}} \\ &= \frac{1}{\operatorname*{Lim}_{x\to c}(\sqrt{x}+\sqrt{c})}\end{aligned}$$

But $\mathrm{Lim}_{x\to c}(\sqrt{x}+\sqrt{c}) = 2\sqrt{c}$ by continuity of the function $g(x) = \sqrt{x} + \sqrt{c}$ at c. Therefore if $f(x) = \sqrt{x}$ and $x > 0$, then $f' = 1/2\sqrt{x}$. However, $f'(0)$ does not exist, because

$$\operatorname*{Lim}_{x\to 0}\frac{f(x)-f(0)}{x-0} = \operatorname*{Lim}_{x\to 0}\frac{\sqrt{x}}{x} = \operatorname*{Lim}_{x\to 0}\frac{1}{\sqrt{x}}$$

does not exist. (Check this.)

The reader no doubt recalls from earlier studies that the derivative $f'(c)$ of f at c is the slope of the line tangent to the graph of f at the point $(c, f(c))$. (See Exercise 6 at the end of this section.) Therefore if the graph does not have a unique nonvertical tangent line at $(c, f(c))$, then it stands to reason that it cannot have a derivative at c. (Thus, the absolute value function does not have a unique tangent line at $(0, 0)$ and, as we have seen, it has no derivative at 0; similarly, the square root function has a vertical tangent line at $(0, 0)$, and it too has no derivative at 0.) Since it is intuitively clear that if f is discontinuous at c, then its graph cannot have a unique tangent line at $(c, f(c))$, this line of reasoning suggests that if f is discontinuous at c, then $f'(c)$ does not exist. This is indeed true, and we begin our study of the properties of the derivative by proving this important fact.

Proposition 8.3.4 Let f be a real-valued function defined on a subset S of $\boldsymbol{R}$, and let $c \in S$ be a limit point of S. If the derivative $f'(c)$ exists, then f is continuous at c.

Proof: If $x \in S$, $x \neq c$, then

$$f(x) - f(c) = \frac{f(x) - f(c)}{x - c}(x - c)$$

Therefore

$$\lim_{x \to c}(f(x) - f(c)) = \lim_{x \to c}\frac{f(x) - f(c)}{x - c}\lim_{x \to c}(x - c)$$

provided that

$$\lim_{x \to c}\frac{f(x) - f(c)}{x - c} \qquad \text{and} \qquad \lim_{x \to c}(x - c)$$

both exist. However

$$\lim_{x \to c}\frac{f(x) - f(c)}{x - c} = f'(c)$$

exists by hypothesis, and

$$\lim_{x \to c}(x - c) = c - c = 0$$

by virtue of the continuity of the linear function $g(x) = x - c$ and Proposition 5.4.4 of Chapter 5. Thus

$$\lim_{x \to c}(f(x) - f(c)) = f'(c) \cdot 0 = 0$$

and hence

$$\lim_{x \to c} f(x) = f(c)$$

which by Proposition 5.4.4 implies that f is continuous at c. □

Proposition 8.3.4 tells us that if f is not continuous at c, then $f'(c)$ cannot exist. It is important to realize that the converse of the proposition is not true: if f is continuous at c, it does not follow that $f'(c)$ exists. Thus the absolute value function is continuous at $c = 0$, but as we have seen, it has no derivative at 0. Indeed, it is not terribly difficult to show that there are functions which are continuous at every real number but which do not have a derivative at any real number. (See Exercise 24 at the end of this section.)

We next state some of the well-known rules of differentiation. The proofs will be left to the reader. (See Exercises 7 and 8 at the end of this section.)

Proposition 8.3.5 Let $c \in \boldsymbol{R}$. If f and g are real-valued functions defined at c, and $f'(c)$ and $g'(c)$ exist, then

1. $(rf)'(c)$ exists and $(rf)'(c) = r \cdot f'(c)$, for all $r \in \boldsymbol{R}$;
2. $(f+g)'(c)$ and $(f-g)'(c)$ exist and

$$(f+g)'(c) = f'(c) + g'(c), \quad (f-g)'(c) = f'(c) - g'(c)$$

3. $(fg)'(c)$ exists and $(fg)'(c) = f(c)g'(c) + f'(c)g(c)$;
4. If $g(c) \neq 0$, $(f/g)'(c)$ exists and

$$\left(\frac{f}{g}\right)'(c) = \frac{f'(c)g(c) - g'(c)f(c)}{[g(c)]^2}$$

Now we are ready to prove the chain rule for derivatives.

Proposition 8.3.6 (Chain Rule) Let $c \in \boldsymbol{R}$. If f is a real-valued function defined at c, g is a real-valued function defined at $f(c)$, and $f'(c)$ and $g'(f(c))$ both exist, then $(g \circ f)'(c)$ exists and

$$(g \circ f)'(c) = g'(f(c))f'(c)$$

Proof: Define a function σ by setting

$$\sigma(x) = \frac{f(x) - f(c)}{x - c} - f'(c)$$

for all x in the domain of f such that $x \neq c$. The following identity holds for all x in the domain of f, including $x = c$:

$$(*) \qquad [\sigma(x) + f'(c)](x - c) = f(x) - f(c)$$

Similarly, we define a function τ by setting

$$\tau(y) = \frac{g(y) - g(f(c))}{y - f(c)} - g'(f(c))$$

for all y in the domain of g such that $y \neq f(c)$. The following identity holds for all y in the domain of g, including $y = f(c)$:

$$(**) \qquad [\tau(y) + g'(f(c))](y - f(c)) = g(y) - g(f(c))$$

Note that

$$\operatorname*{Lim}_{x \to c} \sigma(x) = \operatorname*{Lim}_{x \to c} \frac{f(x) - f(c)}{x - c} - \operatorname*{Lim}_{x \to c} f'(c) = f'(c) - f'(c) = 0$$

and

$$\begin{aligned}\operatorname*{Lim}_{y \to f(c)} \tau(y) &= \operatorname*{Lim}_{y \to f(c)} \frac{g(y) - g(f(c))}{y - f(c)} - \operatorname*{Lim}_{y \to f(c)} g'(f(c)) \\ &= g'(f(c)) - g'(f(c)) = 0\end{aligned}$$

Using the identities $(*)$ and $(**)$, we have

$$\begin{aligned}(g \circ f)(x) - (g \circ f)(c) &= g(f(x)) - g(f(c)) \\ &= [\tau(f(x)) + g'(f(c))](f(x) - f(c)) \\ &= [\tau(f(x)) + g'(f(c))][\sigma(x) + f'(c)](x - c)\end{aligned}$$

Thus if $x \neq c$, then

$$\frac{(g \circ f)(x) - (g \circ f)(c)}{x - c} = [\tau(f(x)) + g'(f(c))][\sigma(x) + f'(c)]$$

Hence

$$\operatorname*{Lim}_{x \to c} \frac{(g \circ f)(x) - (g \circ f)(c)}{x - c} = \operatorname*{Lim}_{x \to c} [\tau(f(x)) + g'(f(c))] \operatorname*{Lim}_{x \to c} [\sigma(x) + f'(c)]$$

provided the limits on the right-hand side of the equation exist. However, $\operatorname{Lim}_{x \to c} \sigma(x) = 0$ implies that $\operatorname{Lim}_{x \to c} [\sigma(x) + f'(c)] = f'(c)$. Furthermore, since $f'(c)$ exists, f is continuous at c, and therefore $\operatorname{Lim}_{x \to c} f(x) = f(c)$; that is, $f(x)$ approaches $f(c)$ as x approaches c. Thus

$$\operatorname*{Lim}_{x \to c} (f(x)) = \operatorname*{Lim}_{f(x) \to f(c)} \tau(f(x))$$

(See Exercise 2 at the end of this section.) But since $\mathrm{Lim}_{y \to f(c)} \tau(y) = 0$, we have $\mathrm{Lim}_{f(x) \to f(c)} \tau(f(x)) = 0$. Therefore

$$\operatorname*{Lim}_{x \to c} [\tau(f(x)) + g'(f(c))] = g'(f(c))$$

and we are done. □

The technique used in the previous proof is one that is often useful when working with derivatives. It consists of writing the difference quotient as the derivative plus an error function which goes to zero as x approaches c, then manipulating the error function to establish the desired result.

Next we are going to consider very briefly the well-known connection between the derivative and relative maxima and minima of a function.

Definition 8.3.7 Let f be a real-valued function defined on the open interval (a, b). The function f has a *relative maximum* at $c \in (a, b)$ if there is some positive real number δ such that

$$f(x) \leqslant f(c) \qquad \text{for all} \qquad x \in (c - \delta, c + \delta)$$

Similarly, f has a *relative minimum* at $c \in (a, b)$ if there is some positive real number δ such that

$$f(x) \geqslant f(c) \qquad \text{for all} \qquad x \in (c - \delta, c + \delta)$$

Proposition 8.3.8 Let f be a real-valued function defined on the open interval (a, b). If f has a relative maximum or a relative minimum at $c \in (a, b)$ and $f'(c)$ exists, then $f'(c) = 0$.

Proof: We prove this for the case in which f has a relative maximum at c. The proof for a relative minimum is similar and is left to the reader. (See Exercise 14 at the end of this section.)

Since f has a relative maximum at $c \in (a, b)$ there is some positive real number γ such that $f(x) \leqslant f(c)$ for all $x \in (c - \gamma, c + \gamma)$. If $f'(c) \neq 0$ there is some positive real number η such that

$$x \in (a, b) \quad \text{and} \quad 0 < |x - c| < \eta \quad \text{imply} \quad \left| \frac{f(x) - f(c)}{x - c} - f'(c) \right| < |f'(c)|$$

Let $\delta = \min\{\gamma, \eta\}$. Then

$$(*) \quad x \in (c - \delta, c + \delta),\ x \neq c \quad \text{implies} \quad \left| \frac{f(x) - f(c)}{x - c} - f'(c) \right| < |f'(c)|$$

Now suppose $f'(c) > 0$. If $x \in (c, c + \delta)$, then

$$\frac{f(x) - f(c)}{x - c} \leqslant 0$$

so that

$$\left| \frac{f(x) - f(c)}{x - c} - f'(c) \right| \geqslant f'(c) = |f'(c)|$$

which contradicts $(*)$ above. Hence $f'(c) > 0$ is impossible. But if $f'(c) < 0$, then $x \in (c - \delta, c)$ implies

$$\frac{f(x) - f(c)}{x - c} \geqslant 0$$

so that

$$\left| \frac{f(x) - f(c)}{x - c} - f'(c) \right| \geqslant -f'(c) = |f'(c)|$$

and thus $f'(c) < 0$ is also impossible, and we are done. □

Theorem 8.3.9 (Rolle's Theorem) Let f be a continuous real-valued function defined on $[a, b]$ and such that $f(a) = f(b) = 0$. If f is differentiable on (a, b), then there is some $c \in (a, b)$ such that $f'(c) = 0$.

Proof: If $f(x) = 0$ for all $x \in [a, b]$ then surely $f'(c) = 0$ for all $c \in (a, b)$ and we are done, so we assume that f is not identically zero on $[a, b]$. Since f is continuous on $[a, b]$, it attains its maximum at some point $c \in [a, b]$ and its minimum at some point $d \in [a, b]$. Since f is not identically zero on $[a, b]$ at least one of $f(c)$, $f(d)$ must be nonzero. For definiteness, suppose $f(c) \neq 0$; then $c \neq a$ and $c \neq b$, so $c \in (a, b)$, and because $f(c) = \max\{f(x) \mid x \in [a, b]\}$, the function f certainly has a relative maximum at c. But f is differentiable on (a, b), so $f'(c)$ exists, and therefore $f'(c) = 0$ by Proposition 8.3.8. □

Corollary 8.3.10 (The Law of the Mean) If f is a real-valued function defined and continuous on $[a, b]$ and differentiable on (a, b), then there is some $c \in (a, b)$ such that $f'(c)(b - a) = f(b) - f(a)$.

Proof: Define

$$g(x) = f(x) - \frac{f(b) - f(a)}{b - a}(x - a) - f(a)$$

for all $x \in [a, b]$ and apply Rolle's Theorem to g. □

Corollary 8.3.11 If f is a real-valued function defined on $[a, b]$ and differentiable on (a, b) and if $f'(x) = 0$ for all $x \in (a, b)$, then f is a constant function on $[a, b]$.

Proof: Apply the Law of the Mean on $[a, x]$, for all $x \in (a, b]$. □

Corollary 8.3.12 Let f and g be continuous real-valued functions defined on $[a, b]$ and differentiable on (a, b). If $f'(x) = g'(x)$ for all $x \in (a, b)$, then there is some $r \in \boldsymbol{R}$ such that $f(x) = g(x) + r$ for all $x \in [a, b]$.

We leave it to the reader to finish the proofs of the preceding corollaries. (See Exercise 17 at the end of this section.)

Now we turn to an examination of the relationship between the derivative and the Riemann integral. The key theorem in this regard is the following one, which says that the derivative of the integral is the integrand.

Theorem 8.3.13 Let f be a real-valued function defined and continuous on $[a, b]$. If F is defined on $[a, b]$ by

$$F(x) = \int_a^x f\,dx$$

for all $x \in [a, b]$, then F is differentiable on $[a, b]$ and $F'(x) = f(x)$ for all $x \in [a, b]$.

Proof: Since f is continuous on $[a, b]$, it is Riemann integrable there, and thus the function F is defined for all $x \in [a, b]$. F is defined at $x = a$ because, by definition,

$$\int_a^a f\,dx = 0$$

Let $c \in [a, b]$, and consider the difference quotient

$$\frac{F(x) - F(c)}{x - c} \qquad \text{for } x \in [a, b],\ x \neq c.$$

If $x > c$, we have

$$\frac{F(x) - F(c)}{x - c} = \frac{1}{x - c}\left[\int_a^x f\,dx - \int_a^c f\,dx\right] = \frac{1}{x - c}\int_c^x f\,dx$$

by Proposition 8.2.5. By Proposition 8.2.8, there is some $y \in (c, x)$ such that

$$\int_c^x f\,dx = f(y)(x - c)$$

Therefore if $x > c$, then

$$\frac{F(x) - F(c)}{x - c} = f(y)$$

for some $y \in (c, x)$. Similarly, if $x < c$, then

$$\frac{F(x) - F(c)}{x - c} = \frac{1}{x - c}\left[-\int_x^c f\,dx\right] = \frac{1}{c - x}\int_x^c f\,dx$$
$$= \frac{1}{c - x} f(y)(c - x) = f(y)$$

for some $y \in (x, c)$. Hence for all $x \in [a, b]$, $x \neq c$, there is some point y between x and c such that

$$\frac{F(x) - F(c)}{x - c} = f(y)$$

Since y is between x and c, y approaches c as x approaches c. Therefore

$$\operatorname*{Lim}_{x \to c} \frac{F(x) - F(c)}{x - c} = \operatorname*{Lim}_{x \to c} f(y) = \operatorname*{Lim}_{y \to c} f(y)$$

But f is continuous on $[a, b]$, so

$$\operatorname*{Lim}_{y \to c} f(y) = f(c)$$

which proves the theorem. □

Corollary 8.3.14 (The Fundamental Theorem of Calculus) Let f be a real-valued function defined and continuous on $[a, b]$. If G is any function defined on $[a, b]$ such that $G'(x) = f(x)$ for all $x \in [a, b]$, then

$$\int_a^b f\,dx = G(b) - G(a)$$

Proof: Let F be as in Theorem 8.3.13. Since $F'(x) = f(x) = G'(x)$ for all $x \in [a, b]$, by Corollary 8.3.12 there is some $r \in \boldsymbol{R}$ such that $F(x) = G(x) + r$ for all $x \in [a, b]$. But then $G(a) + r = F(a) = 0$, so $r = -G(a)$, and hence

$$\int_a^b f\,dx = F(b) = G(b) + r = G(b) - G(a). \ \square$$

The Fundamental Theorem of Calculus is the result which allows us to evaluate certain definite integrals by the familiar process of finding an antiderivative of the integrand and evaluating it at the endpoints of the interval of integration. Thus it is the Fundamental Theorem of Calculus which justifies calculations such as

$$\int_1^2 x^3\,dx = \left.\frac{x^4}{4}\right|_1^2 = \frac{1}{4}(2^4 - 1^4) = \frac{15}{4}$$

However, it is important to realize that not every Riemann integral can be evaluated in this manner. For example, the function f defined by

$$f(x) = \begin{cases} 0 & \text{if } 0 \leqslant x < \dfrac{1}{2} \\ 1 & \text{if } \dfrac{1}{2} \leqslant x \leqslant 1 \end{cases}$$

is Riemann integrable on $[0, 1]$, with

$$\int_0^1 f\,dx = \frac{1}{2}$$

but the Fundamental Theorem of Calculus does not apply here, because f is not continuous on $[0, 1]$ and hence there cannot be any function G such that $G' = f$ on $[0, 1]$.

The Fundamental Theorem of Calculus has several consequences which yield useful techniques of integration. As an example, we state and prove the Integration by Substitution Theorem. For another example, namely the Integration by Parts Theorem, see Exercise 21 at the end of this section.

Proposition 8.3.15 Let g be a real-valued function which is differentiable on $[a, b]$ and suppose that g' is continuous on $[a, b]$. If f is a real-valued function which is continuous on $g([a, b])$, then

$$\int_{g(a)}^{g(b)} f\,dx = \int_a^b (f \circ g)g'\,dx$$

Proof: The continuity of f, g, and g' implies the existence of the integrals. Let F be any function such that $F' = f$ on $g([a, b])$. (Such a function F exists by Theorem 8.3.13.) Let $G = F \circ g$. The function G is defined on $[a, b]$ and by the Chain Rule

$$G' = (F' \circ g)g' = (f \circ g)g'$$

Hence by the Fundamental Theorem of Calculus,

$$\int_{g(a)}^{g(b)} f\,dx = F(g(b)) - F(g(a)) = (F \circ g)(b) - (F \circ g)(a)$$

$$= G(b) - G(a) = \int_a^b (f \circ g)g'\,dx \quad \square$$

Proposition 8.3.15 is known as the Integration by Substitution Theorem because if it is written in the form

$$\int_{g(a)}^{g(b)} f(x)\,dx = \int_a^b f(g(x))g'(x)\,dx$$

it is evident that the right-hand side involves the substitution of $g(x)$ for x in the expression for f.

Example 8.3.16 Proposition 8.3.15 is a useful tool for evaluating integrals, and it may be used in both directions. For instance, suppose we wish to evaluate

$$\int_0^1 x(x^2 + 1)^5\,dx$$

Letting $f(x) = x^5$ and $g(x) = x^2 + 1$ for all $x \in [0, 1]$, we have

$$\int_0^1 x(x^2+1)^5\, dx = \frac{1}{2}\int_0^1 (f \circ g)g'\, dx = \frac{1}{2}\int_{g(0)}^{g(1)} f\, dx = \frac{1}{2}\int_1^2 x^5\, dx$$

But if $F(x) = x^6/6$, then $F'(x) = x^5$ (see Exercise 9 at the end of this section) and hence

$$\int_0^1 x(x^2+1)^5\, dx = \frac{1}{2}\int_1^2 x^5\, dx = \frac{1}{2}\frac{x^6}{6}\bigg|_1^2 = \frac{21}{4}$$

On the other hand, suppose we wish to evaluate

$$\int_0^3 \frac{x}{\sqrt{x+1}}\, dx$$

If we let $f(x) = x/\sqrt{x+1}$ and $g(x) = x^2 - 1$, then

$$\int_0^3 \frac{x}{\sqrt{x+1}}\, dx = \int_{g^{-1}(0)}^{g^{-1}(3)} f\, dx = \int_1^2 (f \circ g)g'\, dx = \int_1^2 \frac{x^2-1}{x}\, 2x\, dx$$

$$= 2\int_1^2 (x^2-1)\, dx = 2\left(\frac{x^3}{3} - x\right)\bigg|_1^2 = \frac{8}{3}$$

We conclude this section with a discussion of the behavior of sequences of differentiable functions. Thus suppose $\{f_n\}$ is a sequence of functions which converges to the function f on $[a, b]$. As we did for integrals, we can ask two questions: if f_n is differentiable on $[a, b]$ for all $n \in \mathbf{N}$, is the limit function f differentiable on $[a, b]$? If f is differentiable on $[a, b]$, is $f' = \operatorname{Lim}_{n \to \infty} f'_n$? We might expect the answers to these questions will be no if the convergence is only pointwise, but yes if it is uniform. Unfortunately, this is not the case: even uniform convergence is not enough to ensure that the limit function is differentiable, or, if it is, that the derivative of the limit is the limit of the derivatives.

Examples 8.3.17 1. This example will show that a sequence of differentiable functions may converge uniformly to a nondifferentiable function. Let f be the absolute value function, restricted to the interval $[-1, 1]$. According to the Weierstrass Approximation Theorem of Section 7.2, Chapter 7, there is a sequence of polynomials which converges uniformly to f. Each polynomial is differentiable on $[-1, 1]$ (see Exercise 7 at the end of this section), but f is not differentiable on $[-1, 1]$ since $f'(0)$ does not exist.

2. This example will show that even if a sequence of differentiable functions converges uniformly to a differentiable function, the derivative of the limit need not be the limit of the derivatives. For each $n \in \boldsymbol{N}$, let

$$f_n(x) = \frac{x^n}{n}$$

for all $x \in [0, 1]$; then for each $n \in \boldsymbol{N}$,

$$f_n(x) = x^{n-1}$$

for all $x \in [0, 1]$ (see Exercise 7 at the end of this section). But the sequence $\{f_n\}$ converges uniformly to the zero function on $[0, 1]$, whereas the sequence of derivatives $\{f'_n\}$ does not converge to the derivative of the zero function (which is again the zero function), because for $x = 1$,

$$\operatorname*{Lim}_{n \to \infty} f_n(1) \neq f'(1) = 0$$

Clearly we will need a stronger condition than uniform convergence of the sequence to ensure that a sequence of differentiable functions converges to a differentiable function. As the following proposition shows, what is needed is the uniform convergence of the sequence of derivatives.

Proposition 8.3.18 Let $\{f_n\}$ be a sequence of functions defined and differentiable on $[a, b]$. If the sequence of derivatives $\{f'_n\}$ converges uniformly on $[a, b]$, if the sequence $\{f_n(c)\}$ converges for some $c \in [a, b]$, and if f'_n is continuous on $[a, b]$ for all $n \in \boldsymbol{N}$, then $\{f_n\}$ converges to a differentiable function f on $[a, b]$, and

$$\operatorname*{Lim}_{n \to \infty} f'_n = f'$$

on $[a, b]$.

Proof: Suppose that $\{f'_n\}$ converges uniformly to the function g on $[a, b]$. By Proposition 8.2.10,

$$\operatorname*{Lim}_{n \to \infty} \int_c^x f'_n \, dx = \int_c^x g \, dx$$

for all $x \in [a, b]$ and the Fundamental Theorem of Calculus implies that

$$\int_c^x f'_n \, dx = f_n(x) - f_n(c)$$

for all $x \in [a, b]$. Hence

$$\lim_{n \to \infty} f_n(x) = \int_c^x g \, dx + \lim_{n \to \infty} f_n(c)$$

for all $x \in [a, b]$. Thus $\text{Lim}_{n \to \infty} f_n(x)$ exists for all $x \in [a, b]$, so the sequence $\{f_n\}$ converges pointwise on $[a, b]$. Let $\text{Lim}_{n \to \infty} f_n = f$; then

$$(*) \qquad f(x) = \lim_{n \to \infty} f_n(x) = \int_c^x g \, dx + f(c)$$

for all $x \in [a, b]$. Since $\{f'_n\}$ is a sequence of continuous functions which converges uniformly to g, g must be continuous, and it follows from Theorem 8.3.13 and $(*)$ above that f is differentiable and $f' = g$ on the closed interval with endpoints c and x, for all $x \in [a, b]$. Hence f is differentiable on $[a, b]$ and

$$f' = g = \lim_{n \to \infty} f'_n$$

on $[a, b]$. □

We should remark that uniform convergence of the sequence of derivatives $\{f'_n\}$ is indeed stronger than uniform convergence of the sequence $\{f_n\}$, for it can be shown that the former implies the latter, whereas we know from Examples 8.3.16 that the latter does not imply the former. Furthermore, Proposition 8.3.17 remains true if we remove the hypothesis that each derivative function f'_n be continuous. We assumed the continuity of the derivatives in order to produce a relatively straightforward proof based on the Fundamental Theorem of Calculus. A proof which does not require the continuity of the derivatives is outlined in Exercise 23 below.

EXERCISES

1. Prove Proposition 8.3.1.
2. Prove that if $\text{Lim}_{x \to c} f(x) = x_0$, then $\text{Lim}_{x \to c} g(f(x)) = \text{Lim}_{z \to x_0} g(z)$, provided $\text{Lim}_{z \to x_o} g(z)$ exists.

3. Let f be a real-valued function defined on an interval I and let $c \in I$. Prove that $f'(c)$ exists if and only if

$$\lim_{h \to 0} \frac{f(c+h) - f(c)}{h} = f'(c)$$

This result is often used as the definition of the derivative of f at c.

4. For each of the following, find the derivative without using any differentiation formulas.
(a) $f(x) = 3x + 2$ for all $x \in \boldsymbol{R}$;
(b) $f(x) = x^3$ for all $x \in \boldsymbol{R}$;
(c) $f(x) = 1/x$ for all $x \in (0, +\infty)$;
(d) $f(x) = 1/x^2$ for all $x \in (0, +\infty)$.

5. (a) Let

$$f(x) = \begin{cases} -1 & \text{if } x < 0 \\ 0 & \text{if } x = 0 \\ 1 & \text{if } x > 0 \end{cases}$$

Show that $f'(0)$ does not exist.
(b) Let $f(x) = x^{2/3}$ for all $x \in \boldsymbol{R}$. Without using any differentiation formulas, prove that

$$f'(x) = \tfrac{2}{3}x^{-1/3}$$

for all $x \in \boldsymbol{R}$, $x \neq 0$, and show that $f'(0)$ does not exist.

6. Let f be a real-valued function defined at $c \in \boldsymbol{R}$. Show that if $f'(c)$ exists, it is the slope of the line tangent to the graph of f at the point $(c, f(c))$ in the Euclidean plane.

7. Prove (1) and (2) of Proposition 8.3.5.

8. Prove (3) and (4) of Proposition 8.3.5. (Hint: for (3),

$$\frac{(fg)(x) - (fg)(c)}{x - c} = \frac{f(x) - f(c)}{x - c} g(x) + \frac{g(x) - g(c)}{x - c} f(c))$$

9. Let $n \in \boldsymbol{N}$ and let $f(x) = x^n$ for all $x \in \boldsymbol{R}$. Prove that

$$f'(x) = nx^{n-1}$$

for all $x \in \boldsymbol{R}$ and then show that every polynomial function on $\boldsymbol{R}$ is differentiable on $\boldsymbol{R}$.

10. Let n be a negative integer and let $f(x) = x^n$ for all $n \in \boldsymbol{R}$, $x \neq 0$. Prove that

$$f'(x) = nx^{n-1}$$

for all $n \in \boldsymbol{R}$, $x \neq 0$.

11. Let $q \in \boldsymbol{Q}$, $q \neq 0$, and let $f(x) = x^q$ for all $x \in (0, +\infty)$. Prove that

$$f'(x) = qx^{q-1}$$

for all $x \in (0, +\infty)$. (Hint: if $q = m/n$, then $[f(x)]^n = x^m$.)

12. Let f be a real-valued function which is differentiable on $[a, b]$. Prove that even though f' need not be continuous on $[a, b]$, it is still true that f' has the intermediate value property; that is, show that f' takes on every value between $f'(a)$ and $f'(b)$. (Hint: if c is between $f'(a)$ and $f'(b)$, consider $g(x) = f(x) - cx$.)

13. Let f be a real-valued one-to-one function defined on $[a, b]$ and let g be its inverse function. Prove that if f is continuous at $c \in [a, b]$ and $g'(f(c))$ exists and is nonzero, then $f'(c)$ exists and

$$f'(c) = \frac{1}{g'(f(c))}$$

14. Prove Proposition 8.3.8 for the case of a relative minimum.

15. Let f be a real-valued function which is continuous on $[a, b]$ and differentiable on (a, b).
    (a) Prove that if $f'(x) \geqslant 0$ for all $x \in (a, b)$, then

$$x \in [a, b],\ y \in [a, b],\ x \leqslant y \qquad \text{imply} \qquad f(x) \leqslant f(y)$$

    (b) Prove that if $f'(x) > 0$ for all $x \in (a, b)$, then

$$x \in [a, b],\ y \in [a, b],\ x < y \qquad \text{imply} \qquad f(x) < f(y)$$

    (c) State and prove results analogous to (a) and (b) if $f'(x) \leqslant 0$ and $f'(x) < 0$ on (a, b)

16. (a) Prove the First Derivative Test for relative maxima:
    Let f be a real-valued function which is continuous at $c \in \boldsymbol{R}$ and suppose there exists a positive real number δ such that f is differentiable on $(c - \delta, c + \delta)$, except possibly at c. If $f'(x) > 0$ for $x \in (c - \delta, c)$ and $f'(x) < 0$ for $x \in (c, c + \delta)$, then f has a relative maximum at c.
    (b) State and prove a First Derivative Test for relative minima.

17. Fill in the details of the proofs for Corollaries 8.3.10, 8.3.11, and 8.3.12.
18. Prove the Cauchy form of the Law of the Mean:

    If f and g are continuous real-valued functions on $[a, b]$ which are differentiable on (a, b), then there is some $c \in (a, b)$ such that

$$f'(c)[g(b) - g(a)] = g'(c)[f(b) - f(a)]$$

19. Prove the following version of L'Hospital's Rule:
    Let f and g be continuous real-valued functions on $[a, b]$ which are differentiable on (a, b). If $f(a) = g(a) = 0$, $g'(x) \neq 0$ for $x \in (a, b)$, and $\operatorname{Lim}_{x \to a} f(x) = \operatorname{Lim}_{x \to a} g(x) = 0$, then

$$\operatorname*{Lim}_{x \to a} \frac{f}{g} = \operatorname*{Lim}_{x \to a} \frac{f'}{g'}$$

    provided the limit on the right-hand side of the equation exists.
20. Prove the following generalization of Theorem 8.3.13:
    Let f be Riemann integrable on $[a, b]$, and define

$$F(x) = \int_a^x f\,dx$$

    for all $x \in [a, b]$. If f is continuous at $c \in [a, b]$, then $F'(c)$ exists and $F'(c) = f(c)$.
21. Prove the Integration by Parts Theorem:
    If f and g are real-valued functions differentiable on $[a, b]$ and f' and g' are continuous on $[a, b]$, then

$$\int_a^b fg'\,dx = f(b)g(b) - f(a)g(a) - \int_a^b f'g\,dx$$

22. Use Proposition 8.3.15 to evaluate the following integrals.

    (a) $$\int_0^1 x^2\sqrt{x^3 + 1}\,dx$$

    (b) $$\int_0^1 \frac{x + 1}{(2x^2 + 4x + 1)^3}\,dx$$

(c) $$\int_0^2 x\sqrt{x+1}\,dx$$

(d) $$\int_2^3 \frac{x}{(x-1)^3}\,dx$$

23. This exercise shows that Proposition 8.3.18 remains true if we drop the hypothesis that the derivatives f'_n be continuous. The reader is asked to fill in the details. (Note:

$$\|f-g\| = \sup\{|f(x)-g(x)| \mid x\in[a,b]\})$$

(a) If $x\in[a,b]$, apply the Law of the Mean to $f_n - f_m$ on $[c,x]$ (or $[x,c]$) and take least upper bounds to conclude that

$$\|f_n - f_m\| \leqslant (b-a)\|f'_n - f'_m\| + |f_n(c) - f_m(c)|$$

Use this inequality, the uniform convergence of $\{f'_n\}$, and the convergence of $\{f_n(c)\}$ to show that $\{f_n\}$ converges uniformly to a function f on $[a,b]$. Note that f is continuous on $[a,b]$.

(b) Let $x\in[a,b]$, $y\in[a,b]$. Apply the Law of the Mean to $f_n - f_m$ on $[y,x]$ (or $[x,y]$) and take least upper bounds to conclude that

$$\left|\frac{f_n(x)-f_n(y)}{x-y} - \frac{f_m(x)-f_m(y)}{x-y}\right| < \|f'_n - f'_m\|$$

Let $\varepsilon > 0$ be given. Because of the uniform convergence of $\{f'_n\}$, there is some $k_1\in \boldsymbol{N}$ such that $n \geqslant k_1$, $m \geqslant k_1$ imply

$$\left|\frac{f_n(x)-f_n(y)}{x-y} - \frac{f_m(x)-f_m(y)}{x-y}\right| < \frac{\varepsilon}{3}$$

from which it follows that if $m \geqslant k_1$, then

$$\left|\frac{f(x)-f(y)}{x-y} - \frac{f_m(x)-f_m(y)}{x-y}\right| \leqslant \frac{\varepsilon}{3}$$

(c) Let $g = \operatorname{Lim}_{n\to\infty} f'_n$ on $[a,b]$. There is some $k_2\in \boldsymbol{N}$ such that

$$m \geqslant k_2 \qquad \text{implies} \qquad |f'_m(y) - g(y)| < \frac{\varepsilon}{3}$$

Let $k = \max\{k_1, k_2\}$. Since $f_k'(y)$ exists, there is some $\delta > 0$ such that

$$0 < |x - y| < \delta \qquad \text{implies} \qquad \left|\frac{f_k(x) - f_k(y)}{x - y} - f_k'(y)\right| < \frac{\varepsilon}{3}$$

Now show that $0 < |x - y| < \delta$ implies that

$$\left|\frac{f(x) - f(y)}{x - y} - g(y)\right| < \varepsilon$$

and hence that $f'(y) = g(y)$.

24. There are many examples of functions which are continuous everywhere yet differentiable nowhere. The particular one we will consider in this exercise can also be found in Chapter 2 of Gelbaum and Olmsted, *Counterexamples in Analysis*, Holden-Day, Inc., San Francisco, 1964. Before we can define the function, we need some preliminary results on series of functions.

If $\{f_n\}$ is a sequence of functions defined on a set $S \subset \boldsymbol{R}$, let

$$s_k = \sum_{n=1}^{k} f_n$$

for all $k \in \boldsymbol{N}$. (Thus s_k is the kth partial sum of the sequence of functions $\{f_n\}$.) If the sequence $\{s_k\}$ converges pointwise (uniformly) to a function f on S, we say that the *series of functions*

$$\sum_{n=1}^{\infty} f_n$$

converges pointwise (uniformly) to the function f on S, and write

$$\sum_{n=1}^{\infty} f_n = f$$

(a) Prove the *Weierstrass M-Test*:

if $\sum_{n=1}^{\infty} M_n$ is a convergent series of nonnegative constants and $|f_n(x)| \leqslant M_n$ for all $n \in \boldsymbol{N}$ and all $x \in S$, then

$$\sum_{n=1}^{\infty} f_n$$

converges uniformly on S.

Now we define a function f which is continuous everywhere and differentiable nowhere. Let

$$f_1(x) = |x| \qquad \text{for} \qquad x \in [-\tfrac{1}{2}, \tfrac{1}{2}]$$

and extend the domain of f_1 to $\boldsymbol{R}$ by requiring that

$$f_1(x+1) = f_1(x)$$

for all $x \in \boldsymbol{R}$. For $n \in \boldsymbol{N}$, $n > 1$, define

$$f_n(x) = 4^{1-n} f_1(4^{n-1}x)$$

for all $x \in \boldsymbol{R}$.

(b) Show that $|f_n(x)| \leqslant \frac{1}{2}(4^{1-n})$ for all $x \in \boldsymbol{R}$ and all $n \in \boldsymbol{N}$. Then show that the series

$$\sum_{n=1}^{\infty} f_n$$

converges uniformly on $\boldsymbol{R}$, and hence that it converges to a function f which is continuous on $\boldsymbol{R}$.

Finally, we prove that the function $f = \Sigma_{n=1}^{\infty} f_n$ is nowhere differentiable.

(c) Let $c \in \boldsymbol{R}$. Show that if

$$x_n = c + 4^{-1-n} \qquad \text{or} \qquad x_n = c - 4^{-1-n}$$

$$|f_m(x_n) - f_m(c)| = \begin{cases} 4^{-1-n} & \text{if } m \leqslant n \\ 0 & \text{if } m > n \end{cases}$$

(d) Show that

$$\frac{f(x_n) - f(c)}{x_n - c}$$

is an even integer if n is even and is an odd integer if n is odd.

(e) Prove that

$$f'(c) = \operatorname*{Lim}_{x \to c} \frac{f(x) - f(c)}{x - c}$$

cannot exist.

Appendix
Cardinal Numbers

In this appendix we continue the study of sets which we initiated in Chapter 1. We begin by proving the important Schroeder–Bernstein Theorem and showing that subsets and countable unions of countable sets are themselves countable sets. We then turn to our main topic in this appendix, cardinal numbers: we define the concept of the cardinal number of a set, examine the properties and arithmetic of cardinal numbers, prove Cantor's Theorem, and conclude with some remarks about the Continuum Hypothesis. The definitions, results, and examples of Section 1.3 of Chapter 1 will form the basis of our work here, and the reader is urged to review the material of Section 1.3 before proceeding.

Recall that we defined a set S to be *equivalent* to a set T if there is a one-to-one function from S onto T. Since the existence of such a function means that the elements of S and T are paired off in a one-to-one manner, it is clear that equivalent sets are the same "size," in the sense that they have the same number of elements.

Suppose we wish to show that a given set S is equivalent to another set T. In order to do this, we would need to find a one-to-one function from S onto T, and this is not always easy to accomplish. The difficulty usually lies in finding a function which is onto. However, it is often the case that we can rather easily find a one-to-one function from S to T which is not onto, and also a one-to-one function from T to S which is not onto. In such a situation S is equivalent to a subset of T and T is equivalent to a subset of S, and intuition tells us that then S and T should be equivalent. This important fact is indeed true; it is known as the Schroeder–Bernstein Theorem.

Theorem A.1 (Schroeder–Bernstein) Let S and T be sets. If S is equivalent to a subset of T and T is equivalent to a subset of S, then S is equivalent to T.

Proof: Let $f: S \to T$ and $g: T \to S$ be one-to-one functions, so that S is equivalent to $f(S) \subset T$ and T is equivalent to $g(T) \subset S$. See Figure A.1. The strategy of the proof is to find a subset X of T such that g^{-1} is defined on $g(X) \subset S$ and T is the disjoint union of X and $f(S - g(X))$ (see Figure A.2), for then we can define a one-to-one function h from S onto T by setting h equal to g^{-1} on $g(X)$ and h equal to f on $S - g(X)$.

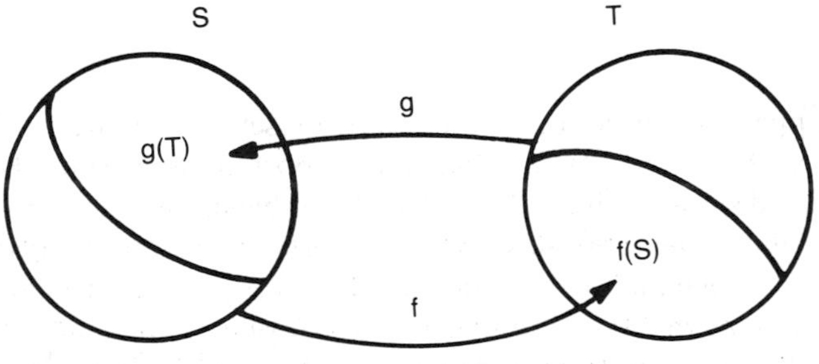

Figure A.1 S is equivalent to $f(S)$, T is equivalent to $g(T)$.

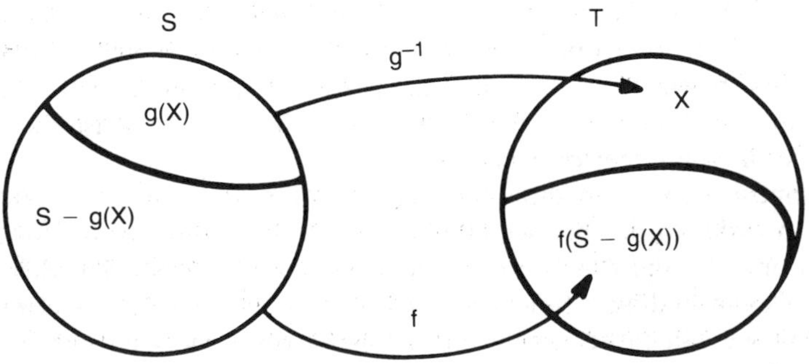

Figure A.2 T is the disjoint union of X and $f(S - g(X))$.

Let $\mathscr{V} = \{V \subset T \mid V \subset T - f(S - g(V))\}$. Thus $\mathscr{V}$ is a set of subsets of T. Since $\emptyset$ is a subset of every set, $\emptyset \subset T - f(S - g(\emptyset))$, and hence $\emptyset \in \mathscr{V}$. Therefore $\mathscr{V}$ is a nonempty set of subsets of T.

Note that

$$(*)\quad A \subset B \subset T \qquad \text{implies} \qquad T - f(S - g(A)) \subset T - f(S - g(B)):$$

if $A \subset B \subset T$, then $g(A) \subset g(B) \subset S$, whence $S - g(B) \subset S - g(A)$, which implies $f(S - g(B)) \subset f(S - g(A))$, which in turn implies

$$T - f(S - g(A)) \subset T - f(S - g(B))$$

Let $X = \bigcup_{V \in \mathscr{V}} V$. For ease of notation, let us set $X' = T - f(S - g(X))$. We will show that $X = X'$. If $V \in \mathscr{V}$, then $V \subset X \subset T$ and thus by virtue of $(*)$ above,

$$T - f(S - g(V)) \subset T - f(S - g(X)) = X'$$

However, $V \in \mathscr{V}$ implies that $V \subset T - f(S - g(V))$, and therefore $V \subset X'$ for all $V \in \mathscr{V}$. But then

$$X = \bigcup_{V \in \mathscr{V}} V \subset X'$$

and hence by $(*)$,

$$X \subset X' \subset T \qquad \text{implies} \qquad X' = T - f(S - g(X)) \subset T - f(S - g(X'))$$

which shows that $X' \in \mathscr{V}$. Therefore $X' \subset \bigcup_{V \in \mathscr{V}} V = X$, and $X \subset X'$, $X' \subset X$ imply that $X = X' = T - f(S - g(X))$.

Since $X = T - f(S - g(X))$, it follows that $T - X = f(S - g(X))$. Define $h\colon S \to T$ by

$$h(s) = \begin{cases} g^{-1}(s) & \text{if } s \in g(X) \\ f(s) & \text{if } s \in S - g(X) \end{cases}$$

The remainder of the proof consists of showing that h is one-to-one and onto.

Suppose s_1 and s_2 are elements of S. If

$$s_1 \in S - g(X) \qquad \text{and} \qquad s_2 \in g(X)$$

then

$$h(s_1) \in f(S - g(X)) = T - X \qquad \text{and} \qquad h(s_2) \in g^{-1}(g(X)) = X$$

and hence $h(s_1) = h(s_2)$ is impossible. Thus if $h(s_1) = h(s_2)$, then either s_1 and s_2 are both elements of $g(X)$, in which case the fact that g^{-1} is one-to-one implies that $s_1 = s_2$, or s_1 and s_2 are both elements of $S - g(X)$, in which case the fact that f is one-to-one implies that $s_1 = s_2$. Therefore h is one-to-one.

Finally, suppose $t \in T$. If $t \in X$, then $g(t) \in g(X)$ and thus

$$t = g^{-1}(g(t)) = h(g(t))$$

If $t \in T - X = f(S - g(X))$, then $t = f(s)$ for some $s \in S - g(X)$, and thus

$$t = f(s) = h(s)$$

Therefore h is onto, and we are done. □

Examples A.2 1. The function f: $\boldsymbol{R} \to (-1, 1)$ defined by

$$f(x) = \frac{x}{1 + |x|}$$

is clearly one-to-one and thus $\boldsymbol{R}$ is equivalent to a subset of $(-1, 1)$. Since $(-1, 1)$ is equivalent to itself, it is equivalent to a subset of $\boldsymbol{R}$. Thus by the Schroeder–Bernstein Theorem, $\boldsymbol{R}$ is equivalent to $(-1, 1)$.

2. By Example 4 of Examples 1.3.2 of Chapter 1, any two closed intervals having more than one point are equivalent; but every open interval contains a closed interval with more than one point, so every closed interval with more than one point is equivalent to a subset of every open interval. Similarly, any two open intervals are equivalent, and every closed interval with more than one point contains an open interval; thus every open interval is equivalent to a subset of every closed interval with more than one point. Therefore by the Schroeder–Bernstein Theorem, every closed interval with more than one point is equivalent to every open interval. Combining this result with that of Example 1 above, we see that $\boldsymbol{R}$ is equivalent to every interval which contains more than one point.

We will use the Schroeder–Bernstein Theorem to give a rigorous proof of the fact that every subset of a countable set is countable.

Proposition A.3 Every subset of a countable set is countable.

Proof: Let S be a countable set. If S is a finite, then by Corollary 1.3.6 every subset of S is finite, hence countable. If S is infinite, then it is denumerable. Let T be a subset of the denumerable set S. If T is finite, it is countable. If T is infinite, then by Corollary 1.3.9 it contains a denumerable subset T', and we have $T' \subset T \subset S$, with S and T' denumerable, hence equivalent. Therefore S is equivalent to a subset of T. But T is equivalent to a subset of S, namely itself, so by the Schroeder–Bernstein Theorem, T is equivalent to S, and thus T is denumerable. $\square$

Next we prove that a countable union of countable sets is countable. The proof is a slight generalization of the argument we gave in Section 1.3 when showing that the set Q of rational numbers is countable.

Proposition A.4 The countable union of countable sets is a countable set.

Proof: Let A be a countable index set and for each $\alpha \in A$ let X_α be a countable set. We must show that $\bigcup_{\alpha \in A} X_\alpha$ is a countable set.

Suppose first that $A = N$, X_n is denumerable for each $n \in N$, and the sets X_n are mutually disjoint. For each $n \in N$ there is a one-to-one function f_n from N onto X_n. For each $m \in N$ let $f_n(m) = x_{nm}$ in X_n. Thus for each $n \in N$ we have $X_n = \{x_{n1}, x_{n2}, \ldots, x_{nm}, \ldots\}$ and it follows that we may write the elements of $\bigcup_{n \in N} X_n$ in the following array:

$$
\begin{array}{cccc}
x_{11} & x_{12} & x_{13} & x_{14} \quad \cdots \\
x_{21} & x_{22} & x_{23} & x_{24} \quad \cdots \\
x_{31} & x_{32} & x_{33} & x_{34} \quad \cdots \\
x_{41} & x_{42} & x_{43} & x_{44} \quad \cdots \\
\vdots & \vdots & \vdots & \vdots
\end{array}
$$

Since the sets X_n are mutually disjoint, the elements in the array are all distinct. Furthermore, every element of $\bigcup_{n \in N} X_n$ appears in the array. We now define a function $g: N \to X_n$ by counting along diagonals of the array:

$$
\begin{array}{cccc}
x_{11} & x_{12} & x_{13} & x_{14} \quad \cdots \\
x_{21} & x_{22} & x_{23} & x_{24} \quad \cdots \\
x_{31} & x_{32} & x_{33} & x_{34} \quad \cdots \\
x_{41} & x_{42} & x_{43} & x_{44} \quad \cdots \\
\vdots & \vdots & \vdots & \vdots \quad \cdots
\end{array}
$$

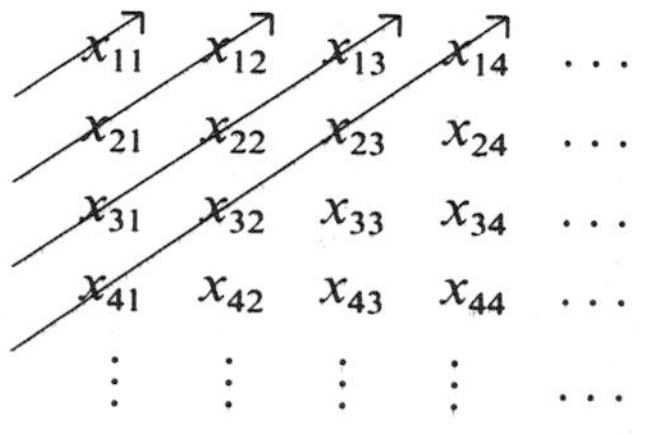

Thus

$$g(1) = x_{11},\ g(2) = x_{21},\ g(3) = x_{12},\ g(4) = x_{31}, \ldots$$

Since the function g is clearly one-to-one and onto, $\bigcup_{n\in N} X_n$ is equivalent to N and hence is countable.

Now we prove the proposition in its full generality. Suppose that A is any countable index set and that Y_α is a countable set for every $\alpha \in A$. We wish to show that $\bigcup_{\alpha\in A} Y_\alpha$ is countable. Since A is countable, and nonempty because it is an index set, A is equivalent to some subset of N. This subset may be a proper subset of N, but in any case each $\alpha \in A$ corresponds to a unique $n \in N$ and we may write $Y_\alpha = Y_n$. But Y_n, being countable, is equivalent to a subset (which may be proper) of the set X_n used above, and thus $\bigcup_{\alpha\in A} Y_\alpha$ is equivalent to a subset of $\bigcup_{n\in N} X_n$. Since we have shown that $\bigcup_{n\in N} X_n$ is countable, $\bigcup_{\alpha\in A} Y_\alpha$ is countable by Proposition A.3. □

Corollary A.5 The countable union of denumerable sets is a denumerable set.

The proof of the corollary is left to the reader.

Corollary A.6 The set Q of rational numbers is denumerable and hence countable.

Proof: Writing $Q = \bigcup_{n\in N} \{m/n \mid m \in Z\}$ shows that it is the countable union of the denumerable sets $\{m/n \mid m \in Z\}$. □

Now we are ready to refine our techniques for comparing sets. We will attach to each set a symbol called its *cardinal number*, then compare sets by manipulating their cardinal numbers according to the rules of cardinal arithmetic.

Definition A.7 Every set has attached to it a symbol called its *cardinal number*. If S is a set, the cardinal number of S is denoted by $|S|$. Order and equality among cardinal numbers are defined as follows: if S and T are sets, then $|S| \leqslant |T|$ if S is equivalent to a subset of T, and $|S| = |T|$ if S is equivalent to T. If $|S| \leqslant |T|$ and $|S| \neq |T|$, we write $|S| < |T|$. Furthermore, we say that a cardinal number is *infinite* if it is the cardinal number of an infinite set, and *finite* if it is the cardinal number of a finite set.

Definition A.7 is rather nebulous: it says that a cardinal number is a symbol attached to a set, but it does not tell us which symbols to attach to which sets. We remedy this in the following examples.

Examples A.8 1. By definition, the cardinal number of the empty set is zero: $|\emptyset| = 0$.

2. Let S be a nonempty finite set. By Corollary 1.3.9, there is some $n \in \boldsymbol{N}$ such that S is equivalent to $\{1, 2, \ldots, n\}$. We define $|S| = n$.

3. We define $|\boldsymbol{N}| = \aleph_0$. (The symbol $\aleph$ is the Hebrew letter aleph; the cardinal number of the set $\boldsymbol{N}$ is thus "aleph-null.") Note that

$$\aleph_0 = |\boldsymbol{N}| = |\boldsymbol{Z}| = |\boldsymbol{Q}|$$

and that in fact, $|S| = \aleph_0$ for any denumerable set S. Furthermore, since every infinite set contains a denumerable subset, $\aleph_0$ is the smallest infinite cardinal number. Note also that $|T| < \aleph_0$ for any finite set T.

4. We define $|\boldsymbol{R}| = c$. (The letter c stands for continuum; the real numbers form a continuum of points.) Note that c is an infinite cardinal number and that $\aleph_0 < c$. (Why?)

Our first result concerning cardinal numbers shows that the order relation $<$ behaves as expected.

Proposition A.9 Let α, β, and γ be cardinal numbers.

1. If $\alpha \leqslant \beta$ and $\beta \leqslant \alpha$, then $\alpha = \beta$.
2. If $\alpha \leqslant \beta$ and $\beta \leqslant \gamma$, then $\alpha \leqslant \gamma$.

Proof: We prove (1), leaving (2) as an exercise. (See Exercise 2 at the end of this appendix.)

Let S and T be sets such that $\alpha = |S|$ and $\beta = |T|$. The inequality $\alpha \leqslant \beta$ implies that S is equivalent to a subset of T, while $\beta \leqslant \alpha$ implies that T is equivalent to a subset of S. Therefore by the Schroeder–Bernstein Theorem, S is equivalent to T, and hence $\alpha = \beta$. □

We begin our consideration of the arithmetic of cardinal numbers by defining addition and multiplication for cardinals.

Definition A.10 Let α and β be cardinal numbers.

1. Let S and T be disjoint sets with $|S| = \alpha$ and $|T| = \beta$. Define

$$\alpha + \beta = |S \cup T|$$

2. Let S and T be sets with $|S| = \alpha$ and $|T| = \beta$. Define

$$\alpha\beta = |S \times T|$$

Examples A.11 1. In this example we show that addition and multiplication of finite cardinal numbers is the same as addition and multiplication of nonnegative integers. For instance, for any cardinal number α, we have $\alpha + 0 = \alpha$ and $\alpha \cdot 0 = 0$ (here 0 is the cardinal number of the empty set), because if S is a set with cardinal number α, then

$$\alpha + 0 = |S \cup \emptyset| = |S| = \alpha$$

and

$$\alpha \cdot 0 = |S \times \emptyset| = |\emptyset| = 0$$

Now suppose that α and β are nonzero finite cardinals; then there exist natural numbers n and m such that

$$\alpha = |\{1, 2, \ldots, n\}| \qquad \text{and} \qquad \beta = |\{1, 2, \ldots, m\}|$$

However, since $\{1, 2, \ldots, m\}$ is clearly equivalent to $\{n + 1, \ldots, n + m\}$, we have

$$\alpha + \beta = |\{1, \ldots, n\} \cup \{n + 1, \ldots, n + m\}| = |\{1, \ldots, n + m\}| = n + m$$

A similar argument shows that $\alpha\beta = nm$. (See Exercise 3 at the end of this appendix.)

2. The arithmetic of infinite cardinals is quite different from that of finite cardinals. For instance, $\aleph_0 + \aleph_0 = \aleph_0$. To see this, let S and T be disjoint denumerable sets; then

$$|S| = \aleph_0, \qquad |T| = \aleph_0, \qquad \text{and} \qquad \aleph_0 + \aleph_0 = |S \cup T|$$

But since S, T are denumerable, so is $S \cup T$, and hence $|S \cup T| = \aleph_0$.

3. Let us show that $\aleph_0 \cdot \aleph_0 = \aleph_0$. Since $\aleph_0 = |\boldsymbol{N}|$, we have $\aleph_0 \cdot \aleph_0 = |\boldsymbol{N} \times \boldsymbol{N}|$. But $|\boldsymbol{N} \times \boldsymbol{N}|$ is obviously denumerable, so $|\boldsymbol{N} \times \boldsymbol{N}| = \aleph_0$.

4. We claim that $c + c = c$ and $c \cdot c = c$. To prove the first of these equalities, recall that every interval in $\boldsymbol{R}$ which has more than one point

is equivalent to $\boldsymbol{R}$. Hence

$$c + c = |[0, 1) \cup [1, 2]| = |[0, 2]| = c$$

To prove the second equality, note that $c = |[0, 1]|$, and hence

$$c \cdot c = |[0, 1] \times [0, 1]|$$

But since $[0, 1]$ is clearly equivalent to the subset $\{(x, 0) \mid x \in [0, 1]\}$ of $[0, 1] \times [0, 1]$, we conclude that $c \leqslant c \cdot c$. On the other hand, we may represent each $x \in [0, 1]$ by a unique decimal expansion, as in the proof of Theorem 1.3.13, and use this to define a function f: $[0, 1] \times [0, 1] \to [0, 1]$ as follows:

$$\text{if } x = 0.a_1a_2 \ldots \text{ and } y = 0.b_1b_2 \ldots \text{ are in } [0, 1]$$

let

$$f(x, y) = 0.a_1b_1a_2b_2 \ldots$$

It is easy to check that f is one-to-one (do so), and this establishes that $[0, 1] \times [0, 1]$ is equivalent to a subset of $[0, 1]$ and thus that $c \cdot c \leqslant c$.

Our next definition shows how to raise a cardinal number to a cardinal number power.

Definition A.12 Let S and T be sets. The *power set* T^S is the set of all functions from S to T. If α and β are cardinal numbers and S and T are sets such that $|S| = \alpha$, $|T| = \beta$, then $\beta^\alpha = |T^S|$.

Example A.13 Let $S = \{a, b\}$ and $T = \{0, 1\}$. There are exactly four functions from S to T, namely f, g, h, and k, defined by

$$f(a) = f(b) = 0$$

$$g(a) = 1, \ g(b) = 0$$

$$h(a) = 0, \ h(b) = 1$$

$$k(a) = k(b) = 1$$

Therefore $T^S = \{f, g, h, k\}$, and since $2 = |S|$ and $2 = |T|$, we have

$$2^2 = |T^S| = |\{f, g, h, k\}| = 4$$

as expected.

Notice that in the preceding example the set S has four distinct subsets, and each of these consists of the elements of S which one and only one of the functions from S to $\{0, 1\}$ carries onto 1. Thus

$$\emptyset = \{x \in S \mid f(x) = 1\}$$
$$\{a\} = \{x \in S \mid g(x) = 1\}$$
$$\{b\} = \{x \in S \mid h(x) = 1\}$$

and

$$S = \{x \in S \mid k(x) = 1\}$$

Clearly, there is a one-to-one correspondence between the power set $\{0, 1\}^S$ and the set of all subsets of S, and therefore the power set $\{0, 1\}^S$ is equivalent to the set of all subsets of S. This is so for every set S: the power set $\{0, 1\}^S$ is equivalent to the set of all subsets of S. Before we prove this it will be convenient to define the characteristic function of a set.

Definition A.14 Let S be a set and let T be a subset of S. The *characteristic function of* T is the function χ_T: $S \to \{0, 1\}$ defined by

$$\chi_T(x) = \begin{cases} 1 & \text{if } x \in T \\ 0 & \text{if } x \in S - T \end{cases}$$

Proposition A.15 Let S be a set. The power set $\{0, 1\}^S$ is equivalent to the set of all subsets of S.

Proof: Let $\mathcal{S}$ denote the set of all subsets of S. For each $T \in \mathcal{S}$, the characteristic function χ_T is an element of $\{0, 1\}^S$, and

$$T\{x \in S \mid \chi_T(x) = 1\}$$

Define a function f: $\mathcal{S} \to \{0, 1\}^S$ as follows:

$$f(T) = \chi_T \qquad \text{for all } T \in \mathcal{S}$$

We will show that f is one-to-one and onto and hence that $\mathcal{S}$ is equivalent to $\{0, 1\}^S$.

Suppose that $f(T) = f(V)$ for subsets T and V of S; then $\chi_T = \chi_V$, and hence $\chi_T(x) = \chi_V(x)$ for all $x \in S$. But then

$$T = \{x \in S \mid \chi_T(x) = 1\} = \{x \in S \mid \chi_V(x) = 1\} = V$$

Therefore f is one-to-one.

Now suppose that g is an element of the power set $\{0, 1\}^S$; then g is a function from S to $\{0, 1\}$. Let $T = \{x \in S \mid g(x) = 1\}$. If $x \in T$, then $g(x) = 1$ and also $\chi_T(x) = 1$, while if $x \in S - T$, then $g(x) = 0$ and $\chi_T(x) = 0$. Thus $f(T) = \chi_T = g$, which shows that f is onto, and we are done. □

Corollary A.16 If S is a set, the set of all subsets of S has cardinal number $2^{|S|}$.

Example A.17 We claim that $2^{\aleph_0} = c$. Let $\mathscr{N}$ denote the set of all subsets of $\boldsymbol{N}$. We know that

$$|\mathscr{N}| = 2^{|\boldsymbol{N}|} = 2^{\aleph_0}$$

so it will suffice to show that $|\mathscr{N}| = c$. We will do this by demonstrating that $\mathscr{N}$ is equivalent to the closed interval $[0, 1]$.

Every $x \in [0, 1]$ has a binary expansion of the form $x = 0.a_1a_2 \ldots$, where $a_n \in \{0, 1\}$ for every $n \in \boldsymbol{N}$. If x has two such binary expansions, we agree to use the one in which all terms after some kth term are zero. With this agreement in force, we may consider that every $x \in [0, 1]$ has a unique binary expansion.

If $x \in [0, 1]$ has the binary expansion $x = 0.a_1a_2 \ldots$, define

$$f(x) = \{n \in \boldsymbol{N} \mid a_n \neq 0\}$$

Clearly f is a function from $[0, 1]$ to $\mathscr{N}$, and if $f(x) = f(y)$, then x and y have the same nonzero terms in their binary expansions and hence have the same binary expansions, so $x = y$. This shows that f is one-to-one and thus that $[0, 1]$ is equivalent to a subset of $\mathscr{N}$.

Now we show that $\mathscr{N}$ is equivalent to a subset of $[0, 1]$. Each $x \in [0, 1]$ has a ternary expansion $x = 0.b_1b_2 \ldots$, where $b_n \in \{0, 1, 2\}$ for all $n \in \boldsymbol{N}$. If x has two such expansions, then in one of them every term after some kth term is a 2. Define a function $g\colon \mathscr{N} \to [0, 1]$ as follows: for every subset M of $\boldsymbol{N}$, $g(M) = 0.b_1b_2 \ldots$, where $0.b_1b_2 \ldots$ is the ternary expansion with

$$b_n = \begin{cases} 1 & \text{if } n \in M \\ 0 & \text{if } n \in \boldsymbol{N} - M \end{cases}$$

If M and M' are subsets of $\mathcal{N}$ such that $M \neq M'$, then the ternary expansions $g(M)$ and $g(M')$ are different in the sense that their terms are not identical. If $g(M)$ and $g(M')$ represent the same real number in $[0, 1]$, then every term in one of them past some kth term must be a 2. But this is impossible, for by the way g is defined, neither $g(M)$ nor $g(M')$ can have a 2 in its ternary expansion. Therefore g is one-to-one, so $\mathcal{N}$ is equivalent to a subset of $[0, 1]$, and the Schroeder–Bernstein Theorem now implies that $\mathcal{N}$ is equivalent to $[0, 1]$. □

If α is a finite cardinal number then of course $\alpha < 2^{\alpha}$. As the preceding example suggests, this inequality remains true for infinite cardinals. This result is known as Cantor's Theorem.

Theorem A.18 (Cantor) For any cardinal number α, $\alpha < 2^{\alpha}$.

Proof: Let S be a set such that $\alpha = |S|$; then 2^{α} is the cardinal of $\mathcal{S}$, the set of all subsets of S. Therefore it will suffice to show that for every set S, S is equivalent to a subset of $\mathcal{S}$, but that S cannot be equivalent to $\mathcal{S}$.

Let $f: S \to \mathcal{S}$ be given by $f(x) = \{x\}$ for all $x \in S$. Since f is obviously one-to-one, S is equivalent to a subset of $\mathcal{S}$.

We show that S cannot be equivalent to $\mathcal{S}$ by proving that if f is any function from S to $\mathcal{S}$, then f cannot be onto. Let f be a function from S to $\mathcal{S}$, so that $f(x)$ is a subset of S for all $x \in S$. Let T be the subset of S defined by $T = \{x \in S \mid x \notin f(x)\}$. The set $T \in \mathcal{S}$, but there is no $x \in S$ such that $f(x) = T$. To see this, suppose $f(x) = T$ and consider the cases $x \in T$, $x \notin T$: if $x \in T$, then by the definition of T, $x \notin f(x) = T$, while $x \notin T = f(x)$, then by the definition of T, $x \in T$. Since each possibility leads to a contradiction, f cannot be onto, and we are done. □

Corollary A.19 There is no largest cardinal number.

According to Cantor's Theorem, we can construct a chain of strictly increasing cardinal numbers, thus:

$$\aleph_0 < 2^{\aleph_0} = c < 2^c < 2^{2^c} < \cdots$$

The first three of the cardinals in this chain are cardinals of what might be called "naturally-occurring" sets: $\aleph_0$ is the cardinal of the natural numbers, c the cardinal of the reals, and 2^c the cardinal of the set of all graphs which can be drawn in the Euclidean plane. (See Exercise 15 at the end of this appendix.) To our knowledge, no one has ever given an example of a naturally-occurring set which has a cardinal greater than 2^c. For this reason it is sometimes said that as far as infinite cardinal numbers are concerned, we can only count up to three.

Let us again consider the chain

$$\aleph_0 < c < 2^c < \cdots$$

Are there cardinal numbers between the ones in this chain? To be more specific, is there a cardinal number between $\aleph_0$ and c? This leads us to a statement of the famous Continuum Hypothesis:

Continuum Hypothesis. There is no cardinal number α such that

$$\aleph_0 < \alpha < c$$

The Continuum Hypothesis was a famous unsolved problem in mathematics for over 50 years. In order to disprove it, one would have to exhibit a set with cardinal greater than $\aleph_0$ but less than c; in other words, a set "larger" than $\boldsymbol{Q}$ but "smaller" than $\boldsymbol{R}$. Since no such set could be found, the Continuum Hypothesis seemed reasonable, but all efforts to prove it failed. P. Cohen finally settled the matter by showing that in fact the Continuum Hypothesis is independent of the other axioms of set theory. This means that it cannot be proved as a theorem of set theory, nor can it be disproved. If we wish it to be true, we must take it as an axiom of set theory; if we wish it to be false, we must take its denial as an axiom.

We conclude this appendix with a paradox. Let A be the set of all cardinal numbers. For each $\alpha \in A$ let S_α be a set such that $\alpha = |S_\alpha|$, and consider $\mathscr{S} = \bigcup_{\alpha \in A} S_\alpha$. Each set S_α is obviously equivalent to a subset of $\mathscr{S}$ and therefore

$$\alpha = |S_\alpha| \leqslant |\mathscr{S}|$$

for all $\alpha \in A$. Therefore $|\mathscr{S}|$ is the largest cardinal number. But this contradicts Corollary A.19, so we have a paradox in the theory of cardinal numbers. What is wrong? Notice that at the beginning of our construction of the largest cardinal we said, "let A be the set of all cardinal numbers." In fact, the collection of all cardinal numbers is not a set; it is too large to be a set, and it cannot be treated as if it were one. Thus our construction of the largest cardinal is invalid. In order to avoid paradoxes such as this, it is necessary to define the concept of a set quite precisely. We have not done this, for it would have taken us much too far afield. The interested reader can find a good introduction to axiomatic set theory in J. L. Kelley's *General Topology*.

EXERCISES

1. Prove Corollary A.5.
2. Prove part (2) of Proposition A.9.
3. Let α and β be finite cardinals, with

$$\alpha = |\{1, \ldots, n\}| \qquad \text{and} \qquad \beta = |\{1, \ldots, m\}|$$

   Prove that $\alpha\beta = nm$.
4. Prove that if α is any cardinal number, then $\alpha \cdot 1 = \alpha$.
5. Find the cardinal numbers of the following sets:
   (a) The set of all open intervals in $\boldsymbol{R}$;
   (b) The set of all open intervals in $\boldsymbol{R}$ which have rational endpoints;
   (c) The Cantor ternary set;
   (d) The set of all lines in the Euclidean plane $\boldsymbol{R}^2$.
6. Let α, β, γ, and δ be cardinal numbers. Prove that if $\alpha \leqslant \beta$ and $\gamma \leqslant \delta$, then $\alpha + \gamma \leqslant \beta + \delta$.
7. Prove that if α and β are cardinal numbers with $\alpha \leqslant \aleph_0$ and $\beta \leqslant \aleph_0$, then $\alpha + \beta \leqslant \aleph_0$.
8. Let α, β, γ, and δ be cardinal numbers. Prove that if $\alpha \leqslant \beta$ and $\gamma \leqslant \delta$, then $\alpha\gamma \leqslant \beta\delta$.
9. Prove that if α and β are cardinal numbers with $\alpha \leqslant \aleph_0$ and $\beta \leqslant \aleph_0$, then $\alpha\beta \leqslant \aleph_0$.
10. Prove that if α is any cardinal number, then $\alpha^0 = 1$.
11. Let α, β, and γ be cardinal numbers. Prove that $\alpha^\beta\alpha^\gamma = \alpha^{\beta+\gamma}$.
12. Let α, β, and γ be cardinal numbers. Prove that $(\alpha^\beta)^\gamma = \alpha^{\beta\gamma}$.
13. Prove that
    (a) $c^{\aleph_0} = c$;
    (b) $\aleph_0^{\aleph_0} = c$.
14. Prove that $c^c = 2^c$.
15. Let S be the set of all graphs which can be drawn in the Euclidean plane $\boldsymbol{R}^2$. Prove that $|S| = 2^c$.
16. A *polynomial of degree* n is a function $p\colon \boldsymbol{R} \to \boldsymbol{R}$ whose defining equation is of the form

$$p(x) = a_n x^n + a_{n-1}x^{n-1} + \cdots + a_1 x + a_0$$

where $a_k \in \boldsymbol{R}$ for $0 \leqslant k \leqslant n$ and $a_n \neq 0$. The real numbers a_k are called the *coefficients* of the polynomial. A real number r is an *algebraic number* if for some $n \in \boldsymbol{N}$ there exists a polynomial p of degree n having rational coefficients such that $p(r) = 0$. If there is no such polynomial, the real number r is said to be a *transcendental number*. Prove that the set of all algebraic numbers is countable and that the set of all transcendental numbers is uncountable. You may use the fact that for a polynomial p of degree n there are at most n real numbers y such that $p(y) = 0$. (Hint: begin the proof by showing that the set of all polynomials of degree n which have integer coefficients is countable.)

Bibliography

ANALYSIS

Bartle, Robert G. (1964). *The Elements of Real Analysis.* John Wiley and Sons: New York.

Gelbaum, Bernard R. and Olmsted, John M. H. (1964). *Counterexamples in Analysis.* Holden-Day: San Francisco.

Goffman, Casper. (1961). *Real Functions.* Holt, Rinehart and Winston: New York.

Randolph, John F. (1968). *Basic Real and Abstract Analysis.* Academic Press: New York.

Royden, H. L. (1963). *Real Analysis.* The Macmillan Company: New York.

Rudin, Walter. (1964). *Principles of Mathematical Analysis.* McGraw-Hill: New York.

TOPOLOGY

Dugundji, James. (1970). *Topology.* Allyn and Bacon: Boston.

Kelley, John L. (1964). *General Topology.* D. Van Nostrand Company: Princeton, New Jersey.

Moore, Theral O. (1964). *Elementary General Topology.* Prentice-Hall: Englewood Cliffs, New Jersey.

Wilansky, Albert. (1970). *Topology for Analysis*. Ginn: Waltham, Massachusetts.

SET THEORY AND CARDINAL NUMBERS

Fraenkel, A. A. (1966). *Set Theory and Logic*. Addison-Wesley: Reading, Massachusetts.

Halmos, Paul R. (1960). *Naive Set Theory*. D. Van Nostrand Company: Princeton, New Jersey.

Index